Lecture Notes
in Computational Science
and Engineering

57

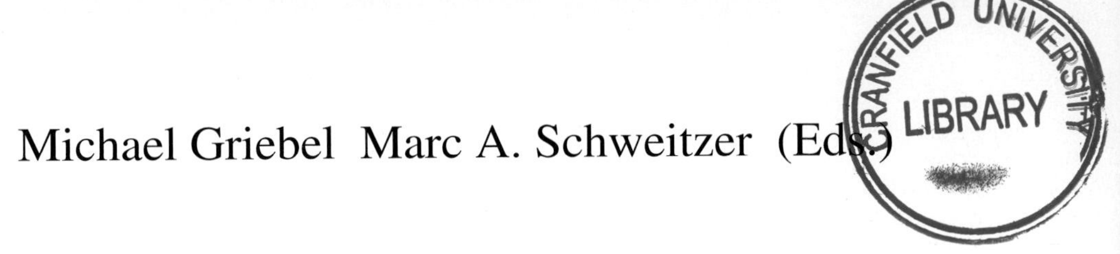

Michael Griebel Marc A. Schweitzer (Eds.)

Meshfree Methods for Partial Differential Equations III

With 137 Figures and 16 Tables

Editors

Michael Griebel
Marc A. Schweitzer

Universität Bonn
Institut für Numerische Simulation
Wegelerstraße 6
53115 Bonn, Germany
email: griebel@ins.uni-bonn.de
schweitzer@ins.uni-bonn.de

Library of Congress Control Number: 2002030507

Mathematics Subject Classification: 65N99, 64M99, 65M12, 6Y99

ISBN-10 3-540-46214-7 Springer Berlin Heidelberg New York
ISBN-13 978-3-540-46214-9 Springer Berlin Heidelberg New York

Springer is a part of Springer Science+Business Media
springer.com

Typesetting: by the authors and techbooks using a Springer LaTeX macro package
Cover design: *design & production* GmbH, Heidelberg

Printed on acid-free paper SPIN: 11887522 46/techbooks 5 4 3 2 1 0

Preface

Meshfree methods for the numerical solution of partial differential equations are becoming more and more mainstream in many areas of applications. Their flexiblity and wide applicability are attracting engineers, scientists, and mathematicians to this very dynamic research area.

Hence, the organizers of the third international workshop on *Meshfree Methods for Partial Differential Equations* held from September 12 to September 15, 2005 in Bonn, Germany aimed at bringing together European, American and Asian researchers working in this exciting field of interdisciplinary research. To this end Ivo Babuška, Ted Belytschko, Michael Griebel, Wing Kam Liu, Helmut Neunzert, and Harry Yserentant invited scientist from twelve countries to Bonn to strengthen the mathematical understanding and analysis of meshfree discretizations but also to promote the exchange of ideas on their implementation and application.

The workshop was again hosted by the Institut für Numerische Simulation at the Rheinische Friedrich–Wilhelms Universität Bonn with the financial support of the Sonderforschungsbereich 611 *Singular Phenomena and Scaling in Mathematical Models* funded by the *Deutsche Forschungsgemeinschaft.*

This volume of LNCSE now comprises selected contributions of attendees of the workshop. Their content ranges from applied mathematics to physics and engineering.

Bonn,
June, 2006

Michael Griebel
Marc Alexander Schweitzer

Contents

Local Maximum-Entropy Approximation Schemes

Marino Arroyo[1] and Michael Ortiz[2]

[1] Universitat Politècnica de Catalunya, Barcelona, Spain
marino.arroyo@upc.edu
[2] California Institute of Technology, Pasadena, USA
ortiz@aero.caltech.edu

Summary. We present a new approach to construct approximation schemes from scattered data on a node set, i.e. in the spirit of meshfree methods. The rational procedure behind these methods is to harmonize the locality of the shape functions and the information-theoretical optimality (entropy maximization) of the scheme, in a sense to be made precise in the paper. As a result, a one-parameter family of methods is defined, which smoothly and seamlessly bridges meshfree-style approximants and Delaunay approximants. Besides an appealing theoretical foundation, the method presents a number of practical advantages when it comes to solving partial differential equations. The non-negativity introduces the well-known monotonicity and variation-diminishing properties of the approximation scheme. Also, these methods satisfy ab initio a weak version of the Kronecker-delta property, which makes essential boundary conditions straightforward. The calculation of the shape functions is both efficient and robust in any spacial dimension. The implementation of a Galerkin method based on local maximum entropy approximants is illustrated by examples.

Key words: Maximum entropy, information theory, Delaunay triangulation, meshfree methods.

1 Introduction

Over the last decade, node-based approximation schemes have experienced a tremendous impetus in the field of numerical methods for PDEs, with the proliferation of meshfree methods (e.g. [14], [4], [13]; see also [11] for a review). The absence of a mesh outstands as a promise of greater flexibility and robustness. The actual realization of these potential advantages is yet to be fully accomplished. Possible reasons include difficulties in the numerical quadrature of the weak form, a topic of intense research (e.g. [7, 6]), and the availability of robust and efficient node-based approximants. The first meshfree approximants were based on the Shepard approximants, while presently most approaches hinge upon Moving Least Squares (MLS) approximants.

We present here a different family of approximation schemes that we call *maximum-entropy* approximation schemes. Here, we present the first-order method [1]. See [2] for a presentation of higher-order methods. In *max-ent* approximation methods, much in the tradition of computational geometric modelling, the shape functions are required to be non-negative. Consequently, owing to the 0th order consistency condition, the approximants can be viewed as discrete probability distributions. Furthermore, the 1st consistency condition renders this class of approximants generalized barycentric coordinates. The local *max-ent* approximation schemes are optimal compromises between two competing objectives: (1) maximum locality of the shape functions (maximum correlation between the approximation and the nodal value at the closest points) and (2) maximum entropy of the scheme. The second objective is a statement of information-theoretical optimality in the sense that it provides the least biased approximation scheme consistent with the reproducing conditions. We prove rigorously that these approximants smoothly and seamlessly bridge Delaunay shape functions and meshfree-style approximants, and that the approximants are smooth, exist and are uniquely defined within the convex hull of the node set. Furthermore, they satisfy ab initio a weak version of the Kronecker-delta property, which ensures that the approximation at each face depends only on the nodes on this particular face. This makes the imposition of essential boundary conditions in the numerical approximation of partial differential equations straightforward.

The formulation of the approximants is presented in Section 2. This section also includes the practical calculation of the shape functions. Section 3 provides a summary of the properties of these approximants. Section 4 gives a number of insightful alternative interpretations of these approximants, and include the notion of relative *max-ent* approximants. The application of the local *max-ent* approximants in a 3D nonlinear elasticity example is provided in Section 5, and the conclusions are collected in Section 6.

2 Formulation

Let $u : \Omega \subset \mathbb{R}^d \to \mathbb{R}$ be a function whose values $\{u_a;\ a = 1, \ldots, N\}$ are known on a node set $X = \{\boldsymbol{x}_a,\ a = 1, \ldots, N\} \subset \mathbb{R}^d$. Without loss of generality, we assume that the affine hull of the node set is $\mathbb{R}^d$. We wish to construct approximations to u of the form

$$u^h(\boldsymbol{x}) = \sum_{a=1}^{N} p_a(\boldsymbol{x}) u_a \tag{2.1}$$

where the functions $p_a : \Omega \to \mathbb{R}$ will be referred to as *shape functions*. A particular choice of shape functions defines an approximation scheme. We shall require the shape functions to satisfy the zeroth and first-order consistency conditions:

$$\sum_{a=1}^{N} p_a(\boldsymbol{x}) = 1, \quad \forall \boldsymbol{x} \in \Omega, \tag{2.2a}$$

$$\sum_{a=1}^{N} p_a(\boldsymbol{x})\, \boldsymbol{x}_a = \boldsymbol{x}, \quad \forall \boldsymbol{x} \in \Omega. \tag{2.2b}$$

These conditions guarantee that affine functions are exactly reproduced by the approximation scheme. In general, the shape functions are not uniquely determined by the consistency conditions when $N > d + 1$.

2.1 Convex Approximants

In addition, we shall require the shape functions be non-negative, i.e.,

$$p_a(\boldsymbol{x}) \geq 0, \quad \forall \boldsymbol{x} \in \Omega, \ a = 1, \ldots, N. \tag{2.3}$$

The positivity of the shape functions, together with the partition of unity property and the 1st order consistency conditions, allows us to interpret the shape functions as generalized barycentric coordinates. This viewpoint is common in geometric modelling, e.g., in Bézier, B-Spline techniques [16], natural neighbor approximations [19], and subdivision approximations [8]. Positive linearly consistent approximants have long been studied in the literature [10]. These methods often present a number of attractive features, such as the related properties of monotonicity, the *variation diminishing property* (the approximation is not more "wiggly" than the data), or smoothness preservation [9], of particular interest in the presence of shocks. Furthermore, they lead to well behaved mass matrices. The positivity restriction is natural in problems where a maximum principle is in force, such as in the heat conduction problem. In the present context, the non-negativity requirement is introduced primarily to enable the interpretation of shape functions as probability distributions (or coefficients of convex combinations). It follows from (2.2a), (2.2b) and (2.3) that the shape functions at $\boldsymbol{x} \in \mathrm{conv}X$ define a convex combination of vertices which evaluates to $\boldsymbol{x}$. In view of this property we shall refer to non-negative and first-order consistent approximation schemes as *convex approximation schemes*.

Our approach to building approximation schemes is to choose selected elements amongst all convex approximation schemes at a point $\boldsymbol{x}$, which we denote by

$$\mathcal{P}_{\boldsymbol{x}}(X) = \left\{ \boldsymbol{p}(\boldsymbol{x}) \in \mathbb{R}_+^N \;\middle|\; \sum_{a=1}^{N} p_a(\boldsymbol{x})\boldsymbol{x}_a = \boldsymbol{x}, \ \sum_{a=1}^{N} p_a(\boldsymbol{x}) = 1 \right\}, \tag{2.4}$$

where $\boldsymbol{p}(\boldsymbol{x})$ denotes the vector of $\mathbb{R}^N$ whose components are $\{p_1(\boldsymbol{x}), \ldots, p_N(\boldsymbol{x})\}$ and $\mathbb{R}_+^N$ is the non-negative orthant.

By direct comparison between the above defined set of convex approximants and the convex hull of the node set

$$\mathrm{conv} X = \left\{ \boldsymbol{x} \in \mathbb{R}^d \;\middle|\; \boldsymbol{x} = \sum_{a=1}^{N} \lambda_a \boldsymbol{x}_a,\ \lambda_a \geq 0,\ \sum_{a=1}^{N} \lambda_a = 1 \right\} \tag{2.5}$$

it follows that

Theorem 1. *The set of convex approximants $\mathcal{P}_{\boldsymbol{x}}(X)$ is non-empty if and only if $\boldsymbol{x} \in \mathrm{conv}\ X$.*

Note carefully that this result does not preclude using convex approximants in non-convex domains, as long as the non-convex domain is a subset of the convex hull of the node set. This restriction does not seem limiting in any reasonable sense. In the following, for simplicity, we shall assume that $\Omega = \mathrm{conv} X$.

We next provide the criteria to select an approximation scheme amongst the convex schemes. One possible rational criterion to select a convex approximation scheme is, based on information-theoretical considerations, to pick the *least biased* convex approximation scheme, i.e. that which maximizes the entropy. Another natural criterion is to select the most local convex scheme, since it most accurately respects the principle that the approximation scheme at a given point should be most influenced by the closest nodes in the node set. As we shall see, these are competing objectives.

2.2 Entropy Maximization

The entropy of a convex approximation scheme is a natural concept since the positivity and partition of unity properties of convex approximants allow us to interpret these approximants at each point $\boldsymbol{x}$ as discrete probability distributions. We remind that the entropy

$$H(\boldsymbol{p}) = -\sum_{a=1}^{N} p_a \log\, p_a$$

is a canonical measure of the uncertainty associated with the probabilities p_a, and measures the lack of information about the system, here, the set of shape functions. Equivalently, the entropy measures the information gained when the random variable is realized. Invoking Jaynes' principle of maximum entropy results in the least biased possible choice of convex scheme, devoid of artifacts or hidden assumptions. In this view, the approximation of a function from scattered data becomes a problem of statistical inference, and is mathematically formulated through the convex program:

$$(ME) \qquad \begin{aligned} \text{Maximize } & H(\boldsymbol{p}) = -\sum_{a=1}^{N} p_a \log p_a \\ \text{subject to } & p_a \geq 0, \quad a = 1, \ldots, N, \\ & \sum_{a=1}^{N} p_a = 1, \\ & \sum_{a=1}^{N} p_a \boldsymbol{x}_a = \boldsymbol{x}. \end{aligned}$$

The solutions of (ME), denoted by $\boldsymbol{p}_0(\boldsymbol{x})$, are referred to as *max-ent* approximants. Owing to the fact that the entropy is strictly concave in the set of convex approximants, we have the following result:

Theorem 2. *The program* (ME) *has a solution iff* $\boldsymbol{x} \in \operatorname{conv} X$, *in which case the solution is unique.*

The *max-ent* approximants, though optimal from an information-theoretic point of view, disregard completely the desirable spacial correlation between the approximation scheme at a given point and the nearby nodal values. Consequently, the shape functions are global. Indeed, the *max-ent* principle tries to find to most uniform distribution consistent with the constraints, here the 1st order consistency condition. Entropy maximization has been independently proposed in [18] as a means to construct C^0 approximants for arbitrary polygonal tesselations. However, the use of strict entropy maximization to define smooth meshfree-style approximants results in global shape functions and poor approximation properties, as illustrated in Figure 1.

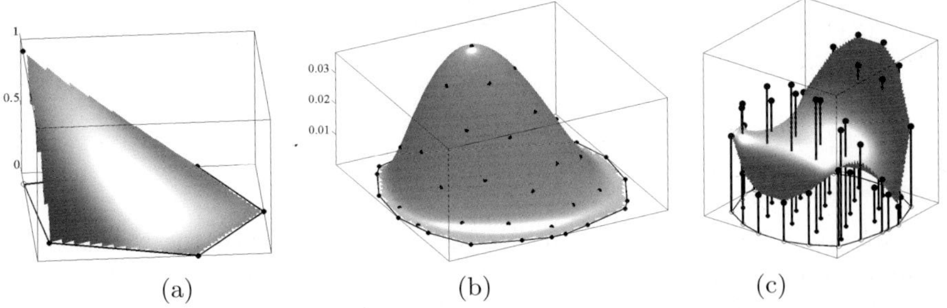

Figure 1. Examples of *max-ent* approximation schemes in the plane. (a) Shape function for the vertex of a pentagon; (b) shape function for an interior node, illustrating the global character of *max-ent* approximation shemes; and (c) *max-ent* approximation, or inference, of a function from scattered data, illustrating the non-interpolating character of *max-ent* approximation schemes.

2.3 Locality Maximization: Delaunay Approximants

A different criterion to select a distinguished approximant in the set of convex approximation schemes is to maximize the locality, or minimize the *width* of the shape functions. Define the width of shape function p_a as

$$w[p_a] = \int_{\Omega} p_a(\boldsymbol{x}) |\boldsymbol{x} - \boldsymbol{x}_a|^2 \, dx, \tag{2.6}$$

i.e. the second moment of p_a about $\boldsymbol{x}_a$. Evidently, other measures of the width of a function can be used instead in order to define alternative approximation schemes [1]. The locality measure presented here is the most natural choice, and emanates from an optimal mass transference theory and the 2−Wasserstein distance [2].

The most local approximation scheme is now that which minimizes the total width

$$W[\boldsymbol{p}] = \sum_{a=1}^{N} w[p_a] = \int_{\Omega} \sum_{a=1}^{N} p_a(\boldsymbol{x}) \; |\boldsymbol{x} - \boldsymbol{x}_a|^2 \, dx, \tag{2.7}$$

subject to the constraints (2.2a), (2.2b) and (2.3). Since the functional (2.7) does not involve shape function derivatives its minimization can be performed pointwise. This results in the linear program:

$$\begin{aligned}
(RAJ) \quad \text{For fixed } \boldsymbol{x} \text{ minimize } \; & U(\boldsymbol{x}, \boldsymbol{p}) = \sum_{a=1}^{N} p_a |\boldsymbol{x} - \boldsymbol{x}_a|^2 \\
\text{subject to } \; & p_a \geq 0, \; a = 1, \ldots, N, \\
& \sum_{a=1}^{N} p_a = 1, \\
& \sum_{a=1}^{N} p_a \boldsymbol{x}_a = \boldsymbol{x}.
\end{aligned}$$

A simple argument shows that the program (RAJ) has solutions if and only if $\boldsymbol{x} \in \text{conv}X$. However, the function $U(\boldsymbol{x}, \cdot)$ is not strictly convex (it is linear) and the solution is not unique in general.

The relationship between the linear program (RAJ) and the Delaunay triangulation has been established. Rajan [17] showed that if the nodes are in general positions (no $(d+1)$ nodes in X are cospherical), then (RAJ) has a unique solution, corresponding to the piecewise affine shape functions supported by the unique Delaunay triangulation associated with the node set X (a Delaunay triangulation verifies that the circumsphere of every simplex contains no point from X in its interior). We shall refer to the convex approximation schemes defined by the solutions $\boldsymbol{p}_{\infty}(\boldsymbol{x})$ of (RAJ) as *Rajan convex approximation schemes*, and to the approximants corresponding to the piecewise affine shape functions supported by a Delaunay triangulation as *Delaunay convex approximants*. Thus, Rajan's result states that for nodes in general positions, the Delaunay convex approximation scheme coincides with the unique Rajan convex approximation scheme, that is optimal in the sense of the width (2.6). When the nodes are not in general positions, the situation is more subtle; the set of Delaunay approximants is obviously non-unique, and the set of Rajan approximants is neither. In addition, the latter contains elements that do not belong to the former. See [2] for a discussion.

2.4 Pareto Optimality between Entropy and Locality Maximization

From the preceeding sections, it is clear that entropy and locality maximization are competing objectives. A standard device to harmonize competing objectives in multi-criterion optimization is to seek for Pareto optimal points [5]. In the present context, the Pareto set is the set of convex approximants such that no other convex approximant is *better* or dominates. An approximant $\boldsymbol{p}$ is *better* than another approximant $\boldsymbol{q}$ if and $\boldsymbol{p}$ meets or beats $\boldsymbol{q}$ on all objectives, and beats it in at least one objective, i.e. $H(\boldsymbol{p}) \geq H(\boldsymbol{q})$, $U(\boldsymbol{x}, \boldsymbol{p}) \leq U(\boldsymbol{x}, \boldsymbol{q})$, and either $H(\boldsymbol{p}) > H(\boldsymbol{q})$ or $U(\boldsymbol{x}, \boldsymbol{p}) < U(\boldsymbol{x}, \boldsymbol{q})$.

In convex multicriterion optimization, the Pareto set can be obtained through scalarization, i.e. considering the solutions of the one-parameter family of convex programs

$$\begin{aligned} (LME)_\beta \quad \text{For fixed } \boldsymbol{x} \text{ minimize } \; & f_\beta(\boldsymbol{x}, \boldsymbol{p}) = \beta U(\boldsymbol{x}, \boldsymbol{p}) - H(\boldsymbol{p}) \\ \text{subject to } \; & p_a \geq 0, \; a = 1, \ldots, N, \\ & \sum_{a=1}^{N} p_a = 1, \\ & \sum_{a=1}^{N} p_a \boldsymbol{x}_a = \boldsymbol{x}. \end{aligned}$$

The *max-ent* and the Rajan programs and approximants are recovered in the limits $\beta = 0$ and $\beta = +\infty$. By the strict convexity of $f_\beta(\boldsymbol{x}, \boldsymbol{p})$ for $\beta \in [0, +\infty)$, the program $(LME)_\beta$ has a solution, $\boldsymbol{p}_\beta(\boldsymbol{x})$, which is unique, if and only if $\boldsymbol{x} \in \text{conv}X$. We shall refer to this family of approximants as local *max-ent* approximation schemes.

It is easily seen [5, 1] that the Pareto set is the set of approximants $\boldsymbol{p}_\beta(\boldsymbol{x})$ for $\beta \in [0, +\infty)$, plus an additional approximation scheme denoted by $\boldsymbol{p}_\infty^{Pareto}(\boldsymbol{x})$. For non-degenerate cases, this additional scheme corresponds to the unique Delaunay triangulation of X. However, for degenerate cases, the selected approximant is the Rajan approximant with maximum entropy,

$$\boldsymbol{p}_\infty^{Pareto}(\boldsymbol{x}) = \arg \max_{\boldsymbol{p} \in S_{\boldsymbol{x}}^{RAJ}(X)} H(\boldsymbol{p}),$$

which is unique by strict concavity of the entropy. In the equation above, $\mathcal{S}_{\boldsymbol{x}}^{RAJ}(X)$ denotes the set of solutions of (RAJ).

2.5 Calculation of the Shape Functions

The practical calculation of the local *max-ent* shape functions $\boldsymbol{p}_\beta(\boldsymbol{x})$ is described next. It relies on standard duality methods, and the proof can be found in [1].

Proposition 1. *Let $\beta \in [0, \infty)$ and let $\boldsymbol{x} \in \text{int}(\text{conv}X)$ be an interior point. Define the* partition function $Z : \mathbb{R}^d \times \mathbb{R}^d \to \mathbb{R}$ *associated with the node set X as*

$$Z(\boldsymbol{x}, \boldsymbol{\lambda}) = \sum_{a=1}^{N} \exp\left[-\beta|\boldsymbol{x} - \boldsymbol{x}_a|^2 + \boldsymbol{\lambda} \cdot (\boldsymbol{x} - \boldsymbol{x}_a)\right]. \tag{2.8}$$

Then, the unique solution of the local max-ent *program $(LME)_\beta$ is given by*

$$p_{\beta a}(\boldsymbol{x}) = \frac{1}{Z(\boldsymbol{x}, \boldsymbol{\lambda}^*(\boldsymbol{x}))} \exp\left[-\beta|\boldsymbol{x} - \boldsymbol{x}_a|^2 + \boldsymbol{\lambda}^*(\boldsymbol{x}) \cdot (\boldsymbol{x} - \boldsymbol{x}_a)\right], \tag{2.9}$$

where $a = 1, ..., N$ and

$$\boldsymbol{\lambda}^*(\boldsymbol{x}) = \arg \min_{\boldsymbol{\lambda} \in \mathbb{R}^d} \log Z(\boldsymbol{x}, \boldsymbol{\lambda}). \tag{2.10}$$

Furthermore, the minimizer $\boldsymbol{\lambda}^*(\boldsymbol{x})$ is unique.

Proposition 2. *Let $\beta \in [0, \infty)$ and let $\boldsymbol{x} \in \text{bd}(\text{conv}X)$ be a boundary point. Let $C(\boldsymbol{x})$ be the contact set of the point $\boldsymbol{x}$ with respect to* conv$X$, *i.e. the smallest face of* conv$X$ *that contains $\boldsymbol{x}$. Then, the local* max-ent *shape functions at $\boldsymbol{x}$ follow from the previous Proposition with a reduced node set $X' = X \cap C(\boldsymbol{x}) - \boldsymbol{x}$, and is formulated in the subspace of $\mathbb{R}^d$ given by the affine hull of the reduced node set $L = \text{aff}X'$.*

The role of the *thermalization parameter* β is clear from Eq. (2.9): it controls the decay or locality of the shape functions. This is expected, since a larger value will give more weight to the locality measure in $f_\beta(\boldsymbol{x}, \boldsymbol{p})$. It is also clear that the shape functions, strictly speaking, have global support. From a numerical perspective, the Gaussian decay leads for all practical purposes to compactly supported functions. This is supported by a detailed numerical study in [1].

We note that in the absence of the 1st order consistency condition, the Lagrange multiplier $\boldsymbol{\lambda}^*(\boldsymbol{x})$ is absent, and the local *max-ent* approximants reduce to the Shepard approximants with Gaussian weight function. In this sense, the Lagrange multipliers introduce a correction so that the approximants satisfy the 1st order consistency condition. We will return later to this discussion.

The above Propositions provide a practical means of calculating the local *max-ent* shape functions at any given point $\boldsymbol{x}$. More schematically, and fixed $\boldsymbol{x}$, the shape functions are computed as follows:

1. Find the proximity index set (relative to a tolerance parameter Tol)

$$I_{\text{Tol}} = \left\{ a \mid \ \|\boldsymbol{x} - \boldsymbol{x}_a| \leq \sqrt{-\log(\text{Tol})/\beta} \right\}$$

2. Solve the minimization problem (2.10) in $\mathbb{R}^d$ using the Newton-Raphson method. In all evaluations of the partition function, the residual and the Jacobian matrix, the sum is only performed over I_{Tol}.
3. Compute the shape functions $p_{\beta a}(\boldsymbol{x}), a \in I_{\text{Tol}}$ according to Eq. (2.9) and set all other shape functions to zero.

The above algorithm is efficient and robust. The smooth minimization problem being solved is guaranteed to have a unique solution by the Kuhn-Tucker theorem and the strict convexity of the function being minimized. The number of unknowns is the space dimension. The expressions for the gradient and the Hessian matrix of the function being minimized are

$$\boldsymbol{r}(\boldsymbol{x}, \boldsymbol{\lambda}) = \partial_{\boldsymbol{\lambda}} \log Z(\boldsymbol{x}, \boldsymbol{\lambda}) = \sum_{a=1}^{N} p_a(\boldsymbol{x}, \boldsymbol{\lambda})(\boldsymbol{x} - \boldsymbol{x}_a), \tag{2.11}$$

and

$$\begin{aligned} \boldsymbol{J}(\boldsymbol{x}, \boldsymbol{\lambda}) &= \partial_{\boldsymbol{\lambda}} \partial_{\boldsymbol{\lambda}} \log Z(\boldsymbol{x}, \boldsymbol{\lambda}) \\ &= \sum_{a=1}^{N} p_a(\boldsymbol{x}, \boldsymbol{\lambda})(\boldsymbol{x} - \boldsymbol{x}_a) \otimes (\boldsymbol{x} - \boldsymbol{x}_a) - \boldsymbol{r}(\boldsymbol{x}, \boldsymbol{\lambda}) \otimes \boldsymbol{r}(\boldsymbol{x}, \boldsymbol{\lambda}). \end{aligned} \tag{2.12}$$

The Newton-Raphson iterations typically converge to reasonable tolerances within 2 or 3 iterations. Note that only the nodes $\boldsymbol{x}_a, a \in I_{\text{Tol}}$ contribute noticeably to the partition function. Restricting all index summations to I_{Tol} greatly reduces the computational cost in the above calculation. For practical applications, $\text{Tol} = 10^{-6}$ is amply sufficient. Changing this tolerance to machine precision does not change the numerical solutions in Galerkin methods based on local *max-ent* approximants.

2.6 Examples

Figure 2 shows the local *max-ent* shape function and its partial derivatives for a node in a two-dimensional node set as a function of the dimensionless parameter $\gamma = \beta h^2$, where h is a measure of the nodal spacing and β is constant over the domain. It can be seen from this figure that the shape functions are smooth and their degree of locality is controlled by the parameter γ. For the maximum value of $\gamma = 6.8$ shown in the figure the shape function ostensibly coincides with the Delaunay shape function. The parameter β can be allowed to depend on position and that dependence can be adjusted adaptively in order to achieve varying degrees of locality [1].

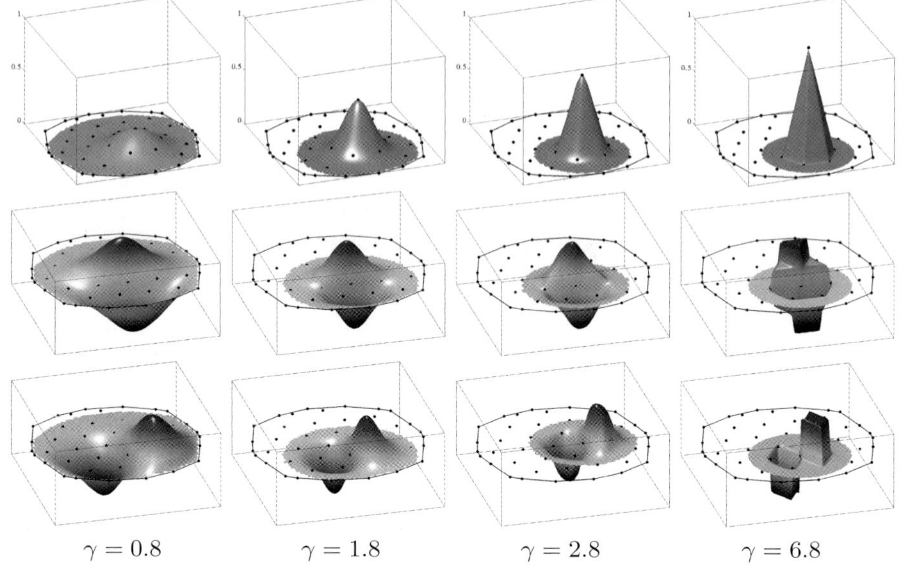

Figure 2. Local *max-ent* shape functions for a two dimensional arrangement of nodes, and spacial derivatives (arbitrary scale) for several values of $\gamma = \beta h^2$.

3 Properties

This section summarizes the basic properties of the above defined family of approximants. Proofs can be found in [1].

3.1 Behavior of Convex Approximants at the Boundary

Property 1. Let $\boldsymbol{p}$ *be a convex scheme with node set* $X = \{\boldsymbol{x}_1, ..., \boldsymbol{x}_N\}$. *Let* F *be a face of the convex polytope* conv$X$. *Then,* $\boldsymbol{x}_a \notin F \quad \Rightarrow \quad p_a = 0$ *on* F.

This property is a weak version of the Kronecker-delta property of interpolating approximants, i.e. that $p_a(\boldsymbol{x}_b) = \delta_{ab}$. This property states that for all convex approximants, and in particular for the local *max-ent* approximants, the approximation at a given face in the boundary depends only on shape functions of nodes in that particular face. Thus, a face is autonomous with regards to approximation, which makes the imposition of Dirichlet boundary conditions in Galerkin methods trivial. If the Dirichlet boundary conditions are affine on a face, then these are exactly imposed by constraining the nodal degrees of freedom to the affine function. If the boundary data is more general, then the imposition of the boundary condition is approximate, though convergent. Also, the shape functions of interior nodes vanish at the boundary, and the extreme points of X verify the strong Kronecker-delta property, in sharp

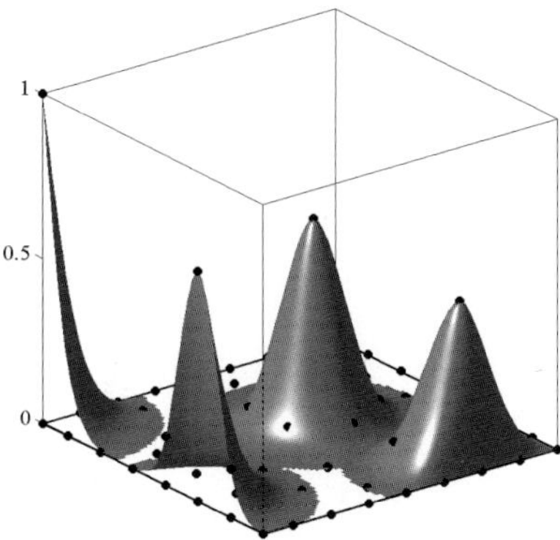

Figure 3. Illustration of the behavior of the local *max-ent* shape functions at the boundary of the domain.

contrast with MLS-based approximation schemes (see Figure 3 for an illustration). We note that, if the domain is a non-convex subset of conv$X$, this property does not hold in non-convex parts of the boundary and the method behaves similarly to MLS approximants. There is an intrinsic difficulty in dealing with non-convex domains in approximation methods that rely in their definition on a notion of distance, such as MLS approximants or the method presented here. The effective treatment of non-convex domains has been extensively studied in the context of MLS-based meshfree methods [15, 3, 12], and this methods are directly applicable in the present context.

3.2 Spatial Smoothness of the Shape Functions

We note that the local *max-ent* shape functions are defined point-wise, so their spacial smoothness must be established. Also, the thermalization parameter can take different values in space, and in general we consider it is given by a function $\beta(\boldsymbol{x})$.

Property 2. Let $\beta : \text{conv}X \rightarrow [0, \infty)$ *be* C^r *in* $\text{int}(\text{conv}X)$. *Then the local* max-ent *shape functions are of class* C^r *in* $\text{int}(\text{conv}X)$.

In particular, when β is constant, the shape functions are C^∞ and the derivative is given by

$$\nabla p_{\beta a}(\boldsymbol{x}) = -p_{\beta a}(\boldsymbol{x}) \boldsymbol{J}(\boldsymbol{x}, \boldsymbol{\lambda}^*(\boldsymbol{x}))^{-1} \ (\boldsymbol{x} - \boldsymbol{x}_a). \tag{3.13}$$

3.3 Smoothness and Limits with Respect to the Thermalization

Property 3. Let $\boldsymbol{x} \in \text{conv}X$. *Then* $\boldsymbol{p}_\beta(\boldsymbol{x})$ *is a* C^∞ *function of* β *in* $(0, +\infty)$. *Furthermore, the limits at* 0 *and* $+\infty$ *exist and are the* max-ent *approximants and the Pareto optimal Rajan approximants respectively*

$$\lim_{\beta\to 0} \boldsymbol{p}_\beta(\boldsymbol{x}) = \boldsymbol{p}_0(\boldsymbol{x}) \quad \textit{and} \quad \lim_{\beta\to +\infty} \boldsymbol{p}_\beta(\boldsymbol{x}) = \boldsymbol{p}_\infty^{Pareto}(\boldsymbol{x}).$$

This result is not surprising, the less obvious part being the limit at $+\infty$ when the node set is degenerate. In this case, the *max-ent* regularization of the Rajan program (RAJ) selects a distinguished element amongst the set of solutions, that does not coincide with any of the multiple Delaunay approximants but rather constitutes a generalized Delaunay approximation scheme for degenerate cases [1].

4 Other Interpretations: Relative Entropy

As noted in [1], the local *max-ent* approximants admit a number of interpretations, which include viewing the new approximants as a regularization of the Delaunay shape functions. A statistical mechanics interpretation is also quite appealing. The regularizing effect of the entropy can also be highlighted by studying the dual problem. In this case, the dual of the Rajan program reduces to the minimization of a polyhedral (non-smooth) convex function, while the dual of the local *max-ent* program is a smooth strictly convex minimization problem. This regularization can be understood through the approximation of the max function by a family of functions based on the log-sum-exp function [1, 5]. Figure 4 illustrates this view, and illustrates how the regularization provides an optimal path that selects a distinguished solution in cases of non-uniqueness, in the same spirit of viscosity solutions of variational problems.

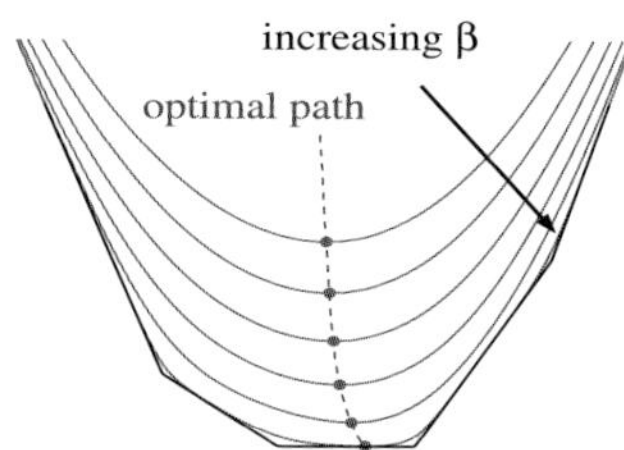

Figure 4. Illustration of the regularization provided by the entropy in the dual problem.

We next develop yet another reformulation of the local *max-ent* approximants, which allows for generalizations of *max-ent* approximation schemes and relies on the concept of *relative entropy*. The formulation of the relative *max-ent* approximants requires the definition of the so-called *Kullback-Leiber distance* between two discrete probability distributions $\boldsymbol{p}$ and $\boldsymbol{q}$

$$D_{\boldsymbol{p}|\boldsymbol{q}} = \sum_a p_a \log \frac{p_a}{q_a}.$$

Its negative is also referred to as *mutual entropy*, *cross entropy*, or *relative entropy*. Although $D_{\boldsymbol{p}|\boldsymbol{p}} = 0$, it is clear that this function is not symmetric in its arguments, and hence it is not strictly a distance. This quantity is used in information theory to measure the amount of information needed to change the description of the system from $\boldsymbol{q}$ to $\boldsymbol{p}$.

Often, the probability distribution $\boldsymbol{q}$ is viewed as *prior* information about the system. Then, some new information may become available, not accounted for by $\boldsymbol{q}$. A natural question in statistical inference is then to determine a new distribution $\boldsymbol{p}$ consistent with the new information, but which is in some sense as close as possible to the prior information. The maximization of the relative entropy between $\boldsymbol{p}$ and $\boldsymbol{q}$ subject to the new information made available provides a means to find such a distribution.

In the present context, suppose that we have a non-negative approximation scheme associated with the node set X, which satisfies the partition of unity property (required for the statistical interpretation) but does not satisfy the 1st order consistency condition. A simple example that comes to mind is a Shepard approximant with weight function $w(\cdot) \geq 0$ given by

$$q_a(\boldsymbol{x}) = \frac{w(\|\boldsymbol{x}_a - \boldsymbol{x}\|)}{\sum_b w(\|\boldsymbol{x}_b - \boldsymbol{x}\|)}.$$

If we regard this approximant as a prior, and wish to construct the closest approximation scheme in the sense of information theory that satisfies the 1st order consistency condition, i.e. the closest convex scheme, we face the convex program

$$\begin{aligned}
(RME)_{\boldsymbol{q}} \quad \text{For fixed } \boldsymbol{x} \text{ maximize} \quad & -\sum_{a=1}^{N} p_a \log \frac{p_a}{q_a} \\
\text{subject to} \quad & p_a \geq 0, \; a = 1, \ldots, N, \\
& \sum_{a=1}^{N} p_a = 1, \\
& \sum_{a=1}^{N} p_a \boldsymbol{x}_a = \boldsymbol{x}.
\end{aligned}$$

The above function could as well be combined with a locality measure, but this is not needed here. It is straightforward to verify that defining the partition function as

$$Z(\boldsymbol{x}, \boldsymbol{\lambda}) = \sum_{b=1}^{N} q_b(\boldsymbol{x}) \exp\left[\boldsymbol{\lambda} \cdot (\boldsymbol{x} - \boldsymbol{x}_b)\right],$$

the relative *max-ent* shape functions follow from

$$p_a(\boldsymbol{x}) = q_a(\boldsymbol{x}) \frac{\exp\left[\boldsymbol{\lambda}^*(\boldsymbol{x}) \cdot (\boldsymbol{x} - \boldsymbol{x}_a)\right]}{Z(\boldsymbol{x}, \boldsymbol{\lambda}^*(\boldsymbol{x}))}, \tag{4.14}$$

where

$$\boldsymbol{\lambda}^*(\boldsymbol{x}) = \arg \min_{\boldsymbol{\lambda} \in \mathbb{R}^d} \log Z(\boldsymbol{x}, \boldsymbol{\lambda}).$$

This procedure transforms any given nonnegative approximant that in general does not satisfy the 1st order consistency conditions and possesses no desirable properties at the boundary into a convex approximant that is 1st order consistent and satisfies the weak Kronecker-delta property at the boundary.

From Eq. (4.14) it is apparent that the relative *max-ent* shape functions inherit the smoothness and support of the prior approximants. Thus, defining relative *max-ent* approximants with compact support is straightforward: it is sufficient to start, for instance, from Shepard approximants based on a weight function with compact support. One should carefully note that the existence and uniqueness properties of the dual problem for the Lagrange multiplier are not granted for all priors, and in particular, there should be enough overlap between the supports of the prior shape functions. The relationship with the local *max-ent* approximants is elucidated by considering Shepard approximants with weight

$$w(d) = \exp(-\beta d^2)$$

as prior. In this case, the relative and the local *max-ent* approximants coincide.

5 Applications in Galerkin Methods

We present here briefly a 3D example of nonlinear elasticity which illustrates the benefits of using smooth shape functions such as the local *max-ent* shape functions in problems with smooth solutions. A Galerkin (or Rayleigh-Ritz in this setting) method is implemented. More details, as well as other benchmark tests of linear elasticity can be found in [1]. Despite the calculation of the local *max-ent* shape functions is computationally more expensive than evaluating the Delaunay shape functions, this example shows how overall, for a give accuracy, remarkable computational saving can be achieved by using the meshfree method.

We consider a hyperelastic compressible Neo-Hookean block subject to a 100% tensile deformation. The calculation is performed with seven uniform node sets of variable resolution. For each node set, two numerical solutions are obtained, one with the local *max-ent* approximants, and another one with the Delaunay approximants (linear simplicial finite elements). Figure 5 shows the dependence of a normalized signed relative error in strain energy (relative to an overkill numerical solution) on the nodal spacing. It is observed from that figure that the accuracy of the local *max-ent* solution is vastly superior to that of the finite element solution. The finest finite element solution has a comparable—albeit slightly larger—error than the second-coarsest local *max-ent* solution. By contrast, the CPU time incurred by the local *max-ent* solution is over a hundred times shorter than that of the finite element solution. This

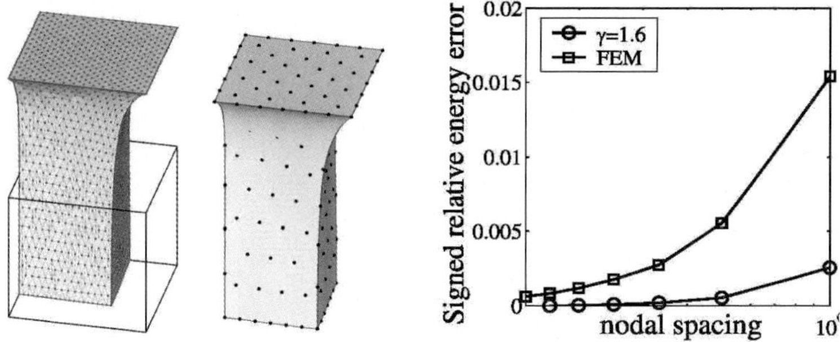

Figure 5. Final deformation for the finest FE mesh and second-coarsest local *max-ent* discretization (left) and signed relative error in strain energy with respect to a reference numerical solution for $\nu_0 = 0.333$ (right).

difference in performance is more pronounced in the nearly incompressible case.

6 Conclusions

We have presented a new approach for building approximation schemes from scattered data using the concept of maximum entropy. This concept not only provides a natural link between approximation and information theory, but also offers a practical computational means to construct a family of approximants that seamlessly and smoothly bridges meshfree-style shape functions and Delaunay shape functions. The theory, computation and properties of these approximants has been provided. Numerical examples emphasize the notable computational savings afforded by smooth approximantion schemes in problems with smooth solutions.

The method presented here allows for extensions that have not been pursued here. These include the approximation in high-dimensional problems, the ability to perform in a very flexible way variational adaptivity, not only of the node positions but also of the thermalization parameter, and the development of higher-order *max-ent* schemes. The latter provide one of the few unstructured non-negative high-order approximants.

References

1. M. Arroyo and M. Ortiz, *Local maximum-entropy approximation schemes: a seamless bridge between finite elements and meshfree methods*, International Journal for Numerical Methods in Engineering **in press** (2005).
2. ———, *Maximum-entropy approximation schemes: higher order methods*, International Journal for Numerical Methods in Engineering **in preparation** (2005).

3. T. Belytschko, Y. Krongauz, D. Organ, M. Fleming, and P. Krysl, *Meshless methods: An overview and recent developments*, Computer Methods in Applied Mechanics and Engineering **139** (1996), no. 1-4, 3–47.
4. T. Belytschko, Y.Y. Lu, and L. Gu, *Element-free Galerkin methods*, International Journal for Numerical Methods in Engineering **37** (1994), no. 2, 229–256.
5. S. Boyd and L. Vandenberghe, *Convex optimization*, Cambridge University Press, Cambridge, UK, 2004.
6. P. Breitkopf, A. Rassineux, J.M. Savignat, and P. Villon, *Integration constraint in diffuse element method*, Computer Methods in Applied Mechanics and Engineering **193** (2004), no. 12-14, 1203–1220.
7. J.S. Chen, C.T. Wu, S. Yoon, and Y. You, *Stabilized conforming nodal integration for Galerkin meshfree methods*, International Journal for Numerical Methods in Engineering **50** (2001), 435466.
8. F. Cirak, M. Ortiz, and P. Schröder, *Subdivision surfaces: a new paradigm for thin-shell finite-element analysis*, International Journal for Numerical Methods in Engineering **47** (2000), no. 12, 2039–2072.
9. C. Cottin, I. Gavrea, H.H. Gonska, D.P. Kacsó, and D.X. Zhou, *Global smoothness preservation and variation-diminishing property*, Journal of Inequalities and Applications **4** (1999), no. 2, 91–114.
10. R.A. DeVore, *The approximation of continuous functions by positive linear operators*, Springer-Verlag, Berlin, 1972.
11. A. Huerta, T. Belytscko, S. Fernández-Méndez, and T. Rabczuk, *Encyclopedia of computational mechanics*, vol. 1, ch. Meshfree methods, pp. 279–309, Wiley, Chichester, 2004.
12. P. Krysl and T. Belytschko, *Element-free galerkin method: Convergence of the continuous and discontinuous shape functions*, Computer Methods in Applied Mechanics and Engineering **148** (1997), no. 3-4, 257–277.
13. W.K. Liu, S. Li, and T. Belytschko, *Moving least square reproducing kernel methods Part I: Methodology and convergence*, Computer Methods in Applied Mechanics and Engineering **143** (1997), no. 1-2, 113–154.
14. B. Nayroles, G. Touzot, and P. Villon, *Generalizing the finite element method: diffuse approximation and diffuse elements*, Computational Mechanics **10** (1992), no. 5, 307–318.
15. D. Organ, M. Fleming, T. Terry, and T. Belytschko, *Continuous meshless approximations for nonconvex bodies by diffraction and transparency*, Computational Mechanics **18** (1996), no. 3, 225–235.
16. H. Prautzch, W. Boehm, and M. Paluszny, *Bézier and B-spline techniques*, Springer-Verlag, Berlin, 2002.
17. V.T. Rajan, *Optimality of the Delaunay triangulation in* R^d, Discrete and Computational Geometry **12** (1994), no. 2, 189–202.
18. N. Sukumar, *Construction of polygonal interpolants: A maximum entropy approach*, International Journal for Numerical Methods in Engineering **61** (2004), no. 12, 2159–2181.
19. N. Sukumar, B. Moran, and T. Belytschko, *The natural element method in solid mechanics*, International Journal for Numerical Methods in Engineering **43** (1998), no. 5, 839–887.

Genetic Algorithms for Meshfree Numerical Integration

Suleiman BaniHani and Suvranu De

Rensselaer Polytechnic Institute, 110 8th st., Troy, NY, USA
{banihs,des}@rpi.edu

Summary. In this paper we present the application of the meshfree method of finite spheres to the solution of thin and thick plates composed of isotropic as well as functionally graded materials. For the solution of such problems it is observed that using Gaussian and adaptive quadrature schemes are computationally inefficient. In this paper a new technique, presented in [26, 21], in which the integration points and weights are generated using genetic algorithms and stored in a lookup table using normalized coordinates as part of an offline preprocessing step, is shown to provide significant reduction of computational time without sacrificing accuracy.

Key words: Meshfree methods, method of finite spheres (MFS), plates, functionally graded material (FGM), genetic algorithms (GA), Gaussian quadrature.

1 Introduction

Due to several advantages over traditional finite element methods, meshfree computational schemes, reviewed in [1], have been proposed. A major advantage of these techniques is that approximation spaces with higher order continuity may be easily generated. Hence it is useful to apply them to the solution of higher order differential equations. In this paper we present an application of the method of finite spheres [2] to the solution of plate problems. Several examples involving thin and thick elastic and functionally graded material plate problems are discussed.

In [4] the moving least squares interpolants were used for the solution of thin elastic plate problems with simply supported and clamped edges. A moving least squares differential quadrature scheme was used in [5] for analyzing thick plates with shear deformation. In [6] radial basis functions were used for solving Kirchhoff plate bending problem with the Hermite collocation approach being used to obtain a system of symmetric and non-singular linear equations.

The hp-clouds method was used in [3] for the analysis of Kirchhoff plates with enrichment being provided by Trefftz functions. Lagrange multipliers were used to apply the essential boundary conditions. The MLPG method was used in [7] for solving the problem of a thin plate in bending with the penalty formulation being used to apply the essential boundary conditions. In [8] the element free Galerkin method was used to solve Kirchhoff plate problems.

In [22] spline functions were used together with a displacement based Galerkin method for the solution of Mindlin-Reissner plate problems. A uniform nodal arrangement was used with nodes lying outside the computational domain.

In [23] the hp-clouds method was employed for the solution of thick plates and it was noticed that using hp-clouds approximations of sufficiently high polynomial degree could control shear locking. However, such approximations are computationally expensive to generate.

In [24] a mesh-independent p-orthotropic enrichment in the generalized finite element method (GFEM) was presented for the solution of the Reissner-Mindlin plate model. In [25] an extended meshfree method was used for solving plate problems involving shear deformations, where the problem was reduced to a homogeneous one. The homogeneous equation was subsequently solved using the reproducing kernel particle method (RKPM).

Functionally graded materials (FGMs) are composites containing volume fractions of two or more materials varying continuously as functions of position along the structure dimensions.

In [29] a theoretical formulation, Navier's solutions of rectangular plates and finite element models based on the third-order shear deformation plate theory, was presented for the analysis of through-thickness functionally graded plates. Numerical results of the linear third-order theory and non-linear first order theory were presented.

In the MLPG method the interpolation of choice is the moving least squares method, which is computationally expensive [9]. Furthermore, to embed derivative information in the interpolation, a "generalized moving least squares" scheme is used, analogous to Hermite interpolations, which results in computationally expensive shape functions.

It is interesting to note that while many meshfree methods have been developed and some have been applied to higher order differential equations, the issue of numerical efficiency has not been given due attention. Most Galerkin-based meshfree methods are notoriously computationally inefficient since rational, nonpolynomial interpolation functions are used and the integration domains are much more complex than in the finite element techniques.

For the solution of higher order differential equations one faces the challenge of efficient numerical integration of the terms in the Galerkin weak form where the integrands are highly peaked in the regions of overlap of the supports of the interpolation functions and many integration points are required unless proper care is taken.

With the goal of achieving computational efficiency in meshfree methods, the method of finite spheres was developed [9]. In this technique, the interpolation functions generated using the partition of unity paradigm [10, 11] are compactly supported on intersecting and overlapping spheres. In [26] we showed that using piece-wise midpoint quadrature, for higher order deferential equations arising in the solution of thin beam and plate problems, is computationally inaccurate and will not converge if h-refinement is performed. Also adaptive numerical integration methods [13, 14], on the other hand, provide much better results but are computationally expensive.

In this paper we employ the genetic algorithm-based lookup table approach, originally proposed in [26, 21], for efficient numerical integration of plate problems. In this technique, the integration points are computed, off line, for a reference nodal configuration using genetic algorithms. The location of these integration points as well as the weights are then nondimensionalized and stored in a lookup table. During computations, the integration points and weights are retrieved from this lookup table. This scheme is used for the solution of several plate problems to demonstrate its accuracy.

In Section 2 we formulate the governing equations for the different plate models followed by a brief review of the MFS discretization scheme. In Section 3 we discuss the genetic algorithm-based lookup table approach. In Section 4 we provide some numerical results demonstrating the effectiveness of the integration scheme.

2 Problem Formulation and Discretization

2.1 Governing Equations

Kirchhoff Plate

We consider a thin isotropic plate as shown in Figure 1 of thickness h and midsurface ($\Omega \in \mathbb{R}^2$), elastic modulus E and Poisson's ratio ν. The boundary of the plate is denoted by Γ. The plate is acted upon by a transverse loading $q(x, y)$.

The corresponding potential energy functional for this problem is [16]

$$\pi_{KP}(w) = U_{\text{int}}(w) - W_{ext}(w) \tag{2.1}$$

where $U_{\text{int}}(w)$ is the internal energy given by

$$U_{\text{int}}(w) = \frac{1}{2}\int\int_{\Omega} (\mathbf{P}w)^{\mathrm{T}}\mathbf{D}(\mathbf{P}w)dxdy \tag{2.2}$$

where $\mathbf{P} = [\partial^2/\partial x^2 \partial^2/\partial y^2 2\partial^2/\partial x\partial y]^{\mathrm{T}}$, and $\mathbf{D} = \frac{Eh^3}{12(1-\nu^2)}\begin{bmatrix} 1 & \nu & 0 \\ \nu & 1 & 0 \\ 0 & 0 & (1-\nu)/2 \end{bmatrix}$. $\Gamma = \Gamma_w \cup \Gamma_\theta \cup \Gamma_M \cup \Gamma_{Q_{eff}}$ are portions of the boundaries where the displacement ($\hat{w}$), rotation ($\hat{\theta}$), moment ($\hat{M}$), and effective shear force ($\hat{Q}_{eff}$)

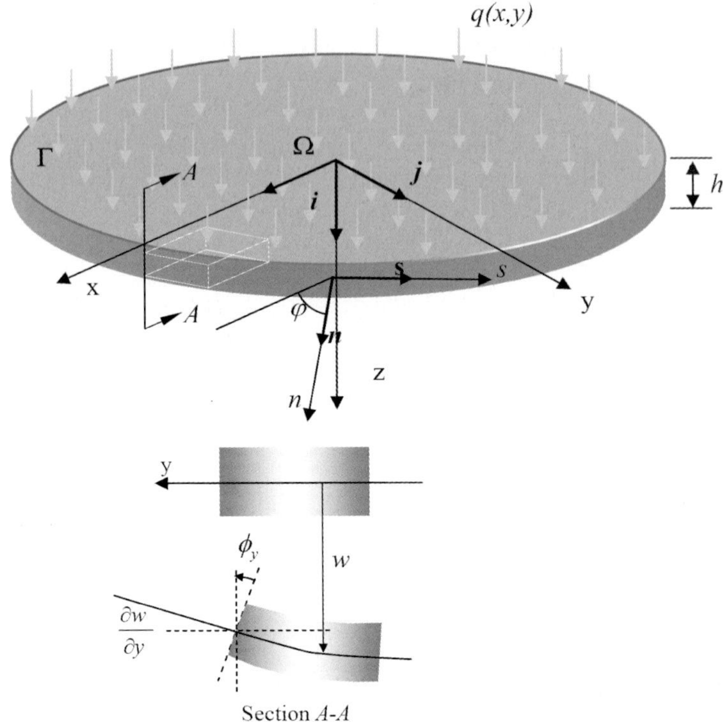

Figure 1. A flat plate in bending. ***(i,j)*** are unit vectors in the global x- and y-directions, respectively. ***(s, n)*** are unit vectors along the tangential and normal directions to the plate boundary, respectively.

are prescribed, respectively. The boundary condition sets (w and Q_{eff}) and (θ and M_n) are disjoint.

The external work $W_{ext}(w)$ has three components due to the applied lateral loads, applied moments and transverse shear, and corner loads

$$W_{ext}(w) = W_{load}(w) + W_{bending}(w) + W_{corners}(w). \tag{2.3}$$

The first two components are

$$W_{load}(w) = \int \int_{\Omega} q(x,y) w dx dy \tag{2.4}$$

$$W_{bending}(w) = \oint_{\Gamma_{Q_{eff}}} \hat{Q}_{eff} w d\Gamma - \oint_{\Gamma_M} \hat{M}_n \frac{\partial w}{\partial n} d\Gamma \tag{2.5}$$

$W_{corners}(w)$ exists when there are N corners where the displacements w_j at the corners are not prescribed

$$W_{corners}(w) = \sum_{j=1}^{N} (\hat{M}_{ns}^{+} - \hat{M}_{ns}^{-}) w_j \tag{2.6}$$

where $\hat{M}_{ns}^{+}, \hat{M}_{ns}^{-}$ are the twisting moments approaching the edge from the right and the left, respectively and w_j is the displacement at the jth corner.

The essential boundary conditions are imposed by defining the augmented functional

$$\pi_{KP}^{*}(w) = \pi_{KP}(w) + \frac{\gamma_1}{2} \int_{\Gamma_w} (w - \hat{w})^2 \, d\Gamma + \frac{\gamma_2}{2} \int_{\Gamma_\theta} \left(\frac{\partial w}{\partial n} - \hat{\theta} \right)^2 d\Gamma \tag{2.7}$$

where γ_1 and γ_2 are penalty parameters.

Mindlin-Reissner Plate

We consider a thick isotropic plate as shown in Figure 1 of thickness h. The corresponding potential energy functional for this problem is [16]

$$\pi_{MP}(w, \theta) = U_{\text{int}}(w, \theta) - W_{ext}(w, \theta) \tag{2.8}$$

where $U_{\text{int}}(w, \theta)$ is the internal energy given by

$$U_{\text{int}}(w, \theta) = \frac{1}{2} \int \int_{\Omega} (\mathbf{L}\theta)^T \mathbf{D} (\mathbf{L}\theta) dx dy + \frac{1}{2} \int \int_{\Omega} (\nabla w - \theta)^T \alpha (\nabla w - \theta) dx dy \tag{2.9}$$

where $\mathbf{L} = \begin{bmatrix} \frac{\partial}{\partial x} & 0 \\ 0 & \frac{\partial}{\partial y} \\ \frac{\partial}{\partial y} & \frac{\partial}{\partial x} \end{bmatrix}$, $\mathbf{D} = \frac{Eh^3}{12(1-\nu^2)} \begin{bmatrix} 1 & \nu & 0 \\ \nu & 1 & 0 \\ 0 & 0 & (1-\nu)/2 \end{bmatrix}$, $\theta = \begin{bmatrix} \phi_x \\ \phi_y \end{bmatrix}$, $\nabla = \begin{bmatrix} \frac{\partial}{\partial x} \\ \frac{\partial}{\partial y} \end{bmatrix}$,
and $\alpha = \begin{bmatrix} kGh & 0 \\ 0 & kGh \end{bmatrix}$. D is the bending rigidity, $\Gamma_w, \Gamma_{\phi_x}, \Gamma_{\phi_y}, \Gamma_{M_n}, \Gamma_{M_{ns}}$, and Γ_{Q_n} are the boundaries where the displacement ($\hat{w}$), rotations ($\hat{\phi_x}$ and $\hat{\phi_y}$), moments ($\hat{M_n}$ and $\hat{M_{ns}}$), and shear force ($\hat{Q_n}$) are prescribed, respectively. The boundary condition sets (w and Q_n), (ϕ_n and M_n), and (ϕ_s and M_{ns}) are disjoint.

The external work $W_{ext}(w, \theta)$ has four components due to the applied lateral loads, applied moments and transverse shear

$$W_{ext}(w, \theta) = \int_{\Omega} q(x, y) w dx dy + \oint_{\Gamma_{M_n}} \hat{M}_n \phi_n d\Gamma + \oint_{\Gamma_{M_{ns}}} \hat{M}_{ns} \phi_s d\Gamma + \oint_{\Gamma_{Q_n}} \hat{Q}_n w d\Gamma. \tag{2.10}$$

The essential boundary conditions are imposed by defining the augmented functional

$$\pi^*_{MP}(w,\theta) = \pi_{MP}(w,\theta) + \frac{\gamma_1}{2}\int_{\Gamma_w} (w-\hat{w})^2\, d\Gamma + \frac{\gamma_2}{2}\int_{\Gamma_{\phi_x}} \left(\phi_x - \hat{\phi}_x\right)^2 d\Gamma + \frac{\gamma_3}{2}\int_{\Gamma_{\phi_y}} \left(\phi_y - \hat{\phi}_y\right)^2 d\Gamma \tag{2.11}$$

where γ_1 and γ_2 are penalty parameters.

Functionally Graded Material Plate Model

We assume that the material property gradation is through the thickness and that the volume fraction is given by the following power law

$$V(z) = (V_t - V_b)\left(\frac{z}{h} + \frac{1}{2}\right)^n + V_b \tag{2.12}$$

where V is the material property, V_t and V_b are the values at the top and bottom faces of the plate, respectively, h is the thickness, and n is a constant that dictates the volume fraction profile through the thickness. The material properties are homogenized using the Mori-Tanaka scheme described in [30]. Here we assume that the modulus E, density ρ, and thermal coefficient of expansion α vary through the thickness, while ν is assumed constant.

The constitutive relations are [29]

$$\begin{Bmatrix} \sigma_{xx} \\ \sigma_{yy} \\ \sigma_{xy} \end{Bmatrix} = \frac{E}{1-\nu^2}\begin{bmatrix} 1 & \nu & 0 \\ \nu & 1 & 0 \\ 0 & 0 & \frac{1-\nu}{2} \end{bmatrix}\left(\begin{Bmatrix} \epsilon_{xx} \\ \epsilon_{yy} \\ \gamma_{xy} \end{Bmatrix} - \begin{Bmatrix} 1 \\ 1 \\ 0 \end{Bmatrix}\alpha\Delta T\right) \tag{2.13}$$

and

$$\begin{Bmatrix} \sigma_{yz} \\ \sigma_{xz} \end{Bmatrix} = \frac{E}{1-\nu^2}\begin{bmatrix} \frac{1-\nu}{2} & 0 \\ 0 & \frac{1-\nu}{2} \end{bmatrix}\begin{Bmatrix} \gamma_{yz} \\ \gamma_{xz} \end{Bmatrix} \tag{2.14}$$

where ΔT is the temperature change, and

$$\begin{Bmatrix} \epsilon_{xx} \\ \epsilon_{yy} \\ \gamma_{xy} \end{Bmatrix} = z\begin{Bmatrix} \frac{\partial \phi_x}{\partial x} \\ \frac{\partial \phi_y}{\partial y} \\ \frac{\partial \phi_x}{\partial y} + \frac{\partial \phi_y}{\partial x} \end{Bmatrix}$$

$$\begin{Bmatrix} \gamma_{yz} \\ \gamma_{xz} \end{Bmatrix} = \begin{Bmatrix} \phi_y + \frac{\partial w}{\partial y} \\ \phi_x + \frac{\partial w}{\partial x} \end{Bmatrix}.$$

The potential energy functional corresponding to this problem can be written as

$$\Pi = U_{int} - W_{ext} \tag{2.15}$$

where U_{int} is the internal energy given by

$$U_{int} = U_b + U_s \tag{2.16}$$

U_b and U_s are the bending and shear energy, respectively, given by

$$U_b = \frac{1}{2}\int_{\Omega} [\epsilon_{xx}\epsilon_{yy}\gamma_{xy}] \begin{bmatrix} \sigma_x \\ \sigma_y \\ \sigma_{xy} \end{bmatrix} d\Omega \tag{2.17}$$

and

$$U_s = \frac{k}{2}\int_{\Omega} [\gamma_{xz}\gamma_{yz}] \begin{bmatrix} \sigma_{xz} \\ \sigma_{yz} \end{bmatrix} d\Omega \tag{2.18}$$

and k is the shear correction factor. W_{ext} is the external work done by the applied force given by

$$W_{ext} = \int_{\Omega} qw d\Omega \tag{2.19}$$

where q is the external load applied to the upper surface of the plate. The corresponding potential energy functional in equation (2.15) can be expressed in terms of the displacements as

$$\begin{aligned} \Pi =& \frac{1}{2}C_1 \int_{\Omega} \left[\frac{\partial \phi_x}{\partial x}^2 + \frac{\partial \phi_y}{\partial y}^2 + 2\nu \frac{\partial \phi_x}{\partial x}\frac{\partial \phi_y}{\partial y} + \frac{1-\nu}{2}(\frac{\partial \phi_x}{\partial y} + \frac{\partial \phi_y}{\partial x})^2 \right] d\Omega \\ & + \frac{k}{2}C_2 \int_{\Omega} \left[(\frac{\partial w}{\partial x} + \phi_x)^2 + (\frac{\partial w}{\partial y} + \phi_y)^2 \right] d\Omega \\ & - \frac{1}{2}\int_{\Omega} (\frac{\partial \phi_x}{\partial x} M^t + \frac{\partial \phi_y}{\partial y} M^t) d\Omega - \int_{\Omega} qw d\Omega \end{aligned} \tag{2.20}$$

where

$$C_1 = \int_{-h/2}^{h/2} \frac{E}{1-\nu^2} z^2 dz$$

and

$$C_2 = \int_{-h/2}^{h/2} \frac{E}{2(1+\nu)} dz$$

M^t is the the thermal moment resultant given by

$$M^t = \frac{1}{1-\nu}\int_{-h/2}^{h/2} \alpha E T z dz. \tag{2.21}$$

The essential boundary conditions are imposed by defining the augmented functional

$$\Pi^* = \Pi + \frac{\gamma_1}{2}\int_{\Gamma_w} (w - \hat{w})^2 d\Gamma + \frac{\gamma_2}{2}\int_{\Gamma_{\phi_x}} (\phi_x - \hat{\phi_x})^2 d\Gamma + \frac{\gamma_3}{2}\int_{\Gamma_{\phi_y}} (\phi_y - \hat{\phi_y})^2 d\Gamma \tag{2.22}$$

where γ_1, γ_2, and γ_3 are penalty parameters. $\Gamma_w, \Gamma_{\phi_x}$, and Γ_{ϕ_y} are the boundaries where $\hat{w}, \hat{\phi_x}$, and $\hat{\phi_y}$ are prescribed, respectively.

Thermal Analysis

In this section we discuss the thermal analysis of functionally graded plates. It is assumed that the top and bottom surfaces of the plate are subjected to constant temperature. The temperature field within the plate depends only on the z-coordinate. The temperature distribution can be obtained by solving the one-dimensional steady state heat transfer equation

$$-\frac{d}{dz}\left(\lambda(z)\frac{dT}{dz}\right)=0 \tag{2.23}$$

with the boundary conditions

$$\begin{array}{l} T=T_t \ at \ \ z=h/2, \\ T=T_b \ at \ z=-h/2 \end{array} \tag{2.24}$$

where λ is the thermal conductivity which varies according to the power law distribution, given by

$$\lambda(z)=(\lambda_t-\lambda_b)\left(\frac{2z+h}{2h}\right)^n+\lambda_b. \tag{2.25}$$

The solution of equation (2.23) with the prescribed boundary conditions is

$$T(z)=T_t-\frac{T_t-T_b}{\int_{-h/2}^{h/2}\frac{dz}{\lambda(z)}}\int_z^{h/2}\frac{d\eta}{\lambda(\eta)}. \tag{2.26}$$

This equation is used for the evaluation of M^t in equation (2.21).

2.2 The Method of Finite Spheres Approximation Scheme

In the method of finite spheres the approximation functions are generated using the partition of unity paradigm [11] and are supported on spheres. In this section we briefly describe this method.

Let $\Omega \in R^d$ (d =1, 2, or 3) be an open bounded domain with Γ as its boundary (Figure 2), and let a set of open spheres $\{B(\mathbf{x}_I,r_I); I=1,2,...,N\}$ cover the entire domain Ω, where $\mathbf{x}_I$ and r_I is the sphere I center and radius, respectively. Each node is placed at the geometric center of a sphere, and the surface of the sphere I is denoted by $S(\mathbf{x}_I,r_I)$.

Since we are interested in solving fourth order differential equations, we define a positive radial weight function $W_I(\mathbf{x})=W(s_I)\in C_0^s(B(\mathbf{x}_I,r_I)), s\geq 1$ with $s_I=||\mathbf{x}-\mathbf{x}_I||_0/r_I$ at each node I which is compactly supported on the sphere at node I. In our work we used a 7th order spline weight functions

$$W_I=\begin{cases}1-35s^4+84s^5-70s^6+20s^7 & 0\leq s<1\\ 0 & s\geq 1\end{cases} \tag{2.27}$$

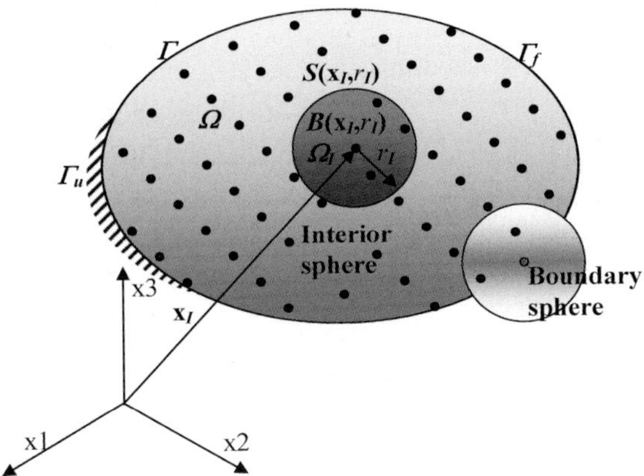

Figure 2. Schematic of the MFS. Discretization of a domain Ω in $\mathbb{R}^2$ by the method of finite spheres, using a set of nodes, for each node I there is a sphere Ω_I . A sphere that lies completely in the domain is an interior sphere, while if the sphere has an intersection with the boundary, Γ, then it is a boundary sphere. The natural boundary conditions are defined on Γ_f, and the essential boundary conditions are defined on $\Gamma_u;\Gamma = \Gamma_u \cup \Gamma_f$ and $\Gamma_u \cap \Gamma_f = 0$.

These functions are utilized to generate the Shepard partition of unity functions

$$\varphi_I^0(\mathbf{x}) = \frac{W_I}{\sum\limits_{J=1}^{N} W_J}, I = 1, 2, ..., N \tag{2.28}$$

which satisfy

1. $\sum\limits_{I=1}^{N} \varphi_I^0(\mathbf{x}) = 1 \ \forall \mathbf{x} \in \Omega$
2. $\varphi_I^0(\mathbf{x}) \in C_0^s(\mathbb{R}^d), \ s \geq 1.$

The Shepard functions $\{\varphi_I^0(\mathbf{x})\}$ satisfy zeroth order consistency. To ensure higher order consistency, a local approximation space $V_I^h = \text{span}_{m\in\Im}\{P_m(\mathbf{x})\}$ is defined at each node I, where $P_m(\mathbf{x})$ is a polynomial and $\Im$ is an index set, and h is a measure of the size of the spheres. For instance,

$$V_I^h = \{1, (x - x_I)/r_I, (y - y_I)/r_I\}$$

may be used to generate a linearly accurate displacement field.

The global approximation space V_h is generated by multiplying the partition of unity functions with the local basis functions at each node

$$V_h = \sum_{I=1}^{N} \varphi_I^0 V_I^h. \tag{2.29}$$

Therefore, any function $w_h \in V_h$ can be written as

$$w_h(\mathbf{x}) = \sum_{I=1}^{N} \sum_{m \in \Im} h_{\mathrm{Im}}(\mathbf{x}) \alpha_{\mathrm{Im}} \tag{2.30}$$

where

$$h_{\mathrm{Im}} = \varphi_I^0(\mathbf{x}) P_m(\mathbf{x})$$

is the shape function associated with the mth degree of freedom $\alpha_{Im}$of node I. It is important to note that

$$h_{\mathrm{Im}}(\mathbf{x}) \in C_0^s(\mathbb{R}^d), s \geq 1.$$

2.3 Discretized Equations

Kirchhoff Plate

For the Kirchhoff plate, using the discretization

$$w_h(x, y) = \sum_{I=1}^{N} \sum_{m \in \Im} h_{\mathrm{Im}}(x, y) \alpha_{\mathrm{Im}} = bf H(x, y) \alpha \tag{2.31}$$

the augmented potential energy functional (2.1) is approximated by

$$\begin{aligned} \pi_{KP}^*(\alpha) = \tfrac{1}{2}\alpha^{\mathrm{T}}\mathbf{K}\alpha - \alpha^{\mathrm{T}}\mathbf{f} + \tfrac{\gamma_1}{2} \int\limits_{\Gamma_w} (\mathbf{H}\alpha - \hat{w})^T (\mathbf{H}\alpha - \hat{w})\, d\Gamma \\ + \tfrac{\gamma_2}{2} \int\limits_{\Gamma_\theta} \left(\tfrac{\partial \mathbf{H}}{\partial n}\alpha - \hat{\theta}\right)^T \left(\tfrac{\partial \mathbf{H}}{\partial n}\alpha - \hat{\theta}\right) d\Gamma \end{aligned} \tag{2.32}$$

where

$$\mathbf{K} = \int\int_{\Omega} \mathbf{B}^{\mathrm{T}} \mathbf{D} \mathbf{B} dx dy \tag{2.33}$$

$$\mathbf{f} = \int\int_{\Omega} \mathbf{H}^{\mathrm{T}} q(x, y) dx dy - \int\limits_{\Gamma_M} \hat{M}_n \frac{\partial \mathbf{H}^{\mathrm{T}}}{\partial n} d\Gamma + \int\limits_{\Gamma_{Q_{eff}}} \hat{Q}_{eff} \mathbf{H}^{\mathrm{T}} d\Gamma + \sum_{j=1}^{N} R_c \mathbf{H}_j^{\mathrm{T}} \tag{2.34}$$

where $\mathbf{B} = \mathbf{PH}$.

Minimizing the augmented potential energy functional with respect to the nodal unknowns results in the following set of linear algebraic equations

$$(\mathbf{K} + \mathbf{G}_1 + \mathbf{G}_2)\, \alpha = \mathbf{f} + \mathbf{f}_{s_1} + \mathbf{f}_{s_2} \tag{2.35}$$

where

$$\mathbf{G}_1 = \gamma_1 \int\limits_{\Gamma_w} (\mathbf{H}^{\mathrm{T}} \mathbf{H})\, d\Gamma$$

$$\mathbf{G}_2 = \gamma_2 \int\limits_{\Gamma_\theta} \left(\frac{\partial \mathbf{H}^{\mathrm{T}}}{\partial n} \frac{\partial \mathbf{H}}{\partial n} \right) d\Gamma$$

$$\mathbf{f}_{s_1} = \gamma_1 \int\limits_{\Gamma_w} \left(\hat{w} \mathbf{H}^{\mathrm{T}} \right) d\Gamma$$

and

$$\mathbf{f}_{s_2} = \gamma_2 \int\limits_{\Gamma_\theta} \left(\hat{\theta} \frac{\partial \mathbf{H}^{\mathrm{T}}}{\partial n} \right) d\Gamma.$$

Mindlin-Reissner Plate

For the Mindlin-Reissner plate, using the method of finite spheres discretization

$$\begin{aligned} w_h(x, y) &= \sum_{I=1}^{N} \sum_{m \in \Im} h_{\mathrm{Im}}(x, y) \tilde{w}_{\mathrm{Im}} = H_w(x, y) \tilde{w} \\ \theta_h(x, y) &= \sum_{I=1}^{N} \sum_{m \in \Im} N_{\mathrm{Im}}(x, y) \tilde{\theta}_{\mathrm{Im}} = H_\theta(x, y) \tilde{\theta} \end{aligned} \tag{2.36}$$

where $H_\theta = \begin{bmatrix} H_{\phi_x} & 0 \\ 0 & H_{\phi y} \end{bmatrix}$, and $\tilde{\theta}^T = \left[\tilde{\phi}_{\mathrm{x}}, \tilde{\phi}_{\mathrm{y}} \right]$. Using (2.34) we may discretize the augmented potential energy functional (2.15) as

$$\begin{aligned} \pi^*_{MP}(a) = \tfrac{1}{2} a^{\mathrm{T}} \mathbf{K} a - a^{\mathrm{T}} \mathbf{f} &+ \tfrac{\gamma_1}{2} \int\limits_{\Gamma_w} \left(H_w \tilde{w} - \hat{w} \right)^T \left(H_w \tilde{w} - \hat{w} \right) d\Gamma \\ &+ \tfrac{\gamma_2}{2} \int\limits_{\Gamma_{\phi x}} \left(H_{\phi_x} \tilde{\phi}_x - \hat{\phi}_x \right)^T \left(H_{\phi_x} \tilde{\phi}_x - \hat{\phi}_x \right) d\Gamma \\ &+ \tfrac{\gamma_3}{2} \int\limits_{\Gamma_{\phi y}} \left(H_{\phi_y} \tilde{\phi}_y - \hat{\phi}_y \right)^T \left(H_{\phi_y} \tilde{\phi}_y - \hat{\phi}_y \right) d\Gamma \end{aligned} \tag{2.37}$$

where

$$\begin{aligned} \mathbf{K} &= \mathbf{K}_b + \mathbf{K}_s \\ a^T &= \left[\tilde{w}, \tilde{\theta} \right] \end{aligned}$$

and

$$\mathbf{f} = \begin{bmatrix} \mathbf{f}_w \\ \mathbf{f}_\theta \end{bmatrix}.$$

The matrices are defined as

$$\mathbf{K}_b = \begin{bmatrix} 0 & 0 \\ 0 & \mathbf{K}^b_{\theta\theta} \end{bmatrix} \tag{2.38}$$

$$\mathbf{K}_s = \begin{bmatrix} \mathbf{K}^s_{ww} & \mathbf{K}^s_{w\theta} \\ \mathbf{K}^s_{\theta w} & \mathbf{K}^s_{\theta\theta} \end{bmatrix} \tag{2.39}$$

with

$$\begin{aligned}
\mathbf{K}^s_{ww} &= \int\int_{\Omega} (\nabla H_w)^{\mathrm{T}} \alpha \nabla H_w dxdy \\
\mathbf{K}^s_{\theta w} &= -\int\int_{\Omega} H_\theta{}^{\mathrm{T}} \alpha \nabla H_w dxdy = (\mathbf{K}^s_{w\theta})^{\mathrm{T}} \\
\mathbf{K}^s_{\theta\theta} &= \int\int_{\Omega} H_\theta{}^{\mathrm{T}} \alpha H_\theta dxdy \\
\mathbf{K}^b_{\theta\theta} &= \int\int_{\Omega} (LH_\theta)^{\mathrm{T}} DLH_\theta dxdy \\
\mathbf{f}_w &= \int\int_{\Omega} \mathbf{H}^{\mathrm{T}}_w q(x,y) dxdy + \int_{\Gamma_{Q_n}} \hat{Q}_n \mathbf{H}^{\mathrm{T}}_w d\Gamma \\
\mathbf{f}_\theta &= \int_{\Gamma_M} \mathbf{H}^{\mathrm{T}}_\theta \hat{M} d\Gamma
\end{aligned}$$

where $\hat{M}^{\mathrm{T}} = \left[\hat{M}_{\mathrm{n}}, \hat{M}_{\mathrm{ns}}\right]$.

Minimizing the augmented potential energy functional with respect to the nodal unknowns results in the following set of linear algebraic equations

$$(\mathbf{K} + \mathbf{G})\,\mathbf{a} = \mathbf{f} + \mathbf{f}_s \tag{2.40}$$

where

$$\mathbf{G} = \begin{bmatrix} \mathbf{G}_w & 0 \\ 0 & \mathbf{G}_\theta \end{bmatrix}$$

and $\mathbf{f}_s = \begin{bmatrix} \mathbf{f}_{s_w} \\ \mathbf{f}_{s_\theta} \end{bmatrix}$ with

$$\mathbf{G}_w = \gamma_1 \int_{\Gamma_w} \left(\mathbf{H}^{\mathrm{T}}\mathbf{H}\right) d\Gamma \tag{2.41}$$

$$\mathbf{G}_\theta = \begin{bmatrix} \mathbf{G}_{\phi_x} & 0 \\ 0 & \mathbf{G}_{\phi_y} \end{bmatrix} \tag{2.42}$$

where

$$\begin{aligned}
\mathbf{G}_{\phi_x} &= \gamma_2 \int_{\Gamma_{\phi x}} \left(H^{\mathrm{T}}_{\phi_x} H_{\phi_x}\right) d\Gamma \\
\mathbf{G}_{\phi_y} &= \gamma_3 \int_{\Gamma_{\phi y}} \left(H^{\mathrm{T}}_{\phi_y} H_{\phi_y}\right) d\Gamma \\
\mathbf{f}_{s_w} &= \gamma_1 \int_{\Gamma_w} \left(\hat{w}\mathbf{H}^{\mathrm{T}}\right) d\Gamma \\
\mathbf{f}_{s_\theta} &= \begin{bmatrix} \mathbf{f}_{s\phi_x} \\ \mathbf{f}_{s\phi_y} \end{bmatrix} \\
\mathbf{f}_{s\phi_x} &= \gamma_2 \int_{\Gamma_{\phi x}} \left(\hat{\phi}_x H^{\mathrm{T}}_{\phi_x}\right) d\Gamma
\end{aligned}$$

and $\mathbf{f}_{s\phi_y} = \gamma_3 \int_{\Gamma_{\phi y}} \left(\hat{\phi}_y H^{\mathrm{T}}_{\phi_y}\right) d\Gamma$.

Functionally Graded Material Plate Model

The method of finite spheres gives the following discretized equations

$$\phi_x^h(x,y) = \sum_{I=1}^{N} \sum_{m \in \mathcal{I}} h_{Im}(x,y)\tilde{a}_{Im} = \mathbf{Ha} \tag{2.43a}$$

$$\phi_y^h(x,y) = \sum_{I=1}^{N} \sum_{m \in \mathcal{I}} h_{Im}(x,y)\tilde{b}_{Im} = \mathbf{Hb} \tag{2.43b}$$

$$w^h(x,y) = \sum_{I=1}^{N} \sum_{m \in \mathcal{I}} h_{Im}(x,y)\tilde{c}_{Im} = \mathbf{Hc}. \tag{2.43c}$$

Minimizing the potential functional (2.22) with respect to the nodal unknowns results in the following set of linear algebraic equations

$$\mathbf{Kd} = \mathbf{F} \tag{2.44}$$

where $\mathbf{K}$, $\mathbf{d}$, and $\mathbf{F}$ are the stiffness matrix, nodal unknowns, and force matrix, respectively. $\mathbf{d}$ and $\mathbf{F}$ are defined as

$$\mathbf{d} = \begin{bmatrix} \mathbf{a} \\ \mathbf{b} \\ \mathbf{c} \end{bmatrix}, \mathbf{F} = \begin{bmatrix} \mathbf{fx} \\ \mathbf{fy} \\ \mathbf{fz} \end{bmatrix} \tag{2.45}$$

where

$$\mathbf{fx} = \int_\Omega \mathbf{B}_x M^t d\Omega, \ \mathbf{fy} = \int_\Omega \mathbf{B}_y M^t d\Omega \quad \text{and} \quad \mathbf{fz} = \int_\Omega \mathbf{H} q d\Omega,$$

where $\mathbf{B}_x$ and $\mathbf{B}_y$ are the derivatives of $\mathbf{H}$ with respect to x and y, respectively.

The stiffness matrix $\mathbf{K}$ is defined as

$$\mathbf{K} = \begin{bmatrix} \mathbf{DA}_{xx} + \frac{1-\nu}{2}\mathbf{A}_{yy} + \mathbf{QA} & \nu\mathbf{DA}_{xy} + \frac{1-\nu}{2}\mathbf{A}_{yx} & \mathbf{QA}_x \\ & \mathbf{DA}_{yy} + \frac{1-\nu}{2}\mathbf{A}_{xx} + \mathbf{QA} & \mathbf{QA}_y \\ \textit{symmetric} & & \mathbf{Q}(\mathbf{A}_{xx} + \mathbf{A}_{yy}) \end{bmatrix} \tag{2.46}$$

where

$$\begin{aligned} &\mathbf{A} = \int_\Omega \mathbf{H}^{\mathrm{T}}\mathbf{H} d\Omega, \\ &\mathbf{A}_x = \int_\Omega \mathbf{H}^{\mathrm{T}}\mathbf{B}_x d\Omega, \quad \mathbf{A}_y = \int_\Omega \mathbf{H}^{\mathrm{T}}\mathbf{B}_y d\Omega, \\ &\mathbf{A}_{xx} = \int_\Omega \mathbf{B}_x^{\mathrm{T}}\mathbf{B}_x d\Omega, \ \mathbf{A}_{yy} = \int_\Omega \mathbf{B}_y^{\mathrm{T}}\mathbf{B}_y d\Omega, \\ &\mathbf{A}_{xy} = \int_\Omega \mathbf{B}_x^{\mathrm{T}}\mathbf{B}_y d\Omega, \ \mathbf{A}_{yx} = \int_\Omega \mathbf{B}_y^{\mathrm{T}}\mathbf{B}_x d\Omega. \end{aligned} \tag{2.47}$$

and

$$\mathbf{D} = \int_{-h/2}^{h/2} \frac{E}{1-\nu^2} \begin{bmatrix} 1 & \nu & 0 \\ \nu & 1 & 0 \\ 0 & 0 & \frac{1-\nu}{2} \end{bmatrix} z^2 dz, \ \mathbf{Q} = \int_{-h/2}^{h/2} \frac{E}{1-\nu^2} \begin{bmatrix} \frac{1-\nu}{2} & 0 \\ 0 & \frac{1-\nu}{2} \end{bmatrix} dz\,. \tag{2.48}$$

3 Numerical Integration

As in most other meshfree methods, the method of finite spheres requires specialized numerical integration techniques to evaluate the integrals in the weak form where the integrands are nonpolynomial rational functions and the integration domains are much more complex than in traditional finite elements. In [12, 1] several computationally efficient integration schemes have been proposed for the method of finite spheres applied to two-dimensional elasto-static problems and it has been reported that piece-wise midpoint quadrature rules are more effective than higher order Gaussian quadrature.

However, for the solution of fourth order problems such as plates, the integrands in the weak form contain higher order derivatives that are more ill behaved than the integrands encountered in second order problems. Hence, in [26, 21] we have developed a novel integration technique based on a lookup table generated using genetic algorithms. This method has much promise since it provides accurate solutions with a reasonable number of integration points. In this section we will briefly describe how genetic algorithms may be used in generating numerical integration schemes and then we will describe the lookup table approach. Details can be found in [26, 21].

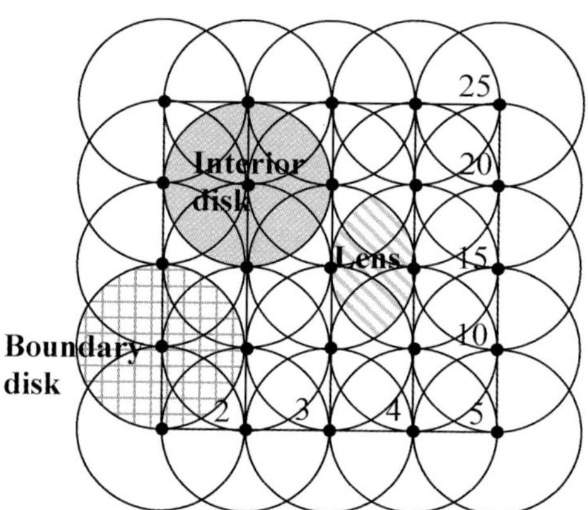

Figure 3. Interior (filled), lens (cross-hatched) and boundary disks (striped) on a two-dimensional domain.

In what follows we will consider the following integral

$$Q_{2D} = \int_{\Omega_I} f(x, y) dx dy \tag{3.49}$$

where Ω_I is a general subdomain in $\mathbb{R}^2$. For the method of finite spheres we will consider three types of subdomains (see Figure 3):

1. ***Interior disk***; defined as a disk that does not intersection the domain boundary.
2. ***Lens***; defined as the region of intersection of two spheres.
3. ***Boundary disk***; defined as a disk with a nonzero intersection with the domain boundary.

3.1 Numerical Integration using Genetic Algorithm

Genetic algorithms [17, 18, 19] are a class of biologically inspired optimization algorithms in which a population of potential solutions is chosen at random. Individuals (population members) compete according to their fitness for selection as parents to the next generation. The genetic material of the parents is modified to produce offsprings (new population) using genetic operators such as mutation and crossover. The process is continued until the maximum fitness in the population has not reached a certain preset value or the number of generations has not reached the maximum limit. This general algorithm is adapted to numerical quadrature, the two dimensional case is described below.

The algorithm starts by choosing μ partitions (β) randomly, the approximate integral Q of each partition is evaluated as well as the relative error E using two different quadrature rules, one more accurate than the other (e.g., we have used one and three point Gauss quadrature rules). For each subinterval the difference between the two integrals is the relative error E_i, while the integration value Q_i corresponds to the more accurate rule.

In the two-dimensional case each subdomain type (interior, lens or boundary disk) has a different partition. An interior disk (of radius R) is divided into m subdomains using concentric circles of radii $\{0 = r_0 < r_1 < \ldots < r_m = R\}$, and each subregion bounded by r_i and r_{i+1} is then divided by n radial lines $\{\theta_{i,0}, \ldots, \theta_{i,(n-1)} | \theta_{ij} \in [0, 2\pi]\}$ and $\theta_{i,j} < \theta_{i,j+1}$, see Figure 4.

The partition β for this case is a matrix of the form

$$\beta = \begin{bmatrix} 0 = r_0, & < r_1, < \ldots, & < r_m = R \\ \theta_{0,0}, & \ldots & , \theta_{m0} \\ \vdots & & \vdots \\ \theta_{0,(n-1)}, & \ldots & , \theta_{m,(n-1)} \end{bmatrix}, \text{ such that } r_0 = 0, \text{ and } r_i < r_{i+1}. \tag{3.50}$$

The integral in equation (3.49) may be written as

$$Q = \sum_{i=0}^{m-1} \sum_{j=0}^{n-1} Q_{i,j}$$

with

$$Q_{i,j} = \int_{r=r_i}^{r_{i+1}} \int_{\theta=\theta_{i,j}}^{\theta_{i,(j+1)}} f(r, \theta) r dr d\theta \approx \sum_{k=1}^{N_r} \sum_{l=1}^{N_\theta} W_{kl} f(\bar{r}_k, \bar{\theta}_l) \tag{3.51}$$

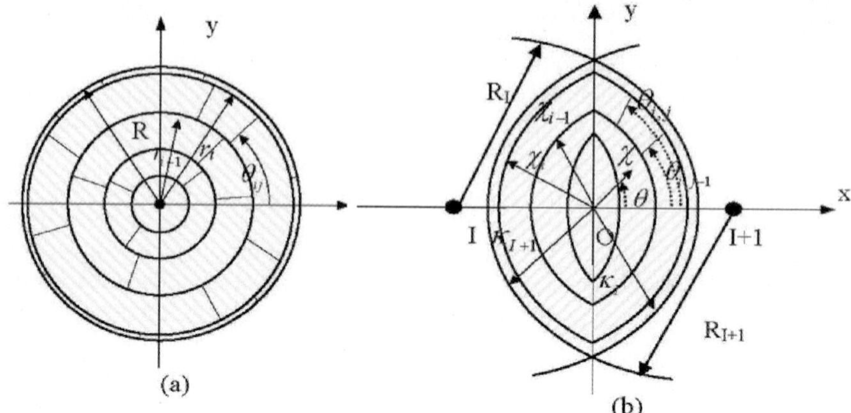

Figure 4. Partition of (a) an interior disk and (b) a lens into integration subdomains.

where N_r, and N_θ are the number of the radii and angles used in the approximation rule, W_{kl} are the weights, e.g., for a mid-point Gauss quadrature

$$W_{kl} = \frac{1}{2}(\theta_{l+1} - \theta_l)(r_k^2 - r_{k-1}^2),$$

and $(\bar{r}_k, \bar{\theta}_l)$ is the centroid of the subdomain.

The relative error is the sum of the relative errors for each subdomain, i.e.,

$$E = \sum_{i=0}^{m-1} \sum_{j=0}^{n-1} |E_{i,j}|.$$

A lens (Figure 4(b)), on the other hand is partitioned by first randomly generating m concentric lenses, the area between any two of which is then partitioned by n randomly generated straight lines passing through the point O. The partitions have the form

$$\beta = \begin{bmatrix} 0 = \chi_0, & \dots, \chi_m = \Xi(\theta) \\ \theta_{0,0}, & \dots \quad , \theta_{m,0} \\ \vdots & \vdots \\ \theta_{0,(n-1)}, & \dots \quad , \theta_{m,(n-1)} \end{bmatrix}, \tag{3.52}$$

such that $\chi \in [0, \Xi(\theta)], \chi_0 = 0,\ \chi_i < \chi_{i+1}$ and

$$\Xi(\theta) = \begin{cases} \kappa_I(\theta) & \frac{3\pi}{2} < \theta < \frac{\pi}{2} \\ \kappa_{I+1}(\theta) & \frac{\pi}{2} < \theta < \frac{3\pi}{2} \end{cases} \tag{3.53}$$

where $\theta \in [0, 2\pi]$, χ_i is the radius of the concentric lenses and $\Xi(\theta)$ is the radius of the outer lens κ_i which is the distance from O to the circumference of the nearest disk $i \in \{I, I+1\}$. For a boundary disk, the integration domain is either a disk or a lens truncated at the boundary.

For a lens we need to evaluate

$$Q = \int_{\chi=0}^{\Xi(\theta)} \int_{\theta=0}^{2\pi} f(\chi, \theta)\chi d\theta d\chi = \sum_{i=0}^{m-1} \sum_{j=0}^{n-1} Q_{i,j} \tag{3.54}$$

where

$$Q_{i,j} = \int_{\chi=\chi_i}^{\chi_{i+1}} \int_{\theta_{i,j}}^{\theta_{i,(j+1)}} f(\chi, \theta)\chi d\theta d\chi \approx \sum_{k=1}^{N_r} \sum_{l=1}^{N_\theta} W_{kl} f(\bar{\chi}_k, \bar{\theta}_l) \tag{3.55}$$

N_r, and N_θ are the number of the radii and angles used in the approximation rule, respectively. $W_k l$ are the weights, e.g., for a one-point Gauss quadrature rule $W_{kl} = \frac{1}{2}(\theta_{l+1} - \theta_l)(\chi_k^2 - \chi_{k-1}^2)$, and $(\bar{\chi}_k, \bar{\theta}_l)$ is the centroid of the subdomain.

In genetic algorithms the data (genetic material) for each member in the population (individual) is stored in chromosomes. In our case the genetic material in the chromosomes is the partition β. The integration value Q and the relative error E are stored along with the partitions in the chromosomes. We define the fitness of a chromosome as

$$F(\beta) = \frac{1}{E} \tag{3.56}$$

to ensure that chromosomes with higher fitness values will indeed correspond to lower overall relative error.

The chromosomes are ranked in ascending order of their fitness values. Parents are selected from them in a process called "rank selection" [19], where each chromosome's probability of being selected is proportional to its rank. After the selection, the parent's genetic material is modified with genetic operators. In genetic algorithms there are two types of operators: "crossover" and "mutation". In mutation of a real-coded chromosome, where the chromosomes are vectors of numbers instead of binary strings [20], the interior points x_i in the chromosomes are perturbed by a small amount δ, where $\epsilon_- < \delta < \epsilon_+$, $\epsilon_- = x_{i-1} - x_i$ and $\epsilon_+ = x_{i+1} - x_i$. The crossover operator takes two parents, randomly select one or more crossover points, and exchange the genetic information between parents after each of these points. Both crossover and mutation are applied on the rows of radii and the columns of angles separately.

3.2 A Genetic Algorithm-Based Look-up Table Approach

A naive use of the genetic algorithm-based numerical integration technique described in Section 3.1 is computationally inefficient. Instead, we generate a lookup table for normalized integration points and weights assuming a regular arrangement of nodal points using genetic algorithms. The key to generating this lookup table is a set of simple normalization schemes described in the following paragraphs.

In two-dimensions the strategy is to map both the interior disk and the lens to a unit disk. For an interior disk at node I, using the transform

$$\eta = \frac{r}{r_I}$$

we may rewrite the integral in equation (3.49) as

$$Q = \int_{\Omega_I} f(x,y)dxdy = \int_0^{2\pi}\int_0^{r_I} f(r,\theta)rdrd\theta = r_I^2 \int_0^{2\pi}\int_0^{1} f(\eta,\theta)\eta d\eta d\theta. \qquad (3.57)$$

For a lens in two-dimensions, formed by the intersection of two disks centered at nodes I and $I+1$ (Figure 4b), we define

$$\eta = \frac{\chi}{\Xi(\theta)} \qquad (3.58)$$

where we use the same notations as in Section 3.1. Equation (3.49) becomes

$$Q = \int_{\chi=0}^{\Xi(\theta)}\int_{\theta=0}^{2\pi} f(\chi,\theta)\chi d\chi d\theta = \int_0^{2\pi}\int_0^{1} f(\eta,\theta)\Xi(\theta)^2\eta d\eta d\theta. \qquad (3.59)$$

For generating the lookup table, a regular arrangement of nodes with inter-nodal spacing of unity in each coordinate direction is used as shown in Figure 5. The genetic algorithm described in Section 3.1 is used on the normalized domain, producing partitions β in the normalized coordinates (η,θ). Once the partitions are generated for the regular arrangement of nodal points, they are stored in the lookup table using these normalized coordinates.

4 Numerical Examples

In this section we provide several example problems of thin and thick isotropic and functionally graded material plate models.

4.1 Kirchhoff Plate

Clamped Edges

We will consider the problem shown in Figure 6 of a rectangular plate with clamped edges. The boundary conditions for this problem are

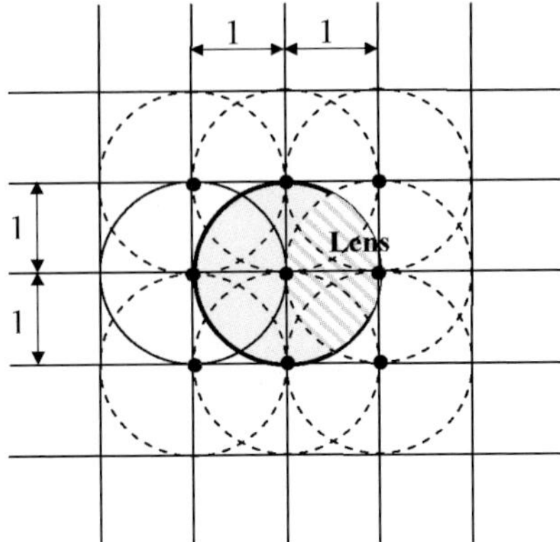

Figure 5. Domains used to generate the integration points and weights in two-dimensional analysis.

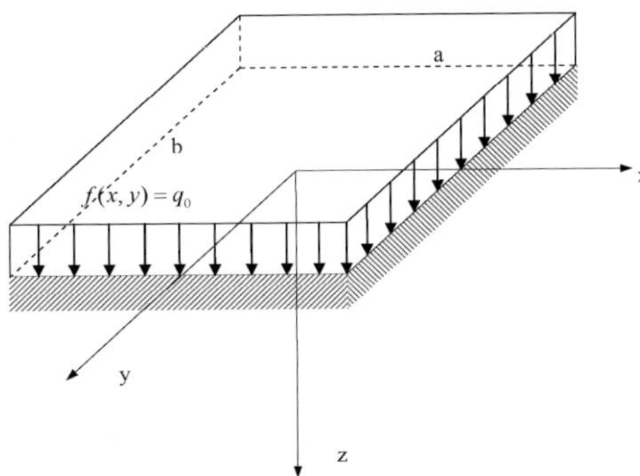

Figure 6. A plate with clamped edges and uniformly distributed load on the top surface.

$$\begin{array}{l} w = \frac{\partial w}{\partial x} = 0 \text{ along } x = \pm a/2, \\ w = \frac{\partial w}{\partial y} = 0 \text{ along } y = \pm b/2. \end{array} \tag{4.60}$$

A distributed constant force $f(x, y) = q_0$ acts on the top surface of the plate as shown in the figure. The values $a = b = 1$, $D = 1$, and $q_0 = 1$ and have been assumed

A comparison between the convergence rate in the solution for the transverse deflection $w(x, y)$ in the H_1 norm, using the method of finite spheres with genetic algorithm-based lookup table for integration and the finite element method is shown in Figure 7, where the exact solution is evaluated as the first 50 terms in the infinite series in [15].

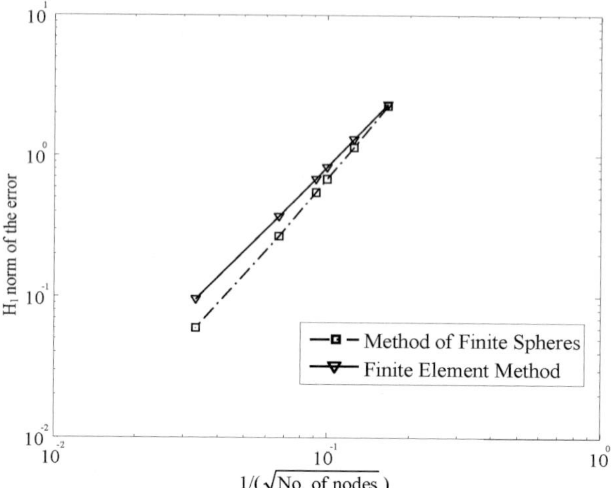

Figure 7. Convergence in the solution for the transverse deflection $w(x, y)$ in the H_1 norm in a rectangular plate with clamped edges obtained using the method of finite spheres and the finite element method.

Mindlin-Reissner Plate

Now we will consider a shear deformable plate [27]

Clamped Edges

We will consider the problem of a rectangular plate with clamped edges. The boundary conditions for this problem are

$$w = \phi_x = \phi_y = 0 \text{ along } \; x = \pm a, y = \pm b\,. \tag{4.61}$$

For simplicity the length of the square plate as well as the rigidity D are assumed to be unity, the Poisson's ratio $\nu = 0.3$, the thickness $h = 0.1$, the shear correction factor $k = 5/6$ and the force $q_0 = 10$. A comparison between the convergence rate in the solution for the transverse deflection $w(x, y)$ in the L_2 norm, using the method of finite spheres with genetic algorithm-based lookup table for integration and the finite element method is shown in Figure 8.

Functionally Graded Material Plate Model

Now we will consider a shear deformable square plate with simply supported edges composed of functionally graded material. The boundary conditions for this problem are

$$w = \phi_x = \phi_y = 0 \quad \text{along } x = \pm a, \quad y = \pm b\,. \tag{4.62}$$

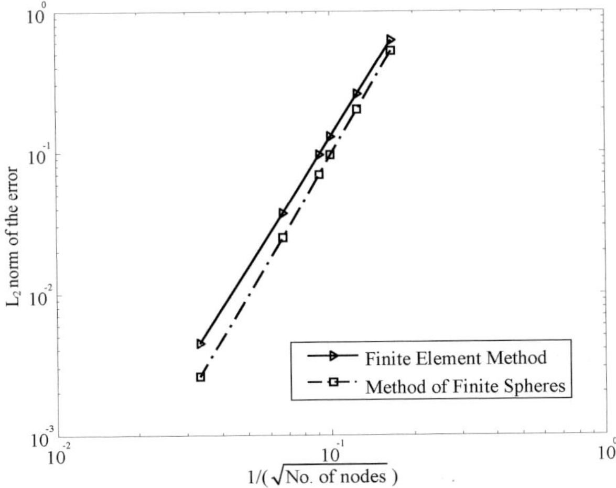

Figure 8. Convergence in the solution for the transverse deflection $w(x, y)$ in a rectangular plate with clamped edges obtained using the method of finite spheres and the finite element method.

A distributed constant load $f(x, y) = q_0$ acts on the top surface of the plate. A temperature difference of $\Delta T = 100°\mathrm{C}$ exists between the top and bottom surfaces; the bottom surface is kept at 0°C and the top surface at 100°C. The lower plate surface is assumed to be aluminum and the top surface is zirconia. The material properties vary as a power law with $n = 1$. The values of the material properties are given in Table 1. The dimensions of the plate are assumed to be unity, while the thickness is $h = 0.3$, and the shear correction factor is $k = 5/6$.

Table 1. Material properties

Material property	Aluminum	Zirconia
Young's modulus	70GPa	151GPa
Poisson's ratio	0.3	0.3
Thermal conductivity	204W/mK	2.09W/mK
Thermal expansion	23×10^{-6}/C	10×10^{-6}/C

Figure 9 shows the thermal solution for the problem presented in Section 2.1.

A comparison between the convergence rate in the solution for the transverse deflection $w(x, y)$ using the method of finite spheres with genetic algorithm-based lookup table for integration and the finite element method is shown in Figure 10. Since the exact solution does not exist in closed form, a finite element solution using a very fine mesh of 100×100 nine-noded quadratic elements was used.

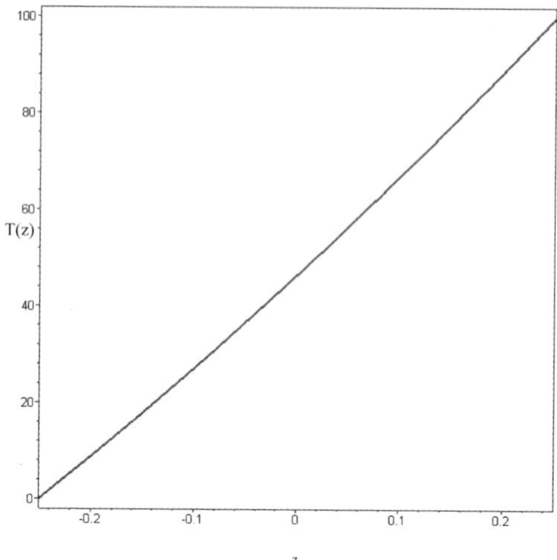

Figure 9. Temperature distribution through the thickness of a functionally graded plate with $n = 1.0$ and thermal conductivity values shown in Table 1. The temperature 0°C and 100°C at the lower and upper surfaces, respectively.

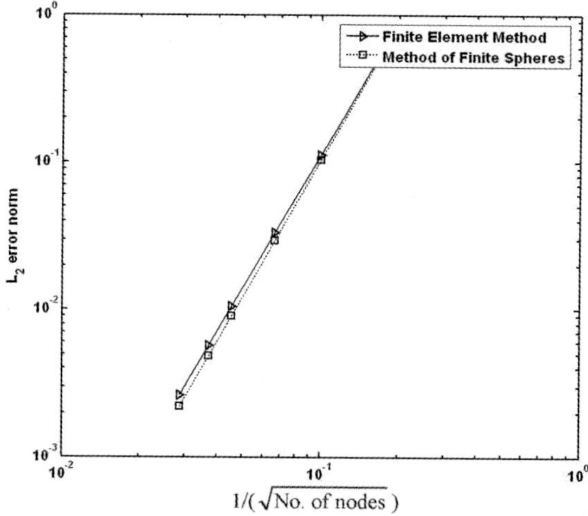

Figure 10. The convergence in the solution for the transverse deflection $w(x, y)$ in the L_2 norm for a circular plate with simply supported edges computed using the method of finite spheres with the genetic algorithm-based lookup table integration scheme.

5 Concluding Remarks

Through the solution of different plate problems, it was confirmed in this study that the genetic algorithm-based lookup table approach provides a computationally efficient scheme to solve higher order differential equations using meshfree methods. The technique is very similar to the use of a lookup table for Gaussian integration points and weights used in traditional finite element methods. The smoothness of the approximation spaces results in excellent solution accuracies as evidenced in the convergence curves. Extension of the integration method to handle nonlinear problems is under development.

References

1. K. J. Bathe and S. De, *The Method of Finite Spheres with Improved Numerical Integration*, Computers and Structures. **79** (2001), pp. 2183–2196.
2. K. J. Bathe and S. De, *The Method of Finite Spheres*, Computational Mechanics. **25** (2000), pp. 329–345.
3. V. Leitão and C. Tiago, *Plate Bending Using hp-clouds and Trefftz Based Enrichment*, 4th International Conference on Boundary Elements Techniques, Granada, Advances in Boundary Element Techniques. **IV** (2003), pp. 341–346.
4. H. A. Mang, J. Sladek, and V. Sladek, *Meshless Formulations for Simply Supported and Clamped Plate Problems*, International Journal for Numerical Methods in Engineering. **55** (2002), pp. 359–375.
5. J. N. Reddy,K. M. Liew, and Y. Q. Huang, *Moving Least Squares Differential Quadrature Method and its Application to the Analysis of Shear Deformable Plates*, International Journal for Numerical Methods in Engineering. **56** (2003), pp. 2331–2351.
6. V. Leitão, *An RBF-Based Hermite Collocation Method for Kirchhoff Plates*, Electronic Journal of Boundary Elements. **1** (2002), pp. 115–127.
7. S. N. Atluri and S. V. Long, *A Meshless Local Petrov-Galerkin Method for Solving the Bending Problem of a Thin Plate*, Computer Modeling in Engineering and Sciences. **3:1** (2002), pp. 11–52.
8. T. Belytschko and P. Krysl, *Analysis of Thin Plates by the Element-Free Galerkin Method*, Computational Mechanics. **17** (1996), pp. 26–35.
9. K. J. Bathe and S. De, *Toward an Efficient Meshless Computational Technique: the Method of Finite Spheres*, Engineering Computations. **18** (2001), pp. 170–192.
10. I. Babuška and J. M. Melenk, *The Partition of Unity Finite Element Method: Basic Theory and Applications*, Computer Methods in Applied Mechanics and Engineering. **V** (1996), pp. 289–314.
11. K. Yosida, *Functional Analysis*, Springer-Verlag: Berlin Heidelberg. **5th Edition** (1978).
12. S. De and M. Macri, *Toward an Automatic Discretization Scheme for the Method of Finite Spheres and its Coupling with the Finite Element Method*, Computers and Structures. **81** (In Press).
13. K. Copps, T. Strouboulis, and I. Babuška, *The Design and Analysis of the Generalized Finite Element Method*, Computer Methods in Applied Mechanics and Engineering. **181** (2000), pp. 43–69.

14. I. Babuška, U. Banerjee, and J. E. Osborn, *Meshless and Generalized Finite Element Methods: A Survey of Some Major Results*, Meshfree Methods for Partial Differential Equations (M. Griebel and M. A. Schweitzer, eds.), Lecture Notes in Computational Science and Engineering, vol. 26, Springer, 2002, pp. 1–20.
15. L. C. Dym and I. H. Shames, *Solid Mechanics: A Variational Approach*, McGraw-Hill, Inc., (1973).
16. S. Timoshenko and S. Woinowsky-Krieger, *Theory of Plates and Shells*, McGraw-Hill, Inc., (1959).
17. G. Goldberg, *Genetic Algorithms in Search, Optimization and Machine Learning*, Addison Wesley., (1989).
18. D. B. Fogel, *Evolutionary Computation*, IEEE press: New York., (1995).
19. Z. Michalewicz, *Genetic Algorithms + Data Structure = Evolution Programs*, Springer-Verlag: Berlin., (1994).
20. A. H. Wright, *Genetic Algorithms for Real Parameter Optimization. Foundations of Genetic Algorithms*, Morgan Kaufmann Publishers: San Mateo, CA. (1991), pp. 205–218.
21. S. BaniHani and S. De, *A Computationally Efficient Technique for the Solution of Timoshenko Beam and Mindlin-Reissner Plate Problems Using the Method of Finite Spheres*, International Journal of Computational Methods., (Submitted).
22. G. R. Liu and Y. T. Gu, *Meshless Methods for Shear-Deformable Beams and Plates*, Computer Methods in Applied Mechanics and Engineering. **152** (1998), pp. 47–72.
23. O. Garcia, E. A. Fancello, C. S. Barcellos, and C. A.Duarte, *hp-Clouds in Mindlin's Thick Plate Model*, International Journal for Numerical Methods in Engineering. **47** (2000), pp. 1381–1400.
24. C. A. Duarte and I. Babuška, *Mesh-Independent p-Orthotropic Enrichment Using the Generalized Finite Element Method*, International Journal for Numerical Methods in Engineering. **55** (2002), pp. 1477–1492.
25. J. S. Chen, D. Wang, and S. B. Dong, *An Extended Meshfree Method for Boundary Value Problems*, Computer Methods in Applied Mechanics and Engineering. **193** (2004), pp. 1085–1103.
26. S. BaniHani and S. De, *Development of a Genetic Algorithm-Based Lookup Table Approach for Efficient Numerical Integration in the Method of Finite Spheres with Application to the Solution of Thin Beam and Plate Problems*, International Journal for Numerical Methods in Engineering., (Submitted).
27. C. M. Wang, J. N. Reddy, and K. H. Lee, *Shear Deformable Beams and Plates: Relationships with Classical Solutions*, Elsevier science Ltd., Oxford, UK. (2000).
28. T. Fuchiyama, N. Noda, T. Tsuji, and Y. Obata, *Analysis of Thermal Stress and Stress Intensity Factor of Functionally Gradient Materials*, Ceramic Transactions, Functionally Gradient Materials. **34** (1993), pp. 425–432.
29. J. N. Reddy, *Analysis of Functionally Graded Plates*, International Journal for Numerical Methods in Engineering. **47** (2000), pp. 663–684.
30. T. Mori and K. Tanaka, *Average Stress in Matrix and Average Elastic Energy of Materials with Misfitting Inclusions*, Acta Metallurgica. **21** (1973), pp. 571–574.

An RBF Meshless Method for Injection Molding Modelling

Francisco Bernal and Manuel Kindelan

Universidad Carlos III de Madrid, Avda. Universidad 30, 28911 Leganes, Spain
{fcoberna,kinde}@ing.uc3m.es

Summary. Recently, very intensive efforts have been devoted to develop *meshless* or *element free* methods that eliminate the need of element connectivity in the solution of PDEs. The motivation is to cut down modelling costs in industrial applications by avoiding the labor intensive step of mesh generation. In addition, these methods are particularly attractive in problems with moving interfaces since no remeshing is necessary.

In this paper, we address the problem of injection molding described as a free boundary problem defined by conservation equations for mass, momentum and energy. This model can be dramatically simplified assuming that the mould is thin so that a Hele-Shaw approximation can be used. In this case the momentum equation is just a 2D elliptic equation whose solution yields the pressure distribution in the filled region of the mould. From this pressure field, the velocity distribution can be computed and the location of the advancing front can be updated.

Therefore, this problem is very well suited to meshfree techniques and, in this paper, we use a radial basis function (RBF) method which is based on the interpolation and collocation of global shape functions, and for simplicity assume a Newtonian fluid so that the model is linear. In particular, we use multiquadric (MQ) RBFs which have been shown to have exponential convergence.

To advance the front we use a level set method which is very efficient for this type of problems because it is very fast and can handle both front collisions and front break-ups without difficulty.

Key words: Meshfree, RBF, Hele-Shaw, Level Set.

1 Model Equations

To model the flow of an injected fluid in a mould, let us start from the conservation equations for mass, momentum and energy,

$$\frac{\partial \rho}{\partial t} + \nabla \cdot (\rho\, \mathbf{v}) = 0$$

$$\rho \left(\frac{\partial \mathbf{v}}{\partial t} + \mathbf{v} \cdot \nabla \mathbf{v} \right) = -\nabla p + \nabla \cdot \tau$$

$$\rho\, C_p \left(\frac{\partial T}{\partial t} + \mathbf{v} \cdot \nabla T \right) = k\, \nabla^2 T + \eta\, \Phi$$

where τ is the extra stress tensor and Φ the dissipation function. Under the assumptions of incompressibility, creeping flow $((\rho_\infty U_\infty L/\eta_\infty)(h/L)^2 << 1)$ and small thickness $(h/L << 1)$, the above equations simplify to,

$$\frac{\partial u}{\partial x} + \frac{\partial v}{\partial y} = 0 \tag{1.1}$$

$$\frac{\partial p}{\partial x} = \frac{\partial}{\partial z}\left(\eta \frac{\partial u}{\partial z}\right), \qquad \frac{\partial p}{\partial y} = \frac{\partial}{\partial z}\left(\eta \frac{\partial v}{\partial z}\right), \qquad \frac{\partial p}{\partial z} = 0 \tag{1.2}$$

$$\rho C_p \left(\frac{\partial T}{\partial t} + u \frac{\partial T}{\partial x} + v \frac{\partial T}{\partial y} \right) = k \frac{\partial^2 T}{\partial z^2} + \eta \Phi \tag{1.3}$$

where the inertia terms in the momentum equations have been neglected in comparison to the viscous terms (creeping flow), and where partial derivatives in the horizontal directions (x, y) have been neglected in comparison to derivatives in the vertical direction (small thickness).

Integrating the momentum equations along the vertical direction from the mid-plane, and taking into account that by symmetry $\frac{\partial u}{\partial z}(z=0) = \frac{\partial v}{\partial z}(z=0) = 0$, results in

$$\frac{\partial u}{\partial z} = -\Lambda_x \frac{z}{\eta} \qquad \frac{\partial v}{\partial z} = -\Lambda_y \frac{z}{\eta}$$

$$\Lambda_x \equiv -\frac{\partial p}{\partial x}, \qquad \Lambda_y \equiv -\frac{\partial p}{\partial y}$$

A new integration in the vertical direction yields,

$$\int_z^b \frac{\partial u}{\partial \tilde{z}}\, d\tilde{z} = \int_z^b \frac{-\Lambda_x \tilde{z}}{\eta}\, d\tilde{z} \qquad \Rightarrow \qquad u = \Lambda_x \int_z^b \frac{\tilde{z}}{\eta}\, d\tilde{z} \tag{1.4}$$

$$\int_z^b \frac{\partial v}{\partial \tilde{z}}\, d\tilde{z} = \int_z^b \frac{-\Lambda_y \tilde{z}}{\eta}\, d\tilde{z} \qquad \Rightarrow \qquad v = \Lambda_y \int_z^b \frac{\tilde{z}}{\eta}\, d\tilde{z} \tag{1.5}$$

where $b(x, y)$ is the half-height and use has been made of the no slip boundary condition $u(z=b) = v(z=b) = 0$. Integrating these equations along the vertical direction results in the following averaged horizontal velocities,

$$\bar{u} = \frac{1}{b} \int_0^b u\, dz = \frac{\Lambda_x}{b} S, \qquad \bar{v} = \frac{1}{b} \int_0^b v\, dz = \frac{\Lambda_y}{b} S \tag{1.6}$$

where

$$S = \int_0^b \frac{z^2}{\eta} dz$$

Thus, the velocity vector is in the direction of $-\nabla p$. Analogously, the continuity equation (1.1) is integrated along the half-height,

$$\int_0^b \left[\frac{\partial u}{\partial x} + \frac{\partial v}{\partial y}\right] dz = \frac{\partial (b\bar{u})}{\partial x} + \frac{\partial (b\bar{v})}{\partial y} \tag{1.7}$$

Introducing the values of the average horizontal velocities (1.6) into the integral form of the continuity equation (1.7) produces,

$$\frac{\partial}{\partial x}\left(S\frac{\partial p}{\partial x}\right) + \frac{\partial}{\partial y}\left(S\frac{\partial p}{\partial y}\right) = 0. \tag{1.8}$$

This is an elliptic, nonlinear equation whose solution yields the pressure field and from (1.4-1.5) the velocity field. Notice that, in general, the viscosity is a function of temperature and shear stress ($\dot{\gamma} = \sqrt{u_z^2 + v_z^2}$). Thus, $\eta = \eta(T, \dot{\gamma})$, and the momentum equation (1.8) is coupled to the energy equation (1.3). For simplicity, we will consider the case of Newtonian fluids for which η is constant and, therefore, the pressure and energy equations are decoupled. The flowability S is simply $S = b^3/(3\eta)$ and equation (1.8) simplifies to,

$$\frac{\partial}{\partial x}\left(b^3\frac{\partial p}{\partial x}\right) + \frac{\partial}{\partial y}\left(b^3\frac{\partial p}{\partial y}\right) = 0. \tag{1.9}$$

This equation is solved with the following boundary conditions,

- inlet pressure (or flow rate) at the mould gate
- $p = 0$ in advancing front
- normal velocity zero at the walls of the mould, $\partial p/\partial n = 0$

The solution of equation (1.8) is used in (1.6) to compute the average velocity field, which is in turn used to advance the front.

This model is the Hele Shaw approximation for injection molding which was introduced by Hieber and Shen [4] and later used by many authors(see for instance [5, 8, 14]).

2 Pressure Field

We use an RBF based method in order to compute, at a certain instant, the pressure distribution within the mould.

Let us consider the rectangular mould shown in Figure 1, where the injected fluid is shown in dark. We consider an injection pressure $p = 1$ and zero pressure in the moving front. In the case of a Newtonian fluid and a mould with constant height (b = const.), equation (1.9) simplifies to the Laplace equation.

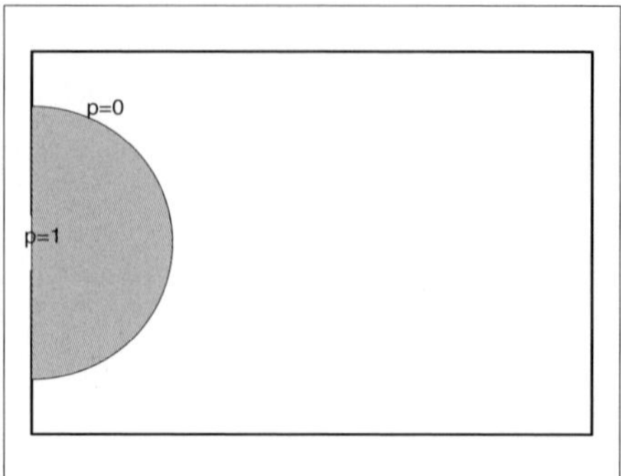

Figure 1. Problem Geometry.

Kansa [6, 7] was the first to use RBF's for the solution of partial differential equations. His method was based on multiquadric (MQ) basis functions,

$$\phi_j\,(\mathbf{x}) \;=\; \sqrt{\|\,\mathbf{x}\;-\;\mathbf{x}_j\,\|^2\;+c_j^2} \tag{2.10}$$

first proposed by Hardy [3]. Using this basis, the solution of equation (1.9) is approximated by

$$p(\mathbf{x}) \;=\; \sum_{k=1}^{N} \alpha_k\,\phi_k(\mathbf{x}) \tag{2.11}$$

and the unknown coefficients $\alpha_k(k = 1, \ldots, N)$ are computed by collocation on N collocation points, $X = \{\mathbf{x}_1, \mathbf{x}_2, \ldots, \mathbf{x}_N\}$ on the domain, Ω, and on its boundary, $\partial\Omega$, where either the PDE (1.9) or the appropriate boundary conditions are imposed.

To specify the MQ basis (2.10), one must specify both the centers $\mathbf{x}_j$ and the shape parameters c_j. Usually, the collocation points are used also as centers for the MQ, although some authors [2] use different data centers and collocation points near the boundaries.

The upper left of Figure 2 shows the location of the RBF centers within the domain. There are N_I interior centers, which are represented by crosses, and N_B boundary centers which are represented by open circles. These N centers ($N = N_I + N_B$) are taken from a finite element mesh.

As a first numerical experiment we consider the solution of equation (1.9) with $b = 1$, using Dirichlet boundary conditions throughout, that is, we replace the null normal velocity at the walls by $p = 0$. Introducing (2.11) in equation (1.9) results in the following linear system of N_I equations in the N unknowns α_k,

$$\sum_{k=1}^{N} \alpha_k\,\Delta\phi_k\,(\mathbf{x}_i) \;=\; 0, \qquad i = 1, 2, \ldots, N_I \tag{2.12}$$

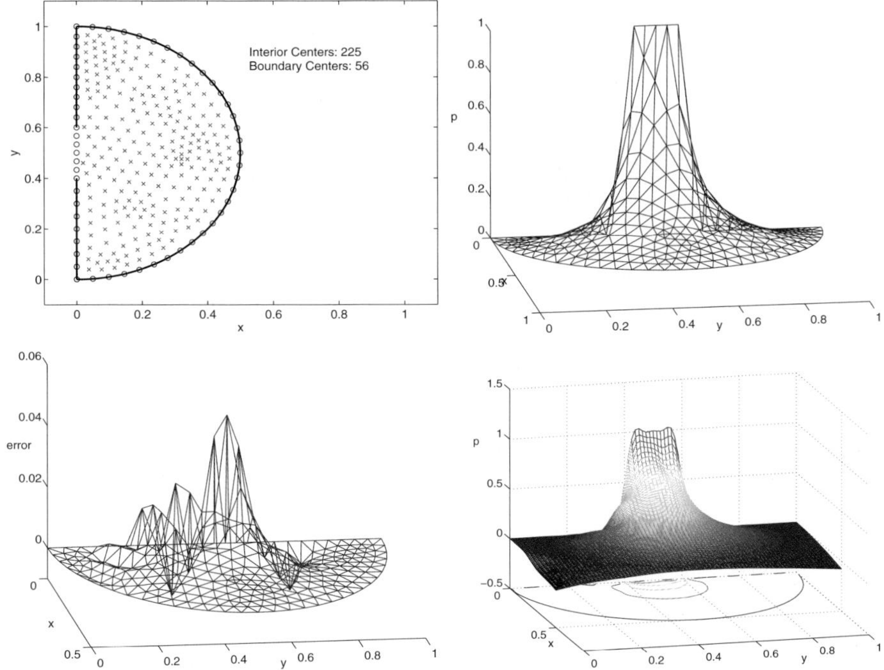

Figure 2. Upper left: RBF centers. Upper right: RBF solution for pressure at the collocation centers ($c_j^2 = 0.02$). Lower left: error at collocation centers. Lower right: RBF solution in a regular mesh.

The remaining equations come from implementing Dirichlet boundary conditions in the N_B boundary centers, namely,

$$\sum_{k=1}^{N} \alpha_k \, \phi_k(\mathbf{x}_i) = p_i, \qquad i = N_I + 1, N_I + 2, \ldots, N_I + N_B \tag{2.13}$$

where $p_i = 0$ except for centers located at the mould entrance where $p_i = 1$. Notice that for multiquadric RBFs

$$\Delta\phi_k\left(\mathbf{x}_i\right) = \frac{1}{\sqrt{\| \mathbf{x}_i - \mathbf{x}_j \|^2 + c_j^2}} + \frac{c_j^2}{\left(\| \mathbf{x}_i - \mathbf{x}_j \|^2 + c_j^2\right)^{3/2}}$$

Equations (2.12, 2.13) can be written in matrix form as,

$$\mathbf{A}\,\alpha = \mathbf{b} \tag{2.14}$$

The upper right of Figure 2 shows the numerical solution at the collocation centers obtained with the RBF method using a constant shape parameter $c_j = 0.1414$ ($c_j^2 = 0.02$). The solution appears smooth and the error relative to the finite element solution is shown in the lower left of Figure 2. Notice that the maximum error at the collocation centers is 0.044 and the mean square error 0.0027. Essentially the error is relevant only in the vicinity of the discontinuity in the boundary condition since we try to approximate a discontinuous solution by a linear combination of smooth and global multiquadric functions. However, if we use the coefficients α_k obtained by RBF using the collocation points shown in Figure 2, to compute the solution in a fine and regular mesh, the solution ceases to be smooth (lower right of Figure 2) and an overshoot near the discontinuity can be clearly seen. Also shown are the isolines corresponding to $p = 0$, $p = 0.25$, $p = 0.5$, $p = 0.75$ and $p = 1$. Notice that the advancing front ($p = 0$) is clearly identified.

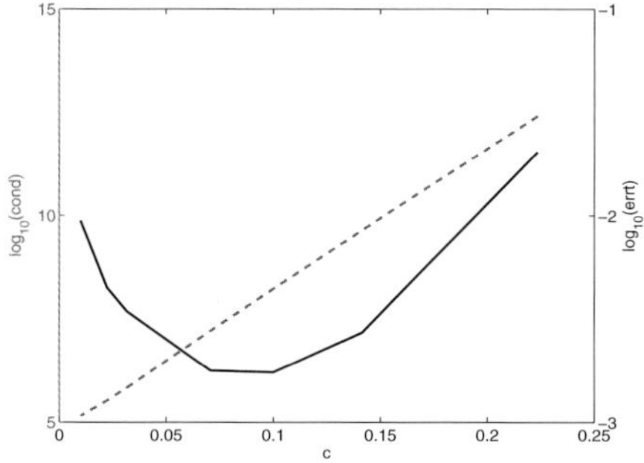

Figure 3. Mean square error at the collocation centers (solid) and condition number (dashed) as a function of the shape parameter c_j.

The accuracy of the solution is strongly dependent on the shape parameter c_j. As the shape parameter increases, accuracy increases, but the condition number of the resulting linear system also increases. Too large or too small shape parameters make the MQ basis too flat or too peaked, and both should be avoided. Thus, an appropriate selection of the shape parameters is crucial but there are no definite rules as to their optimal values. Hardy [3] suggested a constant shape parameter $c_j = c = 0.815\,\bar{d}_j$, where $\bar{d}_j$ is the average distance from each center to its nearest center. Other authors [9] use non constant shape parameters such as $c_j = \theta\, d_j$, where θ is a constant factor. For the mesh shown in the upper left of Figure 2, the average distance between nodes is $\bar{d}_j = \sqrt{\pi/(8\,N)} = 0.0374$. Thus, [3] suggests a value of the shape parameter $c_j = 0.03$.

Figure 3 shows the mean square error and the condition number of matrix **A** as a function of the shape parameter. Notice that the condition number grows exponentially with the shape parameter c_j. On the other hand, the error decreases with the shape parameter until the matrix becomes too ill-conditioned and rounding errors become important. Thus, there is an optimal value of the shape parameter ($c_j \approx 0.1$) which minimizes the average error.

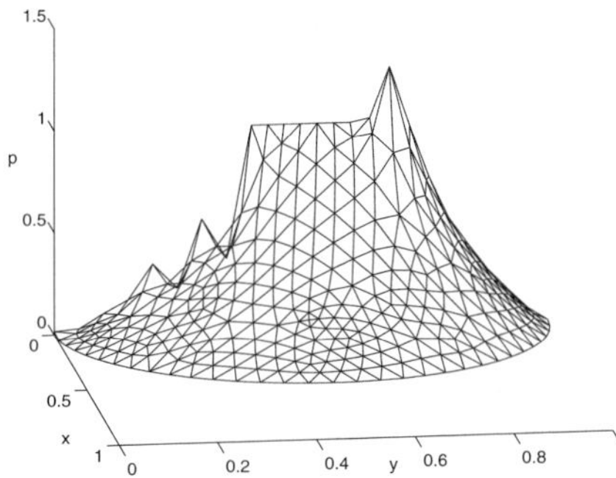

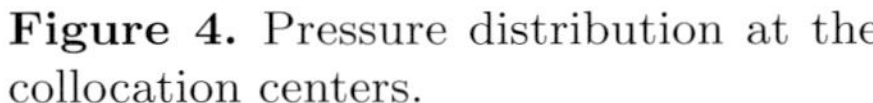

Figure 4. Pressure distribution at the collocation centers.

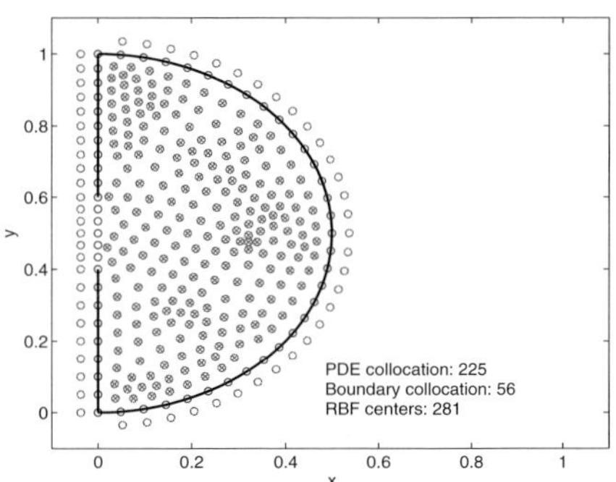

Figure 5. RBF centers (circles) and collocation centers (crosses).

Going back to the initial problem with null normal velocity at the walls ($\partial p/\partial n = 0$), and using the same collocation centers and shape parameters as before, we obtain the solution shown in Figure 4). In this case, large errors occur, specially near the mould entrance and the wall. The maximum error at the collocation centers is 0.778 and the mean square error 0.06. The solution is even worse for other values of the shape parameter.

A well known feature [2] in all RBF approximations is how relatively inaccurate they are at boundaries. As shown by the previous examples, this is specially true in the case of discontinuous boundary conditions. To overcome this problem Fornberg et al. proposed the *Not-a-Knot* method [2], which is based on the idea of using the nodes in the boundary only as collocation points to enforce the boundary conditions, but not as RBF centers for the RBF expansion of the solution.

Figure 5 shows the distribution of collocation centers (circles) and RBF centers (crosses). For interior nodes both sets of centers coincide. However, for each collocation center in the boundary, a new RBF center is introduced which is located in the direction normal to the boundary at a distance equal to the average distance between centers.

The left of Figure 6 shows the pressure distribution obtained with the *Not-a-Knot* method using a constant shape parameter ($c_j = 0.0245$). Notice that the errors in the vicinity of the boundary have disappeared and that the maximum error at the collocation centers has decreased to 0.0435 and the mean square error to 0.006. The right hand side of figure 6 shows the mean square error and the condition number as a function of the shape parameter. In this case the mean square error shows a different behavior that in the case of Dirichlet boundary conditions. In fact, there is a wide range of values of c for which the error remains approximately constant or even decreases with decreasing c.

A different alternative to improve accuracy near the boundaries was proposed by Fedoseyev *et al.* [1]. It is based on the observation that the residual

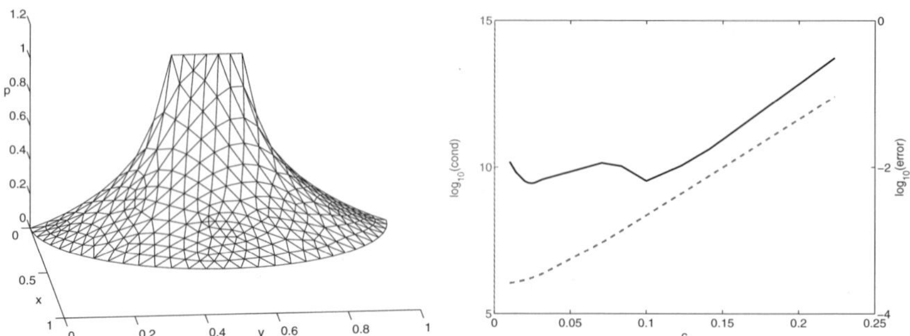

Figure 6. Left: pressure distribution at the collocation centers with the *Not-a-Knot* method. Right: Mean square error and condition number as a function of c.

error is typical largest near the boundaries. Thus, they proposed to enforce collocation of the PDE in boundary nodes so that both the boundary condition and the PDE are imposed in those nodes and the residuals dramatically decrease. However, since the number of equations increases, it is necessary to introduce additional RBF centers to match the number of unknown coefficients α_k. Using Fedoseyev's method with the same additional centers shown in figure 5, we obtain results which have similar accuracy to those obtained with the *Not-a-Knot* method.

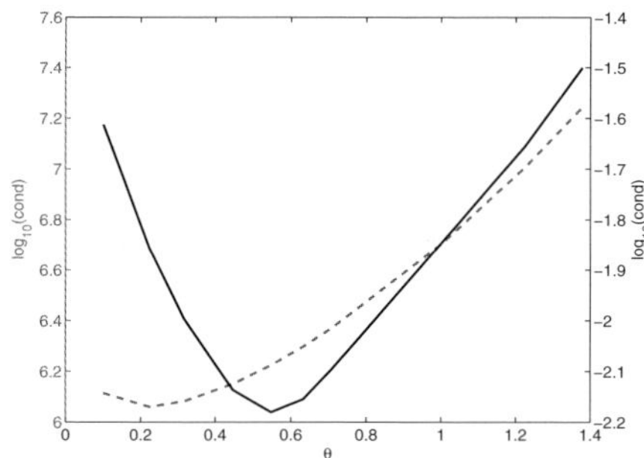

Figure 7. Mean square error and condition number as a function of θ.

Slightly better results can be obtained with a non-constant shape parameter $c_j = \theta\, d_j$. Figure 7 shows the mean square error and the condition number as a function of the parameter θ. These results correspond to the *Not-a-Knot* method using the centers shown in Figure 5. Notice that, similarly to what happened in the case with Dirichlet boundary conditions, the mean square error exhibits a minimum for an optimal value of the parameter θ. It should be pointed out that both the average error and the maximum error are slightly smaller than those obtained with finite elements.

3 Front Displacement

Once the pressure field is known, equation (1.6) provides the average velocity which is shown in Figure 8. Of particular relevance are the velocities in the advancing front which are needed to update its location.

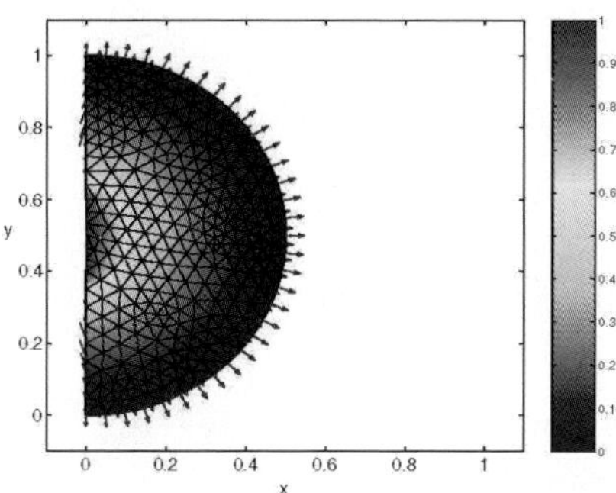

Figure 8. Velocity field at RBF centers (arrows). The absolute value of the velocity is represented in color.

Marker particle techniques [13] and volume-of-fluid techniques [10] are two methods which have been often used to model moving fronts. However, they have severe drawbacks which hamper their use. Since the seminal work of Osher and Sethian [11], level set methods and fast marching methods have been shown to be accurate and numerically efficient techniques, which do not have any of the drawbacks of traditional methods, and which have been successfully applied in a great variety of applications [12, 13].

The level set method considers the moving front as the zero level set of a higher dimensional function $\psi(\mathbf{x}, t)$. The evolution of the front is obtained by evolving the level set function with the simple advection equation,

$$\psi_t + \mathbf{V} \cdot \nabla \psi = 0 \tag{3.15}$$

which defines the motion of the front $\psi(\mathbf{x}, t) = 0$. It is an Eulerian formulation since the front is *captured* by the implicit function ψ. To solve (3.15) requires the velocity field $\mathbf{V}$, and an initial level set function $\psi(\mathbf{x}, 0)$ with the property that its zero level set corresponds to the initial front. Otherwise, the initial level set function is quite arbitrary although a signed distance function is often used.

In order to correctly evolve the front, the velocity field $\mathbf{V}$ has to be defined throughout the domain or, at least, within a narrow band surrounding it. The front, defined by the zero level set of the solution of (3.15), will move correctly provided that the velocity of the front is correct. In the present case, the velocity field is only defined in the filled region of the mould and, therefore, it has to be reasonably *extended* to the whole mould domain.

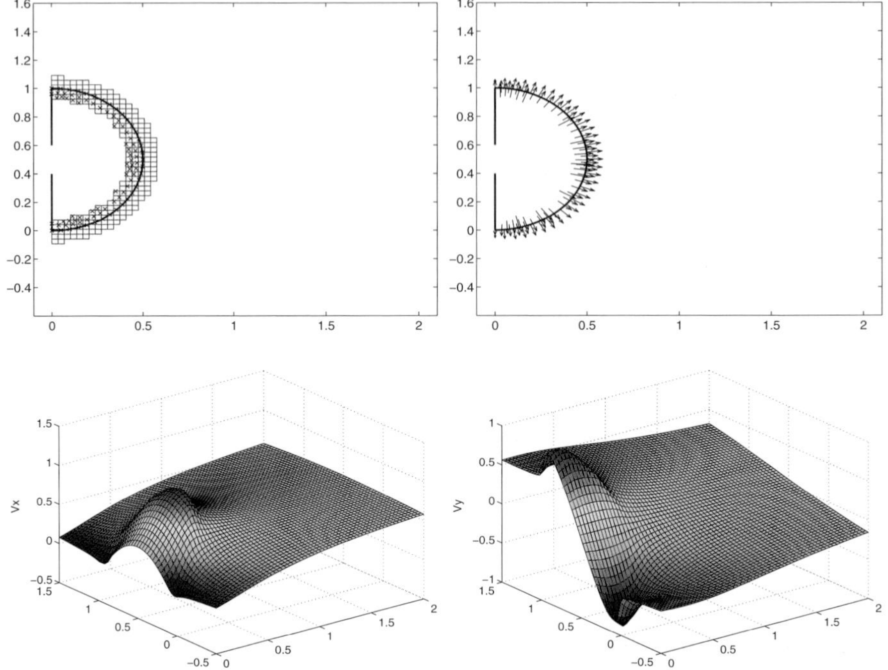

Figure 9. Upper left: narrow band and collocation centers within the narrow band. Upper right: velocity of collocation centers in the narrow band. Lower left: x velocity obtained by RBF interpolation from narrow band. Lower right y velocity obtained by RBF interpolation from narrow band.

In order to correctly extend the velocities we have used the following method:

1. Compute the pressure field by solving equation (2.14)
2. Use equation (1.6) to compute the velocity of collocation centers located in the filled region and within a narrow band surrounding the front (see upper part of figure 9).
3. Extrapolate this velocity field to the rest of the domain using an RBF interpolation technique (see lower part of figure 9).

Notice that the resulting velocities are smooth but not physically correct. However, for the purpose of evolving the front, it is only necessary that the velocity is correct near the front and, therefore, this velocity field is perfectly adequate.

Equation (3.15) is solved in a uniform grid covering the whole domain by means of the upwind method,

$$\psi_{i,j}^{n+1} = \psi_{i,j}^{n} - \nu_x \bar{u}_{ij} \begin{cases} \left(\psi_{i,j}^{n} - \psi_{i-1,j}^{n}\right) & \text{if } \bar{u}_{i,j} \geq 0 \\ \left(\psi_{i+1,j}^{n} - \psi_{i,j}^{n}\right) & \text{if } \bar{u}_{i,j} < 0 \end{cases} - $$

$$- \nu_y \bar{v}_{ij} \begin{cases} \left(\psi_{i,j}^{n} - \psi_{i,j-1}^{n}\right) & \text{if } \bar{v}_{i,j} \geq 0 \\ \left(\psi_{i,j+1}^{n} - \psi_{i,j}^{n}\right) & \text{if } \bar{v}_{i,j} < 0 \end{cases}$$

where $\nu_x = \Delta t/\Delta x$, $\nu_y = \Delta t/\Delta y$, and where subscripts i (j) refer to the horizontal (vertical) discretization and superscript n to the time discretization ($t = n\,\Delta t$). As boundary conditions we use extrapolation of the level set function. Thus, for instance, at the left boundary ($i = 1$), $\psi_{0,j}^{n} = 2\,\psi_{1,j}^{n} - \psi_{2,j}^{n}$ and, therefore,

$$\psi_{1,j}^{n+1} = \psi_{1,j}^{n} - \nu_x \bar{u}_{1j} \left(\psi_{2,j}^{n} - \psi_{1,j}^{n}\right) - \nu_y \bar{v}_{1j} \begin{cases} \left(\psi_{1,j}^{n} - \psi_{1,j-1}^{n}\right) & \text{if } \bar{v}_{1,j} \geq 0 \\ \left(\psi_{1,j+1}^{n} - \psi_{1,j}^{n}\right) & \text{if } \bar{v}_{1,j} < 0 \end{cases}$$

The time step is defined by a CFL condition

$$\Delta t = \frac{0.9}{\max\left\{\dfrac{|\bar{u}_{ij}|}{\Delta x} + \dfrac{|\bar{v}_{ij}|}{\Delta y}\right\}}. \tag{3.16}$$

At each time step, the level set function is updated from (3.15) and a set of points located in the new location of the front are computed by interpolating the level set function in the finite difference mesh. At the next time step, these points become RBF collocation centers, where the boundary condition $p = 0$ is enforced, while the old boundary centers become interior RBF centers. However, a filtering operation is needed to eliminate points in the front that are too much close to each other or too close to some interior point and which would, therefore, make the computation of the pressure very ill conditioned. In any case, as the front advances, the number of RBF centers increases and therefore the size and condition number of matrix A (in equation 2.14) increases.

The upper left of Figure 10 shows the finite difference mesh (30 x 30, $\Delta x = \Delta y = 0.6666$) and the initial set of RBF centers used. The upper right shows the level set function at $t = 0$ (signed distance function) together with its evolution at time step $t = 0.17$. The zero level set of each surface determines the location of the advancing front. The lower left of the figure shows a profile of the initial level set function (solid curve) at $x = 0, 4$ and the same profile at at a later time ($t = 0.17$). The intersection of these curves with the line $\psi = 0$ defines the location of the front at each time. The lower right of the figure shows the zero level set (advancing front) at $t = 0$ and $t = 0.17$.

To test the suitability of the method, we have used the rectangular mould described in Section 3, but with an elliptic inset placed in the center of the mould. The upper left of figure 11 shows the domain, the initial location of the front and the RBF centers. The other figures show the evolution of the front and the RBF centers for $t = 1, 2$ and 4.75 respectively. These centers are used to solve equation (1.9) using the RBF method described in Section 3.

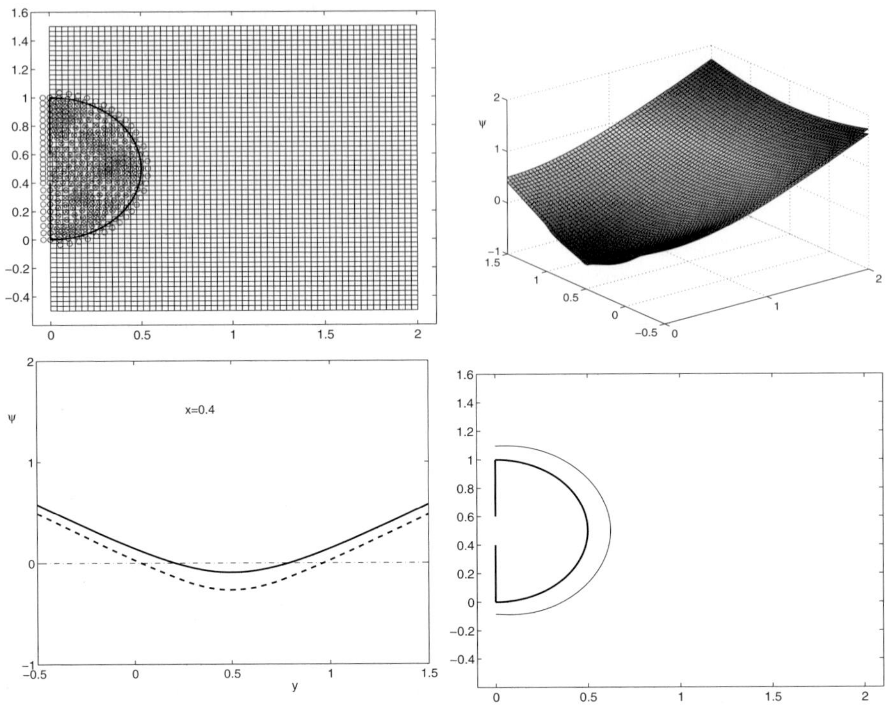

Figure 10. Upper left: uniform finite difference mesh and initial distribution of RBF centers (red circles) and collocation centers (crosses and black and blue circles). Upper right: level set function at $t = 0$ and $t = 0.17$. Lower left: level set function profile $\psi(0.4, y, 0)$ (solid line) and $\psi(0.4, y, 0.17)$ (dashed line). Lower right zero level sets at $t = 0$ and $t = 0.17$; $\psi(\mathbf{x}, 0) = 0$ and $\psi(\mathbf{x}, 0.17) = 0$.

As time progresses, the number of RBF centers increases (from 78 at $t = 0$ to 1051 at $t = 4.75$). Thus, the resulting linear system grows and the computing time per step increases. Notice also how the method is able to handle properly the collision of the two fronts behind the elliptic obstacle. These changes in topology are difficult to handle with other methods.

Figure 12 shows the evolution of the level set function $\psi(\mathbf{x}, t)$ which is obtained by solving (3.15). For $t = 0$ (upper left) this function is initialized to the signed distance from the front. Then this function is advected with the velocity field obtained by RBF interpolation of velocities in the centers located in the narrow band. No re-initialization has been necessary. Notice in the lower right figure that when two fronts collide and produce a change in topology, no difficulty is encountered. The front is always the zero level set of function ψ, and the two fronts coalesce into a new smooth front. However, the level set function has become very flat and, therefore, small errors in its computation may lead to high errors in the location of the front. This can be

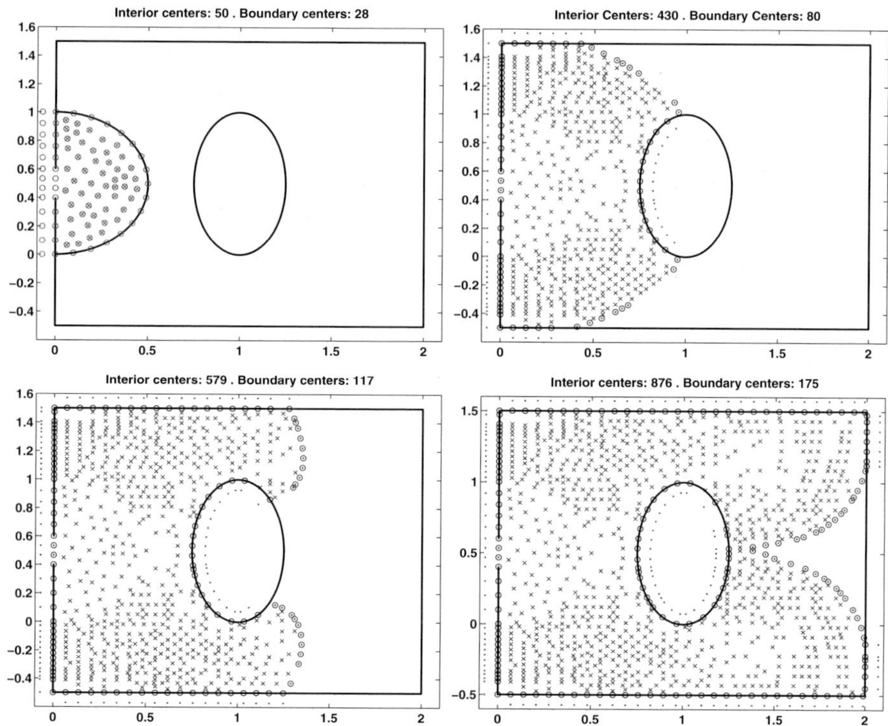

Figure 11. Distribution of RBF centers and collocation centers for $t = 0$ (upper left), $t = 1$ (upper right), $t = 2$ (lower left) and $t = 4.75$ (lower right).

avoided by re-initializing the level set function to a signed distance function, when $\min(\nabla\psi)$ becomes smaller than a predefined value.

Figure 13 shows the evolution of the advancing front as time progresses. For clarity, consequent time steps are plotted in different colors (red, blue). It shows the nodes located in the front for different time steps spaced approximately every 0.5 time units. Notice that initially the velocities are higher and the front advances much faster that in later times. However, the CFL condition (3.16) implies that Δt increases as the velocities decrease and, therefore, the number of time steps to advance a given length is approximately constant. As time progresses, the computational cost for advancing a given length increases because the number of RBF centers increases.

collocation nodes (t=0)	RBF centers (t=0)	cond(A)	time solution (2.14)	RBF centers final step	total time simulation
78	106	$1.461 \cdot 10^7$	0.014	1057	126.45
281	337	$5.818 \cdot 10^{11}$	0.117	1794	386.97
446	531	$9.650 \cdot 10^{13}$	0.332	2109	596.83

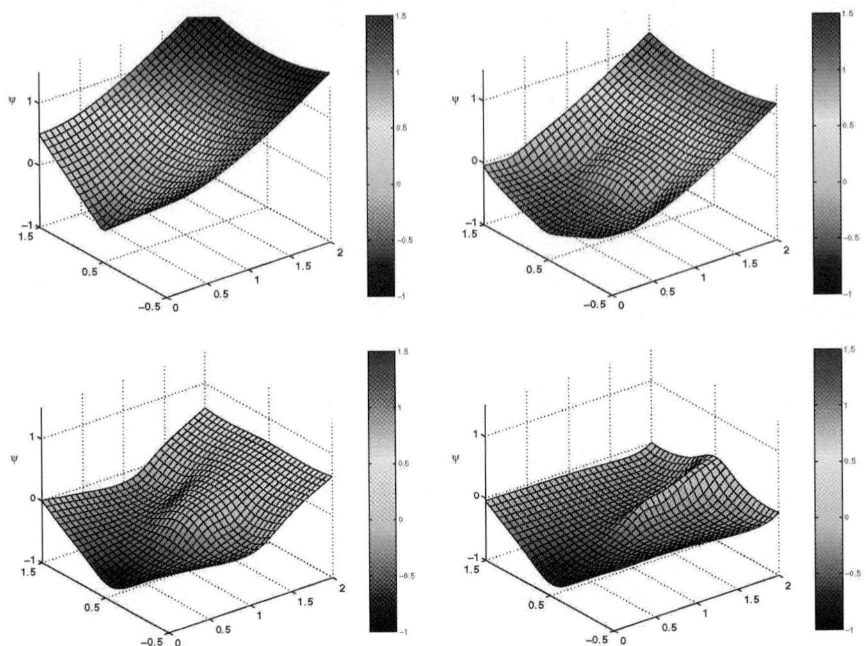

Figure 12. Level set function evolution: $t = 0$ (upper left), $t = 1$ (upper right), $t = 2$ (lower left) and $t = 4.75$ (lower right).

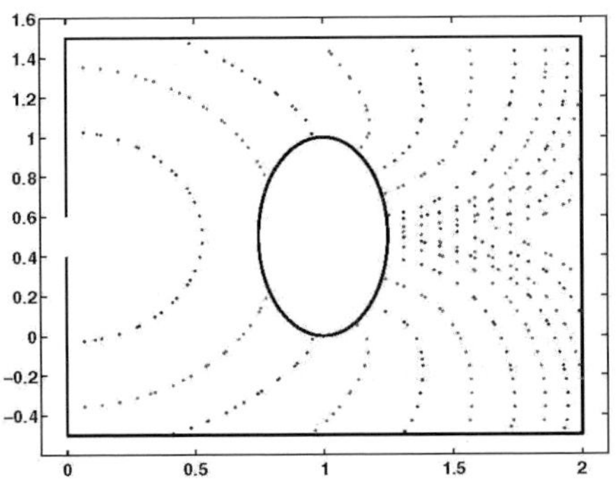

Figure 13. Front evolution. From left to right: $t = 0, 0.4947, 0.9794, 1.4753, 1.9839, 2.4905, 3.0533, 3.5662, 3.98898, 4.4296, 5.0233, 5.5171$ respectively.

The preceding table shows how the number of RBF centers affects the overall computational time. The first column lists the number of collocation nodes used in the first time step, while the second column lists the initial number of RBF centers (collocation nodes plus one additional center outside the boundary for each boundary collocation center). The third column shows how the condition number of matrix A for the linear system (2.14) at the first time step increases with N. The 4th column shows the time in seconds needed for the solution of the linear system (2.14). The results for $N = 337$ and $N = 531$ imply that $t \propto N^{2.29}$. For larger N the exponent should approach 3. As the simulation progresses and the front advances, the number of RBF centers increases. Column 5th shows the number of RBF centers in the last time step of the simulation. Finally, column 6th lists the total time for the complete simulation. Notice that the total simulation time grows approximately linear with the initial number of RBF centers ($T \propto N^{0.96}$). This is due to the fact that the time spent in solving equation (2.14) is only a fraction of the total simulation step and there are many other processes (computation of matrix A, computation of velocities in narrow band, interpolation of velocities, time stepping of level set function, ...) which contribute significantly to the total computational time.

4 Conclusions

We have used a meshfree method to solve a linear moving boundary problem corresponding to the Hele-Shaw approximation of injection molding. The pressure equation is accurately solved using the RBF collocation method and the front motion is modelled by evolving a signed distance level set function.

The discontinuities in the normal derivatives at the boundaries result in large errors when the standard RBF technique is used. However, this difficulty is eliminated by introducing additional RBF centers outside the domain and using either the *not-a-knot* method or PDE collocation at the boundary. With this modification, the method described is an efficient, fully meshfree, alternative to the traditional numerical method (finite elements and *volume of fluid* method) used in injection molding codes.

Acknowledgement. This work has been supported by the Spanish MEC grant FLUMRES DPI2002-04550-c07-03.

References

1. A. I. Fedoseyev, M. J. Friedman, and E. J. Kansa, *Improved multiquadric method for elliptic partial differential equations via PDE collocation on the Boundary*, Comput. Math. Appl., **43** (2002), 439–455.

2. B. Fornberg, T. A. Driscoll, G. Wright and R. Charles, *Observation of the behavior of radial basis function approximations near boundaries*, Comput. Math. Appl., **43** (2002), 473–490.
3. R. L. Hardy, *Multiquadric equations of topography and other irregular surfaces*, J. Geophys. Res., **176** (1971), 1905–1915.
4. C. A. Hieber and S. F. Shen, *A finite-element/finite difference simulation of the injection-molding filling process*, Journal of Non-Newtonian Fluid Mechanics, **7** (1979), 1–32.
5. E. J. Holm and H. P. Langtangen, *A unified finite element model for the injection molding process*, Comput. Methods Appl. Mech. Engrg., **178** (1999), 413–429.
6. E. J. Kansa, *Multiquadrics - a scattered data approximation scheme with applications to computational fluid-dynamics. I. Surface approximations and partial derivative estimates*, Comput. Math. Appls., **19** (1990), 127–145.
7. E. J. Kansa, *Multiquadrics - a scattered data approximation scheme with applications to computational fluid-dynamics. II. Solutions to parabolic, hyperbolic and elliptic partial differential equations*, Comput. Math. Appls., **19** (1990), 147–161.
8. T. H. Kwon and J. B. Park, *Finite element analysis modeling of powder injection molding filling process including yield stress and slip phenomena*, Polymer Engineering and Science, **35** (1995), 741–753.
9. J. Moody and C. J. Daken, *Fast learning in networks of locally-tuned processing units*, Neural Comput., **1** (1989), 281–294.
10. W. Noh and PC. Woodward, *A simple line interface calculation*, Proc. 5th Int. Conf. on Fluid Dynamics, Eds. A.I. vn de Vooran and P.J. Zandberger, Springer-Verlag (1976).
11. S. Osher and J. A. Sethian, *Fronts propagating with curvature dependent speed: algorithms based on Hamilton-Jacobi formulations*, J. Comput. Phys., **79** (1988), 12–49.
12. S. Osher and R. Fedkiw, *Level set methods and dynamic implicit surfaces*, Springer-Verlag (2003).
13. J. A. Sethian, *Level set methods and fast marching methods. Evolving interfaces in computational geometry, fluid mechanics, computer vision and materials science*, Cambridge University Press (1999).
14. O. Verhoyen and F. Dupret, *A simplified method for introducing the Cross viscosity law in the numerical simulation of Hele Shaw flow*, Journal of Non-Newtonian Fluid Mechanics, **74** (1998), 25–46.
15. I. Babuška, U. Banerjee, and J. E. Osborn, *Meshless and Generalized Finite Element Methods: A Survey of Some Major Results*, Meshfree Methods for Partial Differential Equations (M. Griebel and M. A. Schweitzer, eds.), Lecture Notes in Computational Science and Engineering, vol. 26, Springer, 2002, pp. 1–20.
16. T. Belytschko, Y. Y. Lu, and L. Gu, *Element-free Galerkin methods*, Int. J. Numer. Meth. Engrg. **37** (1994), 229–256.
17. D. Braess, *Finite elements: Theory, fast solvers, and applications in solid mechanics*, Cambridge University Press, 2001.
18. M. Griebel and M. A. Schweitzer (eds.), *Meshfree Methods for Partial Differential Equations*, Lecture Notes in Computational Science and Engineering, vol. 26, Springer, 2002.
19. M. Griebel and M. A. Schweitzer (eds.), *Meshfree Methods for Partial Differential Equations II*, Lecture Notes in Computational Science and Engineering, vol. 43, Springer, 2005.

Strain Smoothing for Stabilization and Regularization of Galerkin Meshfree Methods

J.S. Chen[1], W. Hu[1], M.A. Puso[2], Y. Wu[1], and X. Zhang[1]

[1] Department of Civil & Environmental Engineering, University of California, Los Angeles, Los Angeles, CA 90095, USA
jschen@seas.ucla.edu
[2] Department of Mechanical Engineering, Lawrence Livermore National Laboratory, Livermore CA 94550, USA

Summary. In this paper we introduce various forms of strain smoothing for stabilization and regularization of two types of instability: (1) numerical instability resulting from nodal domain integration of weak form, and (2) material instability due to material strain softening and localization behavior. For numerical spatial instability, we show that the conforming strain smoothing in stabilized conforming nodal integration only suppresses zero energy modes resulting from nodal domain integration. When the spurious nonzero energy modes are excited, additional stabilization is proposed. For problems involving strain softening and localization, regularization of the ill-posed problem is needed. We show that the gradient type regularization method for strain softening and localization can be formulated implicitly by introducing a gradient reproducing kernel strain smoothing. It is also demonstrated that the gradient reproducing kernel strain smoothing also provides a stabilization to the nodally integrated stiffness matrix. For application to modeling of fragment penetration processes, a nonconforming strain smoothing as a simplification of conforming strain smoothing is also introduced.

Key words: Strain smoothing, Stabilization, Regularization, Meshfree, Nodal integration

1 Introduction

Nodal integration offers considerable efficiency in Galerkin type meshfree methods, but it encounters spatial instability due to under integration and vanishing derivatives of meshfree shape functions at nodes. Several methods have been introduced as a correction and stabilization of nodal integration. Beissel et al. [1] proposed a least-squares stabilization technique. Randles et al. [12] introduced stress point method to enhance collocation formulation for SPH. Bonet et al. [2] presented a correction term into the derivative of shape function at nodal point, which is constructed by satisfying a linear patch test.

Chen et al. [5, 6, 13, 14] proposed a conforming strain smoothing in a stabilized conforming nodal integration (SCNI) as a stabilization of rank instability in nodal integration, and as a mechanism to pass linear patch test. Our recent study showed that SCNI produces spurious low energy modes under certain conditions. A modified SCNI is introduced, and its stability in reproducing kernel particle method is examined. For application to modeling of fragment penetration processes, a nonconforming strain smoothing as a simplification of conforming strain smoothing is also introduced.

Material instability refers to the event of strain softening and localization that yields an ill-posed problem. The difficulty in strain localization arises because solutions possess features of measure zero, and as such, characteristic length of the mesh introduces a mesh-size perturbation. The dependence on the discretization is not only with respect to mesh refinement but is also with respect to the mesh alignment. The inability of the classical continuum theory to describe the discontinuous strain fields can be corrected if the discontinuous strain field is regularized (smoothed). A commonly used regularization method in strain localization is the gradient method [8, 9, 10]. The classical gradient type regularization results in a governing equation with higher order differentiation, and thus requires additional nonphysical boundary conditions for a solution. A gradient reproducing kernel strain smoothing has been introduced as a regularization of strain localization problem [7, 4]. This gradient type regularization can be implicitly imbedded in a reproducing kernel approximation of strain, and thus does not require additional nonphysical boundary conditions to solve the regularized governing equation. In this work, we show that this gradient strain smoothing offers a stabilization of nodally integrated stiffness matrix similar to the one-point integrated stiffness with stabilization obtained from Taylor expansion of gradient matrix in the finite element setting [11].

This paper is organized as follows. In Section 2, we review the conforming strain smoothing in the stabilized conforming nodal integration (SCNI) for rank instability. In Section 3, we show that SCNI only eliminates improper zero energy modes, and additional stabilization is introduced to suppress spurious nonzero energy modes. In Section 4, we first demonstrate how gradient type regularization method for strain localization can be formulated under a gradient reproducing kernel strain smoothing. The eigenmodes associated with the regularized weak form integrated by nodal integration are also examined. For application to fragment problems in Section 5, we introduce a nonconforming strain smoothing as a simplification of conforming strain smoothing in SCNI. Concluding remarks are given in Section 6.

2 Conforming Strain Smoothing for Rank Instability

Domain integration of weak form poses considerable complexity in Galerkin meshfree method. Employment of Gauss quadrature rules yields integration error when background grids do not coincide with the covers of shape functions. Nodal integration, on the other hand, results in rank deficiency. Further, both Gauss integration and nodal integration methods do not pass linear patch test for nonuniform point distribution. For demonstration of rank instability, consider the following Poisson problem with Dirichlet boundary condition:

$$\nabla^2 u + Q = 0 \text{ in } \Omega \tag{2.1}$$

By introducing test function v^h, the corresponding Galerkin approximation is

$$a\left(v^h, u^h\right) = \left(v^h, Q\right) \tag{2.2}$$

By introducing nodal integration we have the following discrete bilinear form associated with the differential operator:

$$a^h\left(v^h, u^h\right) = \sum_{L=1}^{N} \nabla v^h\left(\mathbf{x}_L\right) \cdot \nabla u^h\left(\mathbf{x}_L\right) w_L \tag{2.3}$$

where w_L is the weight associated with point L. A rank instability resulting from nodal integration in (2.3) is shown in Figure 1 (a), where the test and trial functions are approximated by reproducing kernel approximation with linear basis. Low convergence rate in nodal integration is also observed in Figure 1 (b).

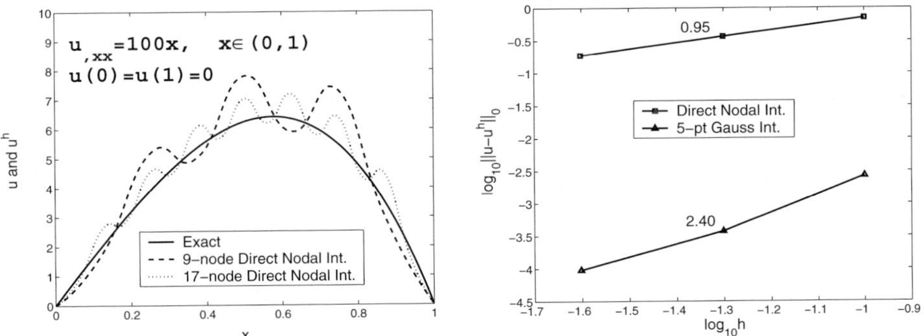

Figure 1. Nodal integration of weak form yields (a) rank instability (b) low convergence rate.

It is shown in Figure 1, even when the approximation of test and trial functions is linearly complete, the first order accuracy is not guaranteed in the Galerkin approximation due to nodal integration. Integration constraints are necessary conditions for linear exactness in the Galerkin approximation [5]. There are two requirements for linear exactness in the Galerkin approximation of second order differential equations. The first condition requires the approximation functions $\Psi_I\left(\mathbf{x}\right)$ of u^h and v^h to possess linear consistency:

$$\sum_{I=1}^{NP} \Psi_I(\mathbf{x}) = 1, \quad \sum_{I=1}^{NP} \Psi_I(\mathbf{x})\mathbf{x}_I = \mathbf{x} \tag{2.4}$$

These conditions are automatically satisfied in the reproducing kernel shape functions $\Psi_I(\mathbf{x})$ if complete linear bases are used. The second condition requires the numerical integration to satisfy the following integration constraint [5]:

$$\hat{\int_{\Omega}} \nabla\Psi_I d\Omega = \hat{\int_{\partial\Omega}} \Psi_I \mathbf{n} d\Gamma \tag{2.5}$$

where $\hat{\int}$ denotes numerical integration. A stabilized conforming nodal integration (SCNI) [5] has been introduced to satisfy linear patch test and to remedy rank instability of nodal integration. In SCNI approach, a conforming smoothed gradient operator at nodal point $\mathbf{x}_L$ is introduced as

$$\bar{\nabla}u^h(\mathbf{x}_L) = \sum_I \bar{\nabla}\Psi_I(\mathbf{x}_L) d_I \tag{2.6}$$

$$\bar{\nabla}\Psi_I(\mathbf{x}_L) = \frac{1}{w_L}\int_{\Omega_L} \nabla\Psi_I d\Omega = \frac{1}{w_L}\int_{\partial\Omega_L} \Psi_I \mathbf{n} d\Gamma, \quad w_L = \int_{\Omega_L} d\Omega \tag{2.7}$$

Here Ω_L is the nodal representative domain, which can be obtained from triangulation or Voronoi cell of a set of discrete points as shown in Figure 2. Note that a divergence theorem has been used in (2.7) to pass linear patch test when the weak form is integrated by nodal integration. It can be easily shown that the conforming smoothed gradient satisfies integration constraint of (2.5) using nodal integration, i.e.,

$$\begin{aligned} \hat{\int_{\Omega}} \bar{\nabla}\Psi_I d\Omega &= \sum_{L=1}^{NP} \bar{\nabla}\Psi_I(\mathbf{x}_L) w_L = \sum_{L=1}^{NP} \left(\frac{1}{w_L} \hat{\int_{\partial\Omega_L}} \Psi_I \mathbf{n} d\Gamma \right) w_L \\ &= \sum_{L=1}^{NP} \hat{\int_{\partial\Omega_L}} \Psi_I \mathbf{n} d\Gamma = \hat{\int_{\partial\Omega}} \Psi_I \mathbf{n} d\Gamma \end{aligned} \tag{2.8}$$

Note that to yield the results of (2.8) we have used the conforming property of nodal representation domain Ω_L as shown in Figure 2. Introducing the smoothed gradient of (2.7) into (2.3) yields the following discrete differential operator:

$$\bar{a}^h(v^h, u^h) = \sum_{L=1}^{NP} \bar{\nabla}v^h(\mathbf{x}_L) \cdot \bar{\nabla}u^h(\mathbf{x}_L) w_L \tag{2.9}$$

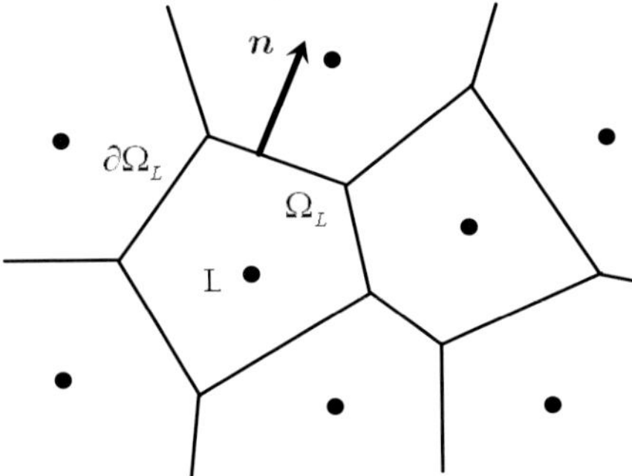

Figure 2. Nodal representative domains.

The corresponding stiffness matrix is

$$\begin{aligned} \mathbf{K}_{IJ} &= \sum_{L=1}^{NP} \bar{\mathbf{B}}_I^T(\mathbf{x}_L)\,\bar{\mathbf{B}}_J(\mathbf{x}_L)\,w_L \\ \bar{\mathbf{B}}_I(\mathbf{x}_L) &= \frac{1}{w_L}\overset{\wedge}{\int_{\partial\Omega_L}} \Psi_I \mathbf{n} d\Gamma, \quad w_L = \int_{\Omega_L} d\Omega \end{aligned} \tag{2.10}$$

The problem in Figure 1 is solved again for comparison of solution of using SCNI, nodal integration, and the fifth order Gauss quadrature rule as shown in Figure 3.

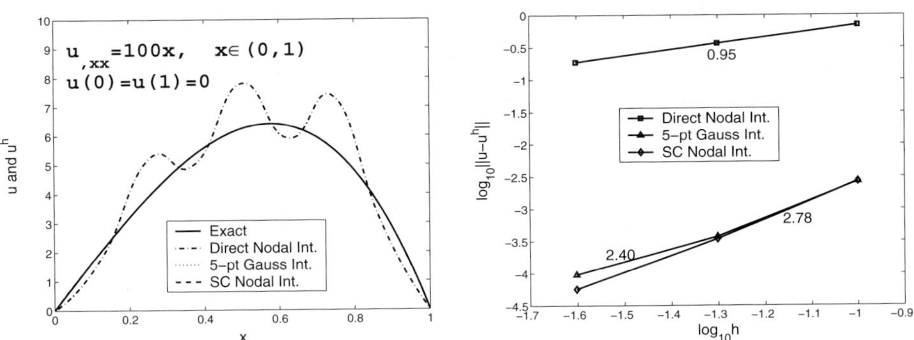

Figure 3. Comparison of solution obtained from various domain integration methods.

3 Additional Stabilization for Spurious Nonzero Energy Modes

The spurious zero energy mode with alternating gradient generates zero smoothed gradient at nodal point, except on the boundary. Thus for a finite domain, zero energy modes cannot propagate as the nonzero smoothed gradient on the boundary precludes the zero energy mode. However, the energy of an oscillatory mode can be very small and stability is not ensured as the mesh is refined. The V_1 coercivity requires the following condition for any non rigid body modes:

$$a^h\left(v^h, u^h\right) \geq \gamma \left\|u^h\right\|_1^2 \tag{3.11}$$

where γ is a mesh independent constant. Consider a uniform discretization of a one-dimensional domain by $N+1$ uniformly distributed points, and let the nodal value of u to take the oscillating pattern $[+1, -1, +1, \cdots +1, -1]$ we have the following bilinear form of SCNI with smoothed gradient:

$$\begin{aligned}\bar{a}^h\left(v^h, u^h\right) &= \frac{h}{2}\left(\frac{u_2 - u_1}{h}\right)^2 + h\left(\frac{u_3 - u_1}{2h}\right)^2 + \cdots + \\ &\quad + h\left(\frac{u_{N+1} - u_{N-1}}{2h}\right)^2 + \frac{h}{2}\left(\frac{u_{N+1} - u_N}{h}\right)^2 \\ &= \frac{4}{h}\end{aligned} \tag{3.12}$$

where h is the nodal distance. It can also be shown

$$\begin{aligned}\left\|u^h\right\|_1^2 &= \int_o^L u^2 dx + \int_o^L u_{,x}^2 dx \\ &= \frac{1}{3}\left(u_1^2 + u_1 u_2 + u_2^2\right) h + \frac{1}{3}\left(u_2^2 + u_2 u_3 + u_3^2\right) h + \cdots + \\ &\quad + \frac{1}{3}\left(u_N^2 + u_N u_{N+1} + u_{N+1}^2\right) h + \left(\frac{u_2 - u_1}{h}\right)^2 h + \\ &\quad + \left(\frac{u_3 - u_2}{h}\right)^2 h + \cdots + \left(\frac{u_{N+1} - u_N}{h}\right)^2 h \\ &= \frac{1}{3} L + \frac{4}{h^2} L \\ &= \frac{h^2 + 12}{3h^2} L\end{aligned} \tag{3.13}$$

It follows $\gamma = \bar{a}^h\left(v^h, u^h\right) / \left\|u^h\right\|_1^2 = \frac{12h}{L(h^2+12)} \rightarrow 0$ as $h \rightarrow 0$ and thus violating coercivity. One way to gain coercivity in SCNI is to consider the following modification:

$$\bar{a}^h\left(v^h,u^h\right) = \sum_{L=1}^{NP}\left\{\underbrace{\bar{\nabla}v^h\left(\mathbf{x}_L\right)\cdot\bar{\nabla}u^h\left(\mathbf{x}_L\right)}_{\text{SCNI}} + \underbrace{\sum_{K\in S_L}\left[c_L^K\left(\left(\nabla v^h\right)_L^K - \bar{\nabla}v^h\left(\mathbf{x}_L\right)\right)\cdot\left(\left(\nabla u^h\right)_L^K - \bar{\nabla}u^h\left(\mathbf{x}_L\right)\right)\right]}_{\text{Additional stabilization}}\right\} w_L \tag{3.14}$$

where $\left(\nabla u^h\right)_L^K$ is ∇u^h evaluated at the centroid of the K-th Delaunay triangles associated with node L as shown in Figure 4, c_L^K is the stabilization parameter, and S_L is the set containing Delaunay triangles associated with node L.

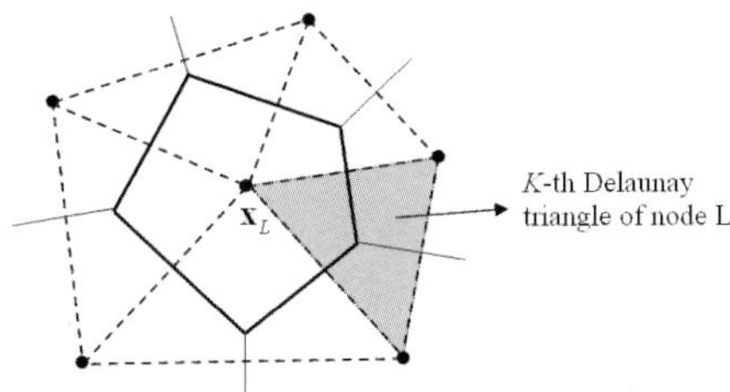

Figure 4. Voronoi cell and Delaunay trangulation.

In elasticity, the Galerkin approximation is

$$a\left(\mathbf{v}^h,\mathbf{u}^h\right) = \left(\mathbf{v}^h,\mathbf{b}\right) + \left(\mathbf{v}^h,\mathbf{h}\right)_{\partial\Omega^{h_i}} \tag{3.15}$$

where $\left(\mathbf{v}^h,\mathbf{b}\right)$ and $\left(\mathbf{v}^h,\mathbf{h}\right)_{\partial\Omega^{h_i}}$ are the standard linear forms in domain Ω and Neumann boundary $\partial\Omega^{h_i}$, respectively, $\mathbf{b}$ is the body force vector, $\mathbf{h}$ is the traction vector, and $a\left(\mathbf{v}^h,\mathbf{u}^h\right)$ in elasticity is

$$a\left(\mathbf{v}^h,\mathbf{u}^h\right) = \int_\Omega \nabla^s\mathbf{v}^h : \mathbf{C} : \nabla^s\mathbf{u}^h d\Omega \equiv \int_\Omega \varepsilon\left(\mathbf{v}^h\right) : \mathbf{C} : \varepsilon\left(\mathbf{u}^h\right) d\Omega \tag{3.16}$$

where

$$\varepsilon\left(\mathbf{u}^h\right)_{ij} \equiv \left(\nabla^s\mathbf{u}^h\right)_{ij} = \frac{1}{2}\left(\frac{\partial u_i^h}{\partial x_j} + \frac{\partial u_j^h}{\partial x_i}\right) \tag{3.17}$$

and $\mathbf{C}$ is the 4-th rank elasticity tensor. By taking a similar modified SCNI procedure with smoothed gradient on ∇^s, we have the following discrete equation:

$$\bar{a}^h\left(\mathbf{v}^h,\mathbf{u}^h\right)=\left(\mathbf{v}^h,\mathbf{b}\right)^h+\left(\mathbf{v}^h,\mathbf{h}\right)^h_{\partial\Omega^{h_i}} \tag{3.18}$$

Here $\left(\mathbf{v}^h,\mathbf{b}\right)^h$ and $\left(\mathbf{v}^h,\mathbf{h}\right)^h_{\partial\Omega^{h_i}}$ are obtained by nodal integration, and

$$\begin{aligned}&\bar{a}^h\left(\mathbf{v}^h,\mathbf{u}^h\right)\\&=\sum_{L=1}^{NP}\left\{\underbrace{\bar{\varepsilon}\left(\mathbf{v}^h\right)_L:\mathbf{C}:\bar{\varepsilon}\left(\mathbf{u}^h\right)_L}_{\text{SCNI}}+\right.\\&\left.+\underbrace{\sum_{K\in S_L}\left[c_L^K\left(\varepsilon\left(\mathbf{v}^h\right)_L^K-\bar{\varepsilon}\left(\mathbf{v}^h\right)_L\right):\mathbf{C}:\left(\varepsilon\left(\mathbf{u}^h\right)_L^K-\bar{\varepsilon}\left(\mathbf{u}^h\right)_L\right)\right]}_{\text{Additional stabilization}}\right\}w_L\end{aligned} \tag{3.19}$$

where

$$\begin{aligned}\left(\bar{\varepsilon}\left(\mathbf{u}^h\right)_L\right)_{ij}&=\frac{1}{w_L}\int\limits_{\Omega_L}^{\wedge}\frac{1}{2}\left(\frac{\partial u_i^h}{\partial x_j}+\frac{\partial u_j^h}{\partial x_i}\right)d\Omega\\&=\frac{1}{w_L}\int\limits_{\partial\Omega_L}^{\wedge}\frac{1}{2}\left(\partial u_i^h n_j+\partial u_j^h n_i\right)d\Gamma\\&=\sum_{I=1}^{NP}\frac{1}{2}\left(\bar{\nabla}_i\Psi_I d_{jI}+\bar{\nabla}_j\Psi_I d_{iI}\right)\end{aligned} \tag{3.20}$$

$$\bar{\nabla}_i\Psi_I=\frac{1}{w_L}\int\limits_{\partial\Omega_L}^{\wedge}\Psi_I n_i d\Gamma, \tag{3.21}$$

and $\varepsilon\left(\mathbf{u}^h\right)_L^K$ is $\varepsilon\left(\mathbf{u}^h\right)$ evaluated at the centroid of the K-th Delaunay triangles associated with node L as shown in Figure 4, and $c_L^K$is the stabilization parameter.

A first nonzero eigenmode associated with the stiffness in two dimensional elasticity ($E=1,\nu=0.499$) integrated using SCNI is shown in Figure 5. The corresponding eigenvalue of this mode is 0.0211. The first nonzero energy eigenmode obtained using modified SCNI (M-SCNI) with stabilization parameter $c_L^K=c=0.01$, which will be used for all the following numerical examples, has shown to provide stabilization.

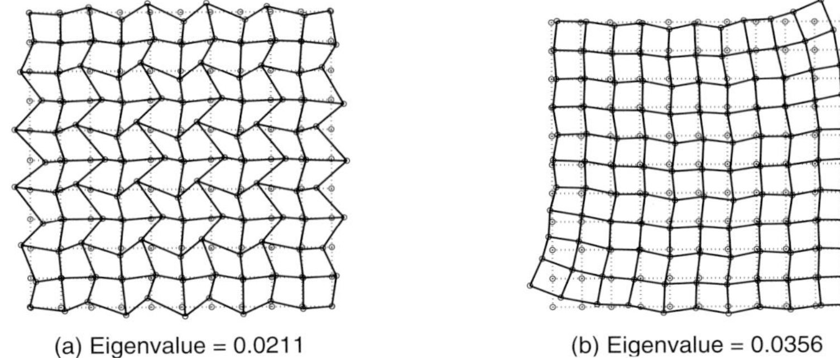

Figure 5. First nonzero energy eigenmode generated using (a) SCNI and (b) M-SCNI with $c = 0.01$.

A cantilever beam shown in Figure 6 is analyzed. Figure 7 compares the results of SCNI and M-SCNI with different stabilization parameters c. It is shown that the use of $c = 0.01$ in M-SCNI that properly suppresses spurious nonzero energy modes in Figure 5 also generates good convergence rate in Figure 7.

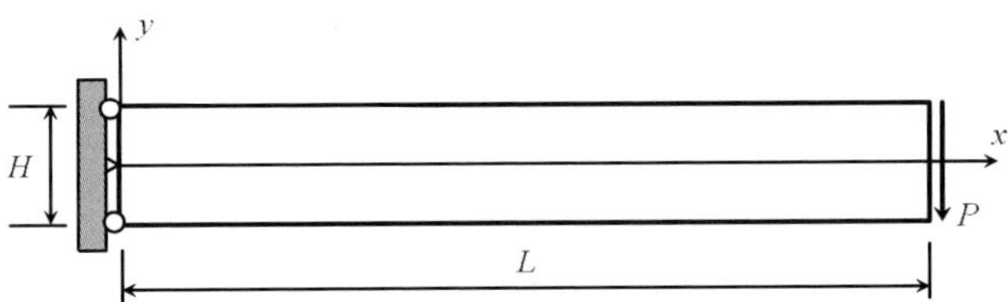

Figure 6. Cantilever beam $(L = 10m, H = 2m, P = 200N, E = 3 \times 10^7 Pa, \nu = 0.25)$.

The numerical studies show that the use of a minimum value in $c_L^K = c = 0.01$ is required for stability. Employment of c_L^K larger than 0.01, on the other hand, yields reduction of convergence rate. It is recommended that $c_L^K = c = 0.01$ be employed in M-SCNI for stability and accuracy.

4 Gradient Strain Smoothing Regularization for Material Instability in Strain Localization

Strain localization is usually a precursor of catastrophic material failure. In the event of strain softing and localization, the change of sign in the tangent moduli yields an ill-posed problem and requires a regularization for unique

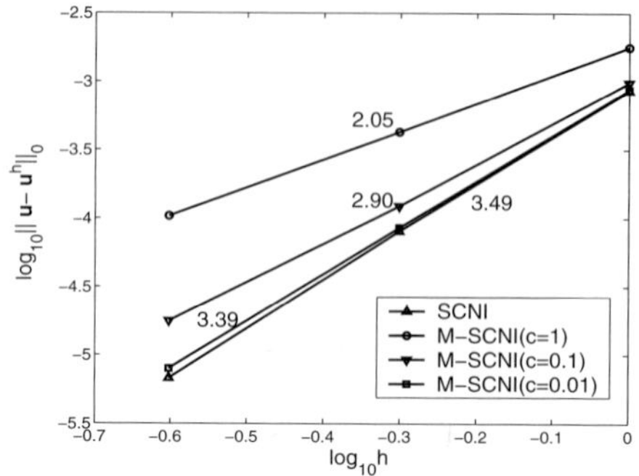

Figure 7. Convergence rate of L_2 error norm.

solution. A commonly used regularization method in strain localization is the gradient method [8, 9, 10], where a "nonlocal strain" $\tilde{\varepsilon}$ is introduced as

$$\tilde{\varepsilon}(\mathbf{x}) = \varepsilon(\mathbf{x}) + \sum_{i+j=1}^{n} \beta_{ij} D_{ij} \varepsilon(\mathbf{x}) \tag{4.22}$$

where ε is any component of strain, and $D_{ij}(\cdot) = \partial^{i+j}(\cdot)/\partial x_1^i \partial x_2^j$. However, introducing strain regularization of (4.22) into equilibrium equation leads to a PDE with higher order differentiation and thus requires additional non-physical boundary conditions to solve the problem. Until now, the physical explanation of these additional boundary conditions is still lacking [9]. The issue of additional boundary conditions has been addressed by introducing a gradient reproducing kernel strain smoothing [4, 7] as follows:

$$\tilde{\varepsilon}(\mathbf{x}) = \int_{\Omega} \tilde{\Phi}_a(\mathbf{x}; \mathbf{x} - \mathbf{s}) \varepsilon(\mathbf{s}) d\mathbf{s} \tag{4.23}$$

We write the discrete counter part of the strain smoothing equation as

$$\tilde{\varepsilon}(\mathbf{x}) = \sum_{I=1}^{NP} \tilde{\Phi}_a(\mathbf{x}; \mathbf{x} - \mathbf{x}_I) \varepsilon(\mathbf{x}_I) \tag{4.24}$$

where $\tilde{\Phi}_a(\mathbf{x}; \mathbf{x} - \mathbf{x}_I)$ is a smoothing function that takes the following reproducing kernel form:

$$\begin{aligned} \tilde{\Phi}_a(\mathbf{x}; \mathbf{x} - \mathbf{x}_I) &= \left[\sum_{i+j=0}^{n} (x_1 - x_{1I})^i (x_2 - x_{2I})^j b_{ij}(\mathbf{x}) \right] \Phi_a(\mathbf{x} - \mathbf{x}_I) \\ &\equiv \mathbf{H}^T(\mathbf{x} - \mathbf{x}_I) \mathbf{b}(\mathbf{x}) \Phi_a(\mathbf{x} - \mathbf{x}_I) \end{aligned} \tag{4.25}$$

The vector $\mathbf{b}(\mathbf{x})$ is obtained by imposing the following gradient reproducing conditions according to the gradient regularization equation (4.22):

$$\sum_{I=1}^{NP} x_{1I}^p x_{2I}^q \tilde{\Phi}_a(\mathbf{x};\mathbf{x}-\mathbf{x}_I) = x_1^p x_2^q + \sum_{i+j=0}^{n} \beta_{ij} D_{ij}(x_1^p x_2^q)$$

$$0 \le p+q \le n \ , \ \beta_{00} = 0 \tag{4.26}$$

It can be shown that the above equation is equivalent to

$$\sum_{I=1}^{NP} (x_1 - x_{1I})^p (x_2 - x_{2I})^q \tilde{\Phi}_a(\mathbf{x};\mathbf{x}-\mathbf{x}_I) = \delta_{p0}\delta_{q0} + \beta_{pq}(-1)^{p+q} p! q!$$

$$0 \le p+q \le n \ , \ \beta_{00} = 0 \tag{4.27}$$

This leads to a system of discrete equations for $\mathbf{b}(\mathbf{x})$

$$\mathbf{M}(\mathbf{x})\mathbf{b}(\mathbf{x}) = \mathbf{g} \tag{4.28}$$

where $\mathbf{g}$ is the vector form of $g_{ij} = \delta_{i0}\delta_{j0} + \beta_{ij}(-1)^{i+j} i! j!$, and

$$\mathbf{M}(\mathbf{x}) = \sum_{I=1}^{NP} \mathbf{H}(\mathbf{x}-\mathbf{x}_I)\mathbf{H}^T(\mathbf{x}-\mathbf{x}_I)\Phi_a(\mathbf{x}-\mathbf{x}_I) \tag{4.29}$$

Substituting $\mathbf{b}(\mathbf{x}) = \mathbf{M}^{-1}(\mathbf{x})\mathbf{g}$ into (4.25), we obtain the following strain smoothing function

$$\tilde{\Phi}_a(\mathbf{x};\mathbf{x}-\mathbf{x}_I) = \mathbf{H}(\mathbf{x}-\mathbf{x}_I)\mathbf{M}^{-1}(\mathbf{x})\mathbf{g}\Phi_a(\mathbf{x}-\mathbf{x}_I) \tag{4.30}$$

Due to the employment of gradient reproducing conditions in the construction of smoothing function $\tilde{\Phi}_a$, we have the following property:

$$\tilde{\varepsilon}(\mathbf{x}) = \sum_{I=1}^{NP} \tilde{\Phi}(\mathbf{x};\mathbf{x}-\mathbf{x}_I)\,\varepsilon(\mathbf{x}_I) \approx \varepsilon(\mathbf{x}) + \sum_{i+j=1}^{n} \beta_{ij} D_{ij}\varepsilon(\mathbf{x}) \tag{4.31}$$

For example, in one-dimension, if 2nd order basis and $\mathbf{g}^T = [1, 0, 2\beta]$ are used in the construction of $\tilde{\Phi}_a$ in (4.30), we have the following property:

$$\tilde{\varepsilon}(\mathbf{x}) = \sum_{I=1}^{NP} \tilde{\Phi}(\mathbf{x};\mathbf{x}-\mathbf{x}_I)\varepsilon(\mathbf{x}_I) \approx \varepsilon(\mathbf{x}) + \beta\varepsilon_{,xx}(\mathbf{x}) \tag{4.32}$$

Recalling strain smoothing equation (4.24), and introducing approximation of displacements, we obtain

$$\begin{aligned}
\tilde{\varepsilon}(\mathbf{x}) &= \sum_{I=1}^{NP} \tilde{\Phi}_a(\mathbf{x};\mathbf{x}-\mathbf{x}_I)\varepsilon(\mathbf{x}_I) \\
&= \sum_{I=1}^{NP} \tilde{\Phi}_a(\mathbf{x};\mathbf{x}-\mathbf{x}_I)\left\{\frac{1}{2}\left[\frac{\partial u_i^h(\mathbf{x}_I)}{\partial x_j} + \frac{\partial u_j^h(\mathbf{x}_I)}{\partial x_i}\right]\right\} \\
&= \sum_{J=1}^{NP} \left(\tilde{B}_{iJ} d_{jJ} + \tilde{B}_{jJ} d_{iJ}\right)
\end{aligned} \tag{4.33}$$

where

$$\tilde{B}_{iJ} = \sum_{I=1}^{NP} \frac{1}{2} \tilde{\Phi}_a(\mathbf{x}; \mathbf{x} - \mathbf{x}_I) \frac{\partial \Psi_J(\mathbf{x}_I)}{\partial x_i} \tag{4.34}$$

A one-dimensional damage mechanics problem as shown in Figure 8 is analyzed. A rod with imperfection in the middle of the bar is subjected to a uniaxial tensile deformation as shown in Figure 8. The equilibrium equation of this problem is

$$\begin{aligned} \left[E(1-D)\varepsilon\right]_{,x} &= 0, \quad 0 < x < L \\ u(0) &= 0 \\ u(L) &= g \end{aligned} \tag{4.35}$$

where D is the damage function. Here, we consider the following damage function:

$$D(\varepsilon) = \begin{cases} \frac{\varepsilon_c(\varepsilon - \varepsilon_i)}{\varepsilon(\varepsilon_c - \varepsilon_i)} & \text{if } \varepsilon_i \leq \varepsilon \leq \varepsilon_c \\ 1 & \text{if } \quad \varepsilon > \varepsilon_c \end{cases} \tag{4.36}$$

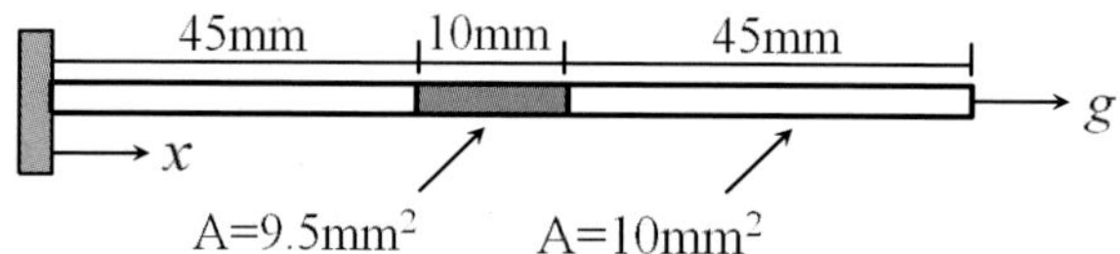

Figure 8. One-dimensional bar model.

Equations (4.35) and (4.36) represent a linear softening behavior as shown in Figure 9.

The following parameters are used: $\varepsilon_i = 1.0 \times 10^{-4}$, $\varepsilon_c = 6.25 \times 10^{-3}$, and $E = 2 \times 10^6 N/mm^2$. An imperfection of the cross sectional area between $x = 30$ mm and $x = 40$ mm is introduced to initiate bifurcation from a homogeneous state of deformation. It is expected that the strain will localize in the imperfection zone, while the rest of the structure will relax elastically. We consider a second order gradient regularization $\tilde{\varepsilon} = \varepsilon + \beta\varepsilon_{,xx}$, with $\beta = 0.0408$, and third order basis functions are employed in the smoothing function. The force-displacement curves obtained using four spatial discretizations regularized with the second order gradient method shown in Figure 10 demonstrate a mesh-independent results using the proposed method.

The regularized strain approximation in (4.33)-(4.34) also provides a stabilization to the nodally integrated discrete differential operator. Consider the weak form with the smoothed strain as follows:

$$\tilde{a}\left(\mathbf{v}^h, \mathbf{u}^h\right) = \left(\mathbf{v}^h, \mathbf{b}\right) + \left(\mathbf{v}^h, \mathbf{h}\right)_{\partial\Omega^{h_i}} \tag{4.37}$$

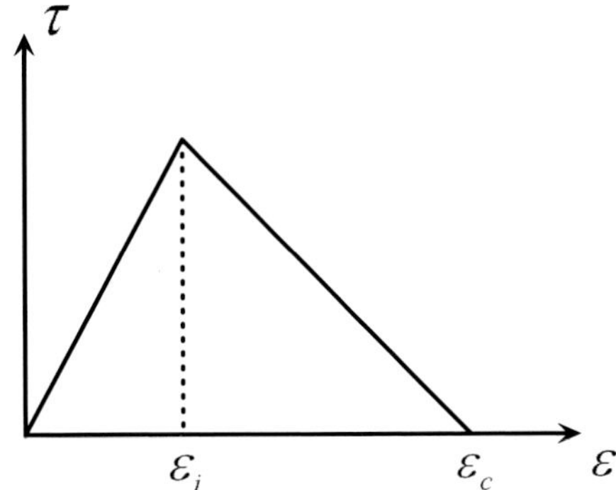

Figure 9. Linear softening stress-strain relation of 1-D elastic damage model.

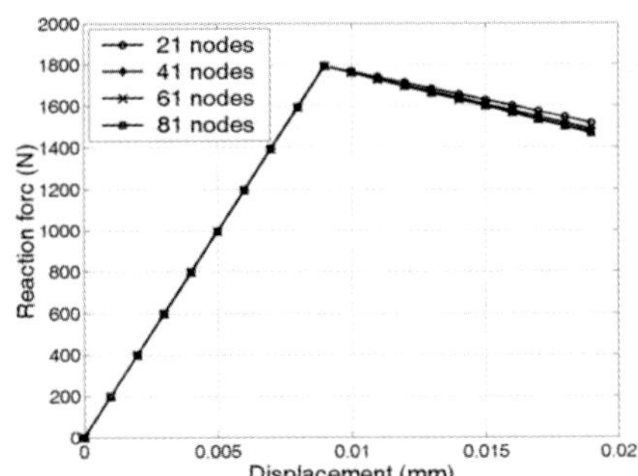

Figure 10. Force-displacement curves obtained by second order strain gradient method.

where

$$\tilde{a}\left(\mathbf{v}^h, \mathbf{u}^h\right) = \int_{\Omega} \tilde{\varepsilon}\left(\mathbf{v}^h\right) : \mathbf{C} : \tilde{\varepsilon}\left(\mathbf{u}^h\right) d\Omega \tag{4.38}$$

The nodal integration of (4.38) yields:

$$\tilde{a}^h\left(\mathbf{v}^h, \mathbf{u}^h\right) = \sum_{L=1}^{NP} \tilde{\varepsilon}\left(\mathbf{v}^h\right)_L : \mathbf{C} : \tilde{\varepsilon}\left(\mathbf{u}^h\right)_L w_L \tag{4.39}$$

Note that by using the gradient producing properties in (4.31), we have:

$$\begin{aligned} &\tilde{a}^h\left(\mathbf{v}^h, \mathbf{u}^h\right) \\ &= \sum_{L=1}^{NP} \tilde{\varepsilon}\left(\mathbf{v}^h\right)_L : \mathbf{C} : \tilde{\varepsilon}\left(\mathbf{u}^h\right)_L w_L \\ &\approx \sum_{L=1}^{NP} \left[\varepsilon\left(\mathbf{v}^h\right)_L + \sum_{i,j} \beta_{ij} D_{ij} \varepsilon\left(\mathbf{v}^h\right)_L\right] : \mathbf{C} : \left[\varepsilon\left(\mathbf{u}^h\right)_L + \sum_{i,j} \beta_{ij} D_{ij} \varepsilon\left(\mathbf{u}^h\right)_L\right] w_L \end{aligned} \tag{4.40}$$

This is analogous to the use of Taylor expansion of gradient matrix in the stiffness matrix as a stabilization of the one-point integrated stiffness matrix [11] in finite element.

Figure 11 shows the comparison of the first nonzero eigenmodes of a nodally integrated stiffness matrix of 2-dimensional elasticity obtained using SCNI from (2.9), modified SCNI (M-SCNI) from (3.19), and gradient SCNI (G-SCNI) from (4.40), and no spurious oscillation is observed in M-SCNI and G-SCNI.

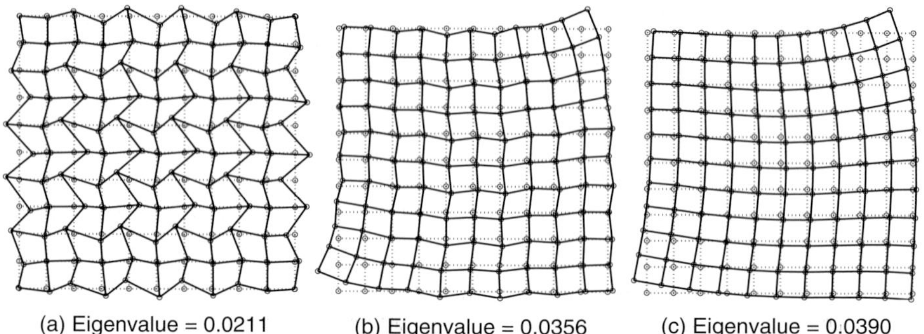

Figure 11. The first nonzero energy eigenmodes of nodally integrated stiffness using (a) SCNI, (b) M-SCNI, and (c) G-SCNI.

5 Application to Fragment Penetration Problems

5.1 Stabilized Nonconforming Nodal Integration (SNNI)

In transient large deformation problems, consider the following governing equation:

$$\rho \ddot{\mathbf{u}} = \nabla \cdot \boldsymbol{\tau} + \mathbf{b} \text{ in } \Omega_{\mathbf{x}} \tag{5.41}$$

with boundary conditions $u_i = g_i$ on $\Gamma_{\mathbf{x}}^{g}$ and $\tau_{ij} n_j = h_i$ on $\Gamma_{\mathbf{x}}^{h_i}$, where $\Omega_{\mathbf{x}}$, $\Gamma_{\mathbf{x}}^{g_i}$, and $\Gamma_{\mathbf{x}}^{h_i}$are the domain, Dirichlet boundary, and Neumann boundary at the current configuration, respectively, ρ is density, $\boldsymbol{\tau}$ is Cauchy stress, $\mathbf{b}$ is the body force vector, h_i is the traction. The corresponding Gakerkin approximation is

$$\left(\mathbf{v}^h, \rho \mathbf{u}^h\right) + a\left(\mathbf{v}^h, \mathbf{u}^h\right) = \left(\mathbf{v}^h, \mathbf{b}\right) + \left(\mathbf{v}^h, \mathbf{h}\right)_{\partial \Omega^{h_i}} \tag{5.42}$$

For fragment problems, updated Lagrangian formulation is employed, and the above weak form is integrated over the domain and boundary at the current configuration, in which $a\left(\mathbf{v}^h, \mathbf{u}^h\right)$ is expressed as:

$$a\left(\mathbf{v}^h, \mathbf{u}^h\right) = \int_{\Omega_x} \nabla^s \mathbf{v}^h : \boldsymbol{\tau}\left(\nabla^s \mathbf{u}^h\right) d\Omega \equiv \int_{\Omega_x} \varepsilon\left(\mathbf{v}^h\right) : \boldsymbol{\tau}\left(\varepsilon\left(\mathbf{u}^h\right)\right) d\Omega \tag{5.43}$$

where Ω_x is the domain at the current configuration, and

$$\varepsilon\left(\mathbf{u}^h\right)_{ij} \equiv \left(\nabla^s \mathbf{u}^h\right)_{ij} = \frac{1}{2}\left(\frac{\partial u_i^h}{\partial x_j} + \frac{\partial u_j^h}{\partial x_i}\right) \tag{5.44}$$

and x_i is the spatial coordinate at the current configuration. In updated Lagrangian formulation, the approximation of $\mathbf{u}^h$ is formulated at the current configuration [3], and constructing Voronoi cell for conforming strain smoothing in SCNI becomes tedious, and sometimes impossible during the fragmentation processes. To simplify the conforming strain smoothing in SCNI, we consider the following stabilized nonconforming nodal integration (SNNI), in which a simple nonconforming circular smoothing domain ϖ_L with boundary $\partial\varpi_L$ is used, as shown in Figure 12. For elasticity, a nonconforming strain smoothing is introduced

$$\bar{\varepsilon}_{ij}^h(\mathbf{x}_L) = \frac{1}{2\pi r^2}\int_{\varpi_L}(u_{i,j}^h + u_{j,i}^h)d\Omega = \frac{1}{2\pi r^2}\int_{\partial\varpi_L}(u_i^h n_j + u_j^h n_i)d\Gamma \quad (5.45)$$

where r is the radius of the circular smoothing domain.

Remark 1. Additional stabilization in the construction of stiffness follows (3.19), where $\varepsilon\left(\mathbf{u}^h\right)_L^K$ can be evaluated at the centroid of the quarter circle associated with node L.

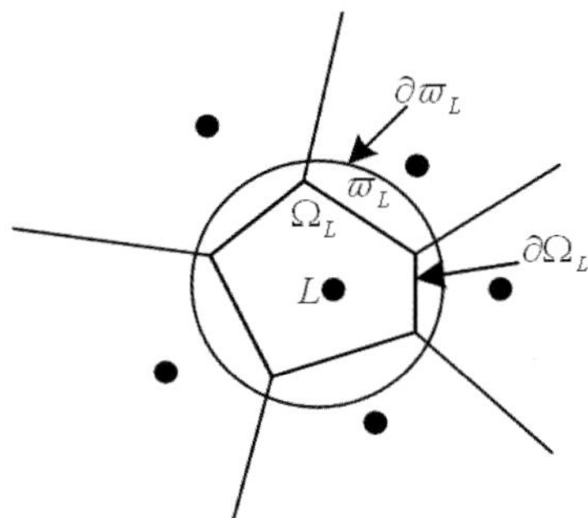

Figure 12. Averaging domain for SNNI.

5.2 Test of Convergence Property of SNNI in Elastostatics

A cantilever beam as shown in Figure 6 is analyzed, where $L = 10m$, $H = 2m$, $P = 1.0 \times 10^4 N$, $E = 2.11 \times 10^7 Pa$, $\nu = 0.3$. The meshfree discretization (of half the beam) and the corresponding Voronoi nodal representative domain for SCNI and smoothing domain for SNNI with $\alpha = 1.5$ are shown in Figure 13(a) and Figure 13(b), where $\alpha = \sqrt{\int_{\varpi_L} d\Omega} / \sqrt{\int_{\Omega_L} d\Omega}$. Since this beam geometry has large surface to volume ratio, we do not add the additional stabilization in SNNI. The tip displacement normalized by the analytical solution is shown in table 1. The comparison of L_2 error norms of the solution obtained with various integration methods are compared in Figure 14. The numerical experiment shows that the smoothing domain with α around 1.0 gives the best convergence rate for SNNI and is used for the following fragment problem. Although SNNI method offers a slight reduction in the solution accuracy

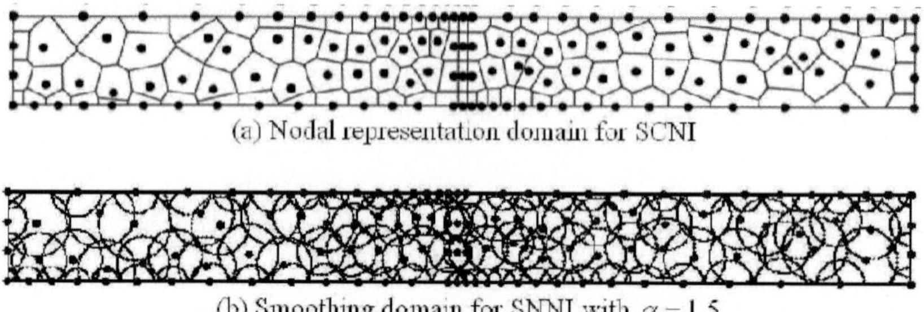

Figure 13. Nodal representative and smoothing domains for SCNI and SNNI.

Table 1. Accuracy of tip displacement of beam problem by different integration techniques.

Integration Method	Accuracy (%)
Gauss 5×5	95.0
Direct Nodal	192.8
SCNI	99.2
SNNI	98.5

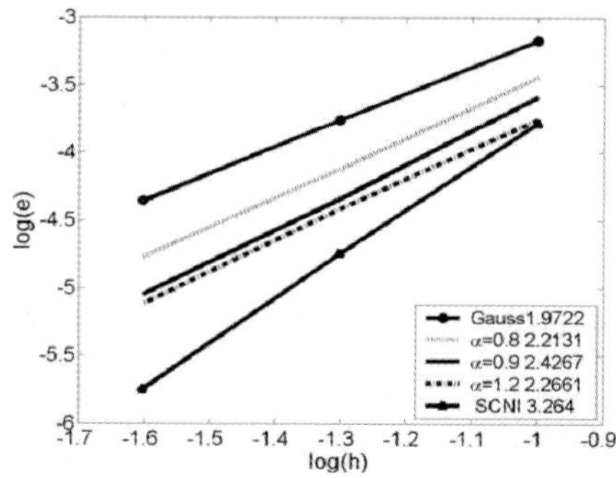

Figure 14. L_2 error norms of solution obtained from SNNI $(e = ||\mathbf{u}^h - \mathbf{u}||_0)$ with various α values, SCNI, and Gauss integration.

compared to that of SCNI, the simplicity in the strain smoothing makes it particularly attractive for fragment problems. For problems that exhibit excessive particle motion and large degree of material separation, SNNI is the obvious choice over SCNI.

5.3 High Velocity Penetration

A Carbon-Tungsten projectile ball is impacting on a $0.4m \times 1m$ concrete wall. Plane strain assumption is assumed. The material constants of the ball are: Young's modulus $E = 630GPa$, Poisson's ratio $\nu = 0.33$, density

$\rho = 8700kg/m^3$, initial yield stress $\sigma_y = 800MPa$ and the initial velocity of the ball is $500m/s$. The ball is discretized into 100 particles. The material constants of the concrete wall are: Young's modulus $E = 30GPa$, Poisson's ratio $\nu = 0.2$, density $\rho = 2400kg/m^3$, initial yield stress $\sigma_y = 40MPa$. The wall is fixed at both vertical ends and uniformly discretized into $41 \times 17 = 697$ particles. The two vertical surfaces of the concrete wall are fixed. The SNNI domain integration is employed. The progressive deformation of the penetration is shown in Figure 15.

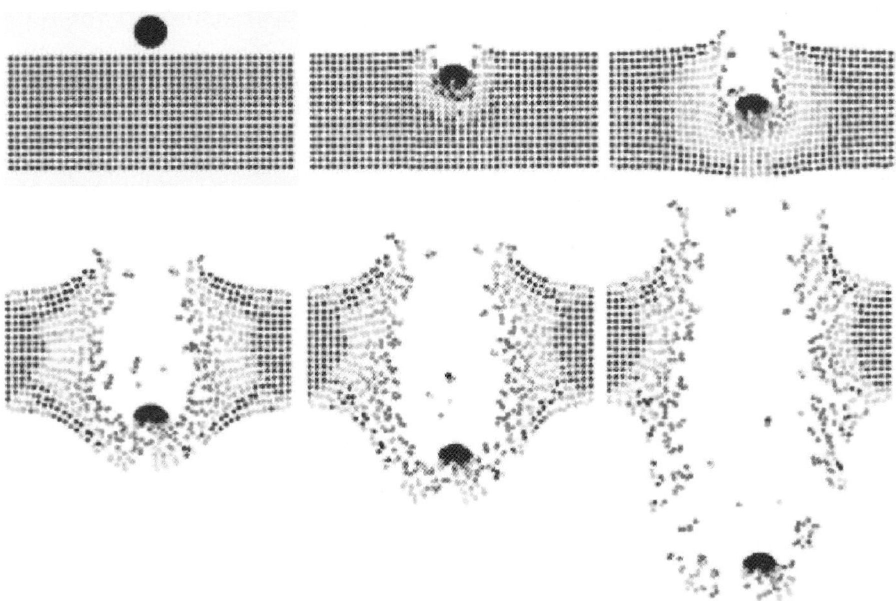

Figure 15. Progressive deformation in high velocity penetration.

6 Conclusions

In this paper, we have introduced several forms of strain smoothing for stabilization and regularization of numerical instability resulting from nodal integration of meshfree Galerkin weak form and material instability due to strain softening and localization. In many large deformation and fragment penetration problems, nodal integration in meshfree methods is highly desirable for the ability to trace complex material motion and for computational efficiency. Nodal integration, however, yields spatial instability. Conforming strain smoothing in stabilized conforming nodal integration (SCNI) has been introduced to stabilize spurious zero energy modes. This study demonstrated that SCNI does not guarantee coercivity. The energy of an oscillatory mode can be very small and spurious nonzero energy modes can exist as the mesh is refined. To recover coercivity, an additional stabilization has been added to the SCNI with a rather simple construction. Test problems have been analyzed to examine the stability and convergence of the modified SCNI method.

In the event of strain softing and localization, the change of sign in the tangent moduli yields an ill-posed problem and requires a regularization for unique solution. The classical gradient type regularization operating on strain results in a governing equation with higher order differentiation, and thus requires additional nonphysical boundary conditions for a solution. This strain gradient regularization can be implicitly imbedded in a gradient reproducing kernel strain smoothing, leading to a naturally regularized weak form without the need of additional nonphysical boundary conditions. It is shown in the numerical test that the imbedded gradient term in the gradient strain smoothing offers a stabilization to the nodally integrated stiffness matrix similar to the stabilization obtained from Taylor series expansion of the one-point evaluated gradient matrix in finite element [11].

For application to fragment and penetration problems, the construction of Voronoi cells in the conforming strain smoothing in SCNI is too tedious and sometimes impossible. We have introduced a stabilized nonconforming nodal integration (SNNI) in which a nonconforming circular strain smoothing zone has been employed. It is shown that although the conforming property is relaxed in SNNI strain smoothing which results in a slight reduction in the solution accuracy compared to that of SCNI, the simplicity in the nonconforming strain smoothing makes it particularly attractive for fragment penetration problems.

References

1. S. Beissel and T. Belytschko, *Nodal integration of the element-free Galerkin method*, Computer Methods in Applied Mechanics and Engineering **139** (1996), 49–74.
2. J. Bonet and S. Kulasegaram, *Correction and stabilization of smooth particle hydrodynamics methods with applications in metal forming simulation*, International Journal for Numerical Methods in Engineering **47** (1999), 1189–1214.
3. J. S. Chen, C. Pan, C. T. Wu, and W. K. Liu, *Reproducing Kernel Particle Methods for Large Deformation Analysis of Nonlinear Structures*, Computer Methods in Applied Mechanics and Engineering **139** (1996), 195–227.
4. J. S. Chen, C. T. Wu, and T. Belytschko, *Regularization of Material Instabilities by Meshfree Approximation with Intrinsic Length Scales*, International Journal for Numerical Methods in Engineering **47** (2000), 1303–1322.
5. J. S. Chen, C. T. Wu, S. Yoon, and Y. You, *A Stabilized Conforming Nodal Integration for Galerkin Meshfree Methods*, International Journal for Numerical Methods in Engineering **50** (2001), 435–466.

6. J. S. Chen, C. T. Wu, S. Yoon, and Y. You, *Nonlinear Version of Stabilized Conforming Nodal Integration for Galerkin Meshfree Methods*, International Journal for Numerical Methods in Engineering **53** (2002), 2587–2615.
7. J. S. Chen, X. Zhang, and T. Belytschko, *An Implicit Gradient Model by a Reproducing Kernel Strain Regularization in Strain Localization Problems*, Computer Methods in Applied Mechanics and Engineering **193** (2004), 2827–2844.
8. R. De Borst, and J. Pamin, *Gradient plasticity in numerical simulation of concrete cracking*, Eur. J. Mech. A/Solids **15** (1996), 295–3202.
9. R. De Borst, R. H. Pamin, R. H.J. Peerlings, and L. J. Sluys, *On Gradient-enhanced Damage and Plasticity Models for Failure in Quasi-brittle and Frictional Materials*, Computational Mechanics **17** (1996), 130–141.
10. D. Lasry, and T. Belytschko, *Localization Limiters in Transient Problems*, Int. J. of Solids and Struct. **24** (1988), 581–597.
11. W. K. Liu, J. S.J. Ong, and R. A. Uras, *Finite element stabilization matrices-a unification approach*, Computer Methods in Applied Mechanics and Engineering **53** (1985), 13–46.
12. P. W. Randles, and L. D. Libersky, *Normalized SPH with stress points*, International Journal for Numerical Methods in Engineering **48** (2000), 1445–1462.
13. D. Wang, and J. S. Chen, *Locking Free Stabilized Conforming Nodal Integration for Meshfree Mindlin-Reissner Plate Formulation*, Computer Methods in Applied Mechanics and Engineering **193** (2004), 1065–1083.
14. J. W. Yoo, B. Moran, and J. S. Chen, *Stabilized Conforming Nodal Integration in the Natural-Element Method*, International Journal for Numerical Methods in Engineering **60** (2004), 861–890.

Fuzzy Grid Method for Lagrangian Gas Dynamics Equations

O.V. Diyankov and I.V. Krasnogorov

NeurokTechsoft LLC, Troitsk, Moscow region, Russia,
diyankov@aconts.com

Summary. This paper addresses the numerical solution of the system of the Lagrangian gas dynamics equation. Usually Finite Difference Methods are applied for the simulation of fluid dynamics. However, for problems with large deformations as the application under consideration they encounter problems with intercrossed numerical grids. This disadvantage can be overcome by meshfree methods which do not require a numerical grid like the Smoothed Particle Hydrodynamics method in principle [1], [2]. We show, however, that this approach does not work properly for the solution of the Lagrangian system for hydrodynamic flow, except for special cases, because of conservation violations.

In order to overcome this problem, we construct a new fully conservative particle method – a Fuzzy grid approach. More precisely a system of Lagrangian partial differential equations for non-radiating inviscid fluid is considered. The presented numerical solution scheme is based on the central difference scheme [3]. In order to fulfill the consistency condition and the corresponding conservation law the use of modified weights is proposed. Extra smoothing is reduced by including an anti-diffusion term. For the determination of the modified weights, a symmetrical and an unsymmetrical approaches are considered. In both cases the weights can be obtained by the solution of a linear system. The performance of the different modifications is demonstrated on numerical examples in one space dimension. According to the numerical results the symmetrical modification leads to better approximations than the unsymmetrical one.

Key words: Meshless, hydrodynamics, central schemes.

1 Smooth Particle Hydrodynamics

Simulation of fluid dynamic problems have been widely performed using traditional Finite Difference Methods (FDM). The main advantage of FDM is simple construction of a difference scheme, but the destructive distinction is necessity of a numerical grid. When modeling problems with large deformation or moving material interfaces this method encounters many difficulties. For instance, if you use Lagrange-based methods for a penetration problem, you perfect simulate material interfaces, but encounter problems with intercrossed numerical grid.

Another alternative method is Smooth Particle Method (SPH). SPH doesn't need a numerical grid and uses smoothed particles as interpolation points to represent materials at discrete locations, so it can easily trace material interfaces and moving boundaries.

In SPH method function $f(\boldsymbol{r})$ is approximated by the following representation [1]

$$\langle f(\boldsymbol{r})\rangle = \int_{\Omega} f(\boldsymbol{\xi}) W\left(\boldsymbol{r}-\boldsymbol{\xi}, h\right) d\boldsymbol{\xi}$$

where $W\left(\boldsymbol{r}-\boldsymbol{\xi}, h\right)$ is the SPH symmetric kernel function, h - smoothed distance, which controls size of the support domain $\Omega \in \mathbb{R}^3$.

The transition from continuum to discrete representation for arbitrary hydrodynamic function is denoted by the following equation:

$$\langle f_i\rangle = \sum_{j\in\Omega_i} \frac{m_j f_j}{\rho_j} W_{ij}\left(\boldsymbol{r}_i - \boldsymbol{r}_j, h\right)$$

where Ω_i is the neighbouring particles within a support domain of the i-th particle, m_j - mass, ρ_j - density, f_j - function value of the j-th particle located at the point $\boldsymbol{r_j}$.

The approximation of spatial derivatives of the function can be obtained from

$$\langle \boldsymbol{\nabla} f_i\rangle = \sum_{j\in\Omega_i} \frac{m_j f_j}{\rho_j} \boldsymbol{\nabla}_i W_{ij} \tag{1.1}$$

$$\boldsymbol{\nabla}_i W_{ij} = \frac{\boldsymbol{r}_i - \boldsymbol{r}_j}{|\boldsymbol{r}_i - \boldsymbol{r}_j|}\frac{\partial W_{ij}}{\partial \boldsymbol{r}}.$$

The system of Lagrangian partial differential equations of hydrodynamics flow are given by

$$\frac{d\rho}{dt} = -\rho\cdot \mathrm{div}\boldsymbol{U}$$

$$\frac{d\boldsymbol{U}}{dt} = -\frac{1}{\rho}\cdot \mathrm{grad}P$$

$$\frac{dE}{dt} = -\frac{P}{\rho}\cdot \mathrm{div}\boldsymbol{U}$$

$$P = (\gamma-1)\rho\left(E - \frac{U^2}{2}\right)$$

where γ denotes the ratio of specific heats and other symbols have their conventional meanings.

The most common SPH approximation for this system can be written

$$\left\langle \frac{d\rho}{dt} \right\rangle_i = \sum_{j\in\Omega_i} m_j \left(\boldsymbol{U}_i - \boldsymbol{U}_j\right) \cdot \boldsymbol{\nabla}_i W_{ij}$$

$$\left\langle \frac{d\boldsymbol{U}}{dt} \right\rangle_i = \sum_{j\in\Omega_i} m_j \frac{P_i + P_j}{\rho_i \rho_j} \cdot \boldsymbol{\nabla}_i W_{ij}$$

$$\left\langle \frac{dE}{dt} \right\rangle_i = \sum_{j\in\Omega_i} m_j \frac{P_i + P_j}{2\rho_i \rho_j} \left(\boldsymbol{U}_i - \boldsymbol{U}_j\right) \cdot \boldsymbol{\nabla}_i W_{ij}.$$

This approach works good only for very special tests, when all distances between the i-th and the j-th particles are the same. For all other tasks this method doesn't work properly, because of conservation violation that follows from inequality $\nabla_i W_{ij} \neq -\nabla_j W_{ji}$.

This fact led us to construct fully conservative particle method.

2 Fuzzy Grid

For our purpose it's more convenient to study the following system of Lagrangian partial differential equations for non-radiating inviscid fluid

$$\frac{dV}{dt} - V\mathrm{div}\boldsymbol{U} = 0$$

$$\frac{d(m\boldsymbol{U})}{dt} + V\mathrm{grad}P = 0 \tag{2.2}$$

$$\frac{d(mE)}{dt} + V\mathrm{div}(P\boldsymbol{U}) = 0.$$

The difference scheme is based on central difference scheme, proposed in [3]

$$\boldsymbol{F}_i^{n+1} = \boldsymbol{F}_i^n + \tau \cdot \left(\boldsymbol{Q}_i - \boldsymbol{G}_i\right) \tag{2.3}$$

$$\boldsymbol{Q}_i = \frac{1}{2} \sum_{j\in\Omega_i} \left(\boldsymbol{F}_i^n - \boldsymbol{F}_j^n\right) \cdot \left(\frac{c_i}{V_i} + \frac{c_j}{V_j}\right) \cdot \max\left(|\widetilde{W}_{ij}|, |\widetilde{W}_{ji}|\right) \tag{2.4}$$

$$\tau_{max} = 0.5 \min_i \left[\sum_{j\in\Omega_i} \left(\frac{c_i}{V_i} + \frac{c_j}{V_j}\right) \cdot \max\left(|\widetilde{W}_{ij}|, |\widetilde{W}_{ji}|\right) \right]^{-1} \tag{2.5}$$

where $\boldsymbol{F} = \{V, m\boldsymbol{U}, mE\}$, $\boldsymbol{G} = \{-V\text{div}\boldsymbol{U}, V\text{grad}P, V\text{div}(P\boldsymbol{U})\}$, $\boldsymbol{Q}$ - additional mathematical viscousity, c_i - local sound speed, $\widetilde{W}_{ij}$ - modified weights (will be discussed later), τ - time step.

Instead of spatial derivatives calculation procedure in the SPH (1.1) we obtain derivatives from

$$\left\langle V\nabla g_i^k \right\rangle = \sum_{j\in\Omega_i} \left(g_i^k - g_j^k\right) \cdot sign\left(x_i - x_j\right) \cdot \widetilde{W}_{ij} \tag{2.6}$$

where k is the component number of vector $\boldsymbol{G}$.

In order to reduce extra smoothing (especially in a depression wave) we include anti-diffusion term. It is based on well-known *MinMod* limiter function

$$MM\left(f_1, f_2, ..., f_n\right) = \begin{cases} \min_j \left\{f_j\right\}, & f_j > 0, \forall j \in 1, ..., n \\ \max_j \left\{f_j\right\}, & f_j < 0, \forall j \in 1, ..., n \\ 0, & otherwise. \end{cases} \tag{2.7}$$

Thus, the time evolution scheme (2.3) can be rewritten

$$\boldsymbol{F}_i^{n+1} = \boldsymbol{F}_i^n + \tau \cdot \left[\left(\boldsymbol{Q}_i - \boldsymbol{M}_i\right) - \boldsymbol{G}_i\right] \tag{2.8}$$

$$\boldsymbol{M}_i = \frac{1}{2}\sum_{j\in\Omega_i}\left(\frac{c_i}{V_i} + \frac{c_j}{V_j}\right)\cdot sign\left(x_j - x_i\right)\cdot MM\left(\nabla\boldsymbol{F}_j^n, \nabla\boldsymbol{F}_i^n\right)\cdot\max\left(|\widetilde{W}_{ij}|, |\widetilde{W}_{ji}|\right). \tag{2.9}$$

3 The Unsymmetrical Approach to Modify Weights

We assume, that

$$\widetilde{W}_{ij} = \begin{cases} \alpha_i^+ W_{ij}, & x_i - x_j < 0 \\ \alpha_i^- W_{ij}, & otherwise. \end{cases} \tag{3.10}$$

Our goal is to achieve two equalities:

1. Approximation law: $\sum_{j\in\Omega_i} \widetilde{W}_{ij} = 1$.
2. Conservation law:

$$\sum_{j\in\Omega_i} \widetilde{W}_{ij} \cdot sign\left(x_i - x_j\right) - \sum_{j\in\Omega_i} \widetilde{W}_{ji} \cdot sign\left(x_j - x_i\right) = 0. \tag{3.11}$$

The approximation equation follows from unity condition that assures the zeroth order consistency of the integral form representation of the continuum function [2].

The conservation equation follows from derivative representation (2.6) that guarantees absence of inner artificial source. To prove this fact looks at the following computations.

Let's consider 1D case and assume that each particle has 4 neighbour particles (2 left and 2 right) except, of course, boundary ones and expand expression for gradient (2.6)

$$
\begin{aligned}
\langle \nabla g_1 \rangle &= \widetilde{W}_{12} \cdot (g_2 - g_1) + \widetilde{W}_{13} \cdot (g_3 - g_1) \\
\langle \nabla g_2 \rangle &= \widetilde{W}_{21} \cdot (g_2 - g_1) + \widetilde{W}_{23} \cdot (g_3 - g_2) + \widetilde{W}_{24} \cdot (g_4 - g_2) \\
\langle \nabla g_3 \rangle &= \widetilde{W}_{31} \cdot (g_3 - g_1) + \widetilde{W}_{32} \cdot (g_3 - g_2) + \widetilde{W}_{34} \cdot (g_4 - g_3) + \widetilde{W}_{35} \cdot (g_5 - g_3) \\
\langle \nabla g_4 \rangle &= \widetilde{W}_{42} \cdot (g_4 - g_2) + \widetilde{W}_{43} \cdot (g_4 - g_3) + \widetilde{W}_{45} \cdot (g_5 - g_4) \\
\langle \nabla g_5 \rangle &= \widetilde{W}_{53} \cdot (g_5 - g_3) + \widetilde{W}_{54} \cdot (g_5 - g_4).
\end{aligned}
\tag{3.12}
$$

Extract coefficients for each $g_i, i \in 1, ..., 5$

$$
\begin{aligned}
g_1 &: \left(-\widetilde{W}_{12} - \widetilde{W}_{13}\right) - \left(\widetilde{W}_{21} + \widetilde{W}_{31}\right) \\
g_2 &: \left(\widetilde{W}_{21} - \widetilde{W}_{23} - \widetilde{W}_{24}\right) - \left(-\widetilde{W}_{12} + \widetilde{W}_{32} + \widetilde{W}_{42}\right) \\
g_3 &: \left(\widetilde{W}_{31} + \widetilde{W}_{32} - \widetilde{W}_{34} - \widetilde{W}_{35}\right) - \left(-\widetilde{W}_{13} - \widetilde{W}_{23} + \widetilde{W}_{43} + \widetilde{W}_{53}\right) \\
g_4 &: \left(\widetilde{W}_{42} + \widetilde{W}_{43} - \widetilde{W}_{45}\right) - \left(-\widetilde{W}_{24} - \widetilde{W}_{34} + \widetilde{W}_{54}\right) \\
g_5 &: \left(\widetilde{W}_{53} + \widetilde{W}_{54}\right) - \left(-\widetilde{W}_{35} - \widetilde{W}_{45}\right).
\end{aligned}
\tag{3.13}
$$

Thus, in order to achieve conservation requirement each expression for g_i must be equal to zero, which was to be proved.

We have $(2n - 2)$ unknowns α^+ and α^- on the one side and we can write $(2n - 2)$ equations from the system (3.11) on the other side, where n is the number of particles. Though we can solve system of linear equations respectively unknowns α^+ and α^-.

The problem is the huge magnitude of resulted α, because large α significantly reduces time step τ (2.5) and leads to extra mathematical viscousity $\boldsymbol{Q}$ (2.4) and smoothing. This fact is obviously seen in figures (7 – 10).

4 The Symmetrical Approach to Modify Weights

We assume, that

$$
\begin{cases}
\widetilde{W}_{ij} = \alpha_{ij} \cdot W_{ij} \\
\widetilde{W}_{ij} = \widetilde{W}_{ji}.
\end{cases}
\tag{4.14}
$$

Our goal is to achieve two equalities:

1. Approximation law: $\sum_{j \in \Omega_i} \widetilde{W}_{ij} = 1$.
2. Conservation law:

$$
\sum_{j \in \Omega_i} \widetilde{W}_{ij} \cdot sign\,(x_i - x_j) = 0. \tag{4.15}
$$

The conservation law equation was simplified because of $\widetilde{W}_{ij}$ symmetry.

At first we need to build matrix Θ that is consisted from equation coefficients of system (4.15) relatively to unknowns α_{ij}. The number of unknowns are equal to number of $W_{ij}, j > i$. So, the system (4.15) is transformed to the equation

$$\Theta \cdot \alpha = B \tag{4.16}$$

where B is the right hand side of the system (4.15).

In order to obtain optimal α we pose the following minimization problem

$$min\left[\alpha^T \cdot \alpha - \lambda^T \cdot (\Theta \cdot \alpha - B)\right] \tag{4.17}$$

where λ is the Lagrange multiplier.

We can minimize (4.17) by setting the partial derivatives of the α to zero, which is a necessary condition for a minimum

$$2\alpha^T - \lambda^T \cdot \Theta = 0$$

$$\alpha = \frac{1}{2}\Theta^T \cdot \lambda. \tag{4.18}$$

Substitute (4.18) into (4.16) we can find λ

$$\frac{1}{2}\Theta\Theta^T \cdot \lambda = B$$

$$\lambda = 2\left(\Theta\Theta^T\right)^{-1} \cdot B. \tag{4.19}$$

And finally compute α

$$\alpha = \Theta^T \cdot \left(\Theta\Theta^T\right)^{-1} \cdot B. \tag{4.20}$$

Corresponding results are presented in figures (3-6, 11-18). With using antidifussion term the depression wave is closer to exact solution. To more accurate simulation of the contact discontinuity it needs to use special algorithm for solving Riemann problem.

5 Numerical Examples

In this section we will present numerical examples which demonstrate performance of our Fuzzy Grid approach. We consider the approximate solution of the Lagrange equations of hydrodynamics flow (2.2)

$$\frac{\mathrm{d}}{\mathrm{d}t}\begin{bmatrix} V \\ mU \\ mE \end{bmatrix} + \begin{bmatrix} -V \\ V \\ V \end{bmatrix} \cdot \frac{\partial}{\partial x}\begin{bmatrix} U \\ P \\ PU \end{bmatrix} = 0, \tag{5.21}$$

$$P = (\gamma - 1)\rho\left(E - \frac{U^2}{2}\right), \quad \rho = \frac{m}{V}.$$

We experiment with the following algorithms:

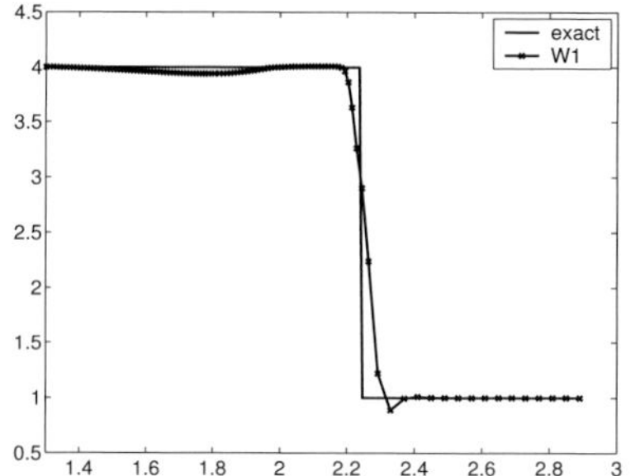

Figure 1. Shock wave - density. t = 1.3

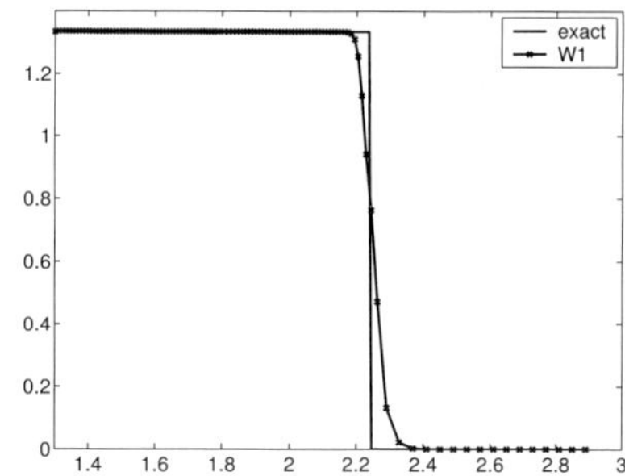

Figure 2. Shock wave - pressure. t = 1.3

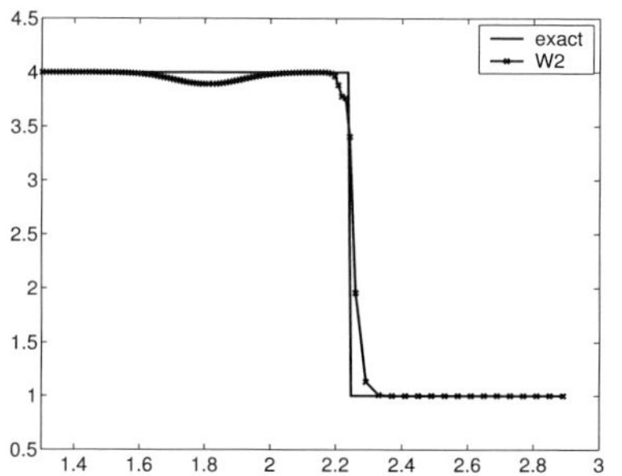

Figure 3. Shock wave - density. t = 1.3

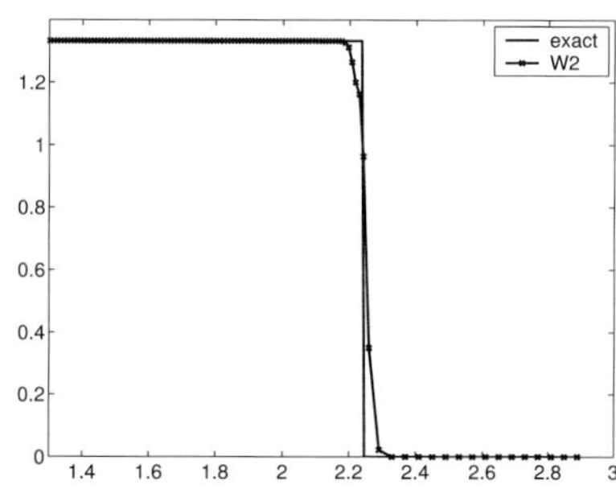

Figure 4. Shock wave - pressure. t = 1.3

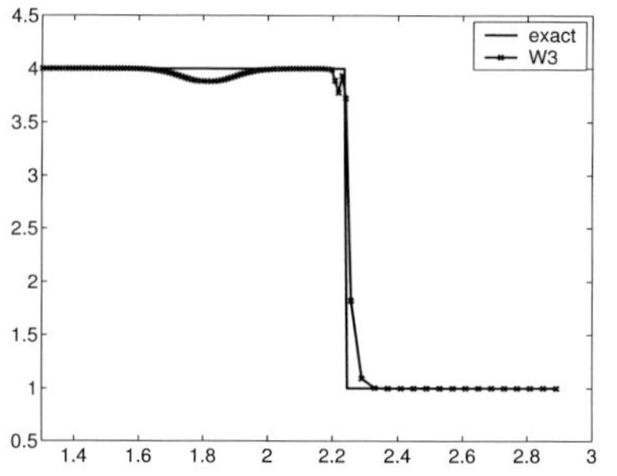

Figure 5. Shock wave - density. t = 1.3

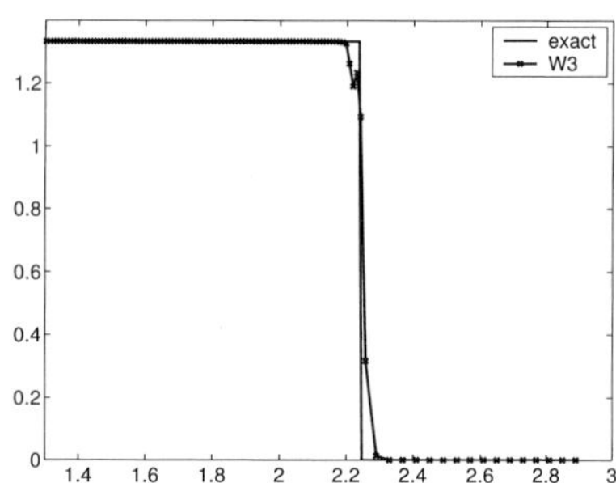

Figure 6. Shock wave - pressure. t = 1.3

1. The unsymmetrical modification, see (3.10). It is referred to as **W1.**
2. The symmetrical modification, see (4.14). It is referred to as **W2.**
3. The symmetrical modification with *MinMod* anti-diffusion terms, see (4.14), (2.8). It is referred to as **W3.**

We solve the system (5.21) with two sets of initial conditions.

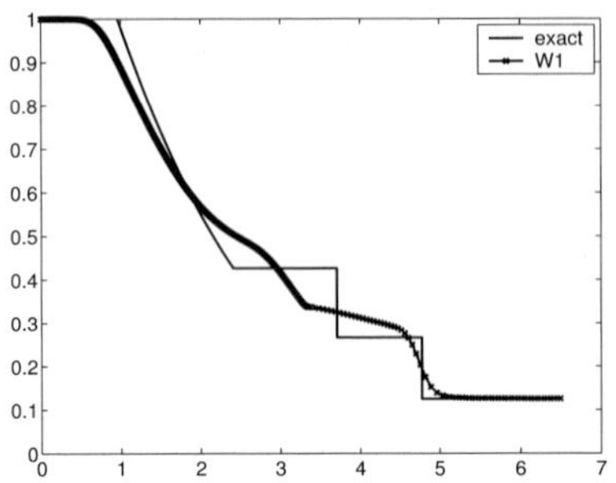

Figure 7. Sod problem - density. t = 1.3

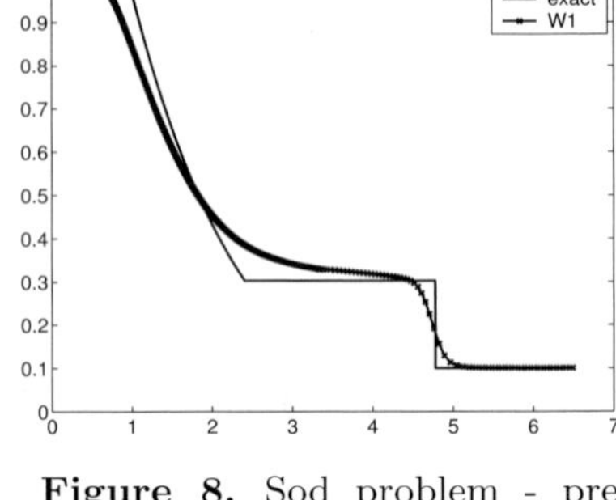

Figure 8. Sod problem - pressure. t = 1.3

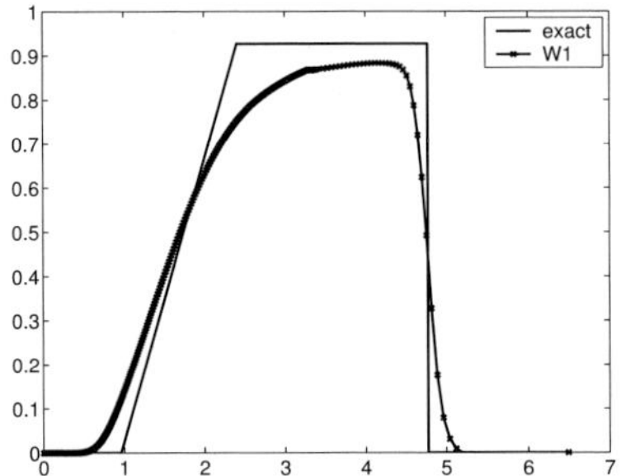

Figure 9. Sod problem - velocity. t = 1.3

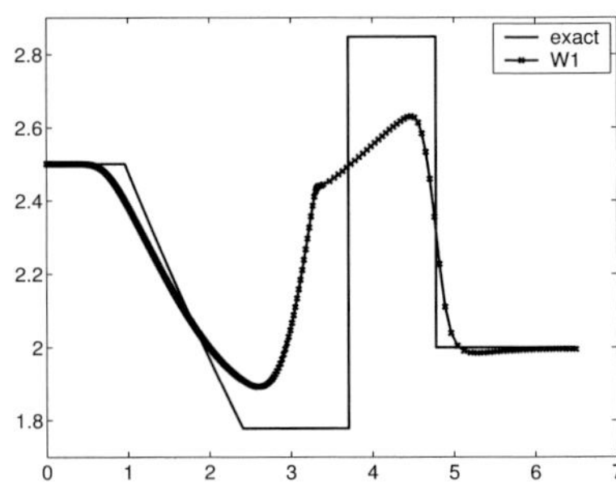

Figure 10. Sod problem - energy. t = 1.3

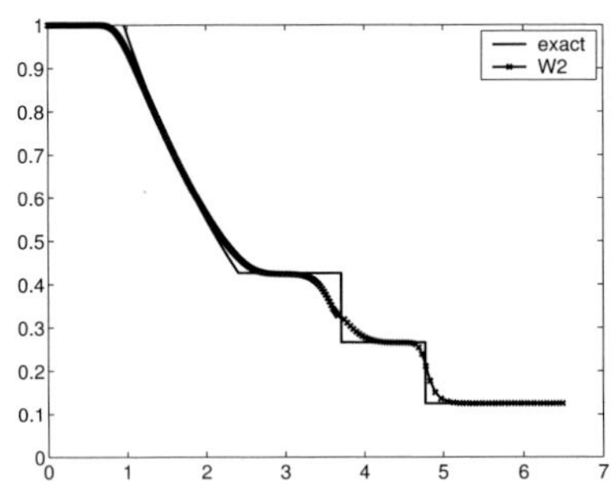

Figure 11. Sod problem - density. t = 1.3

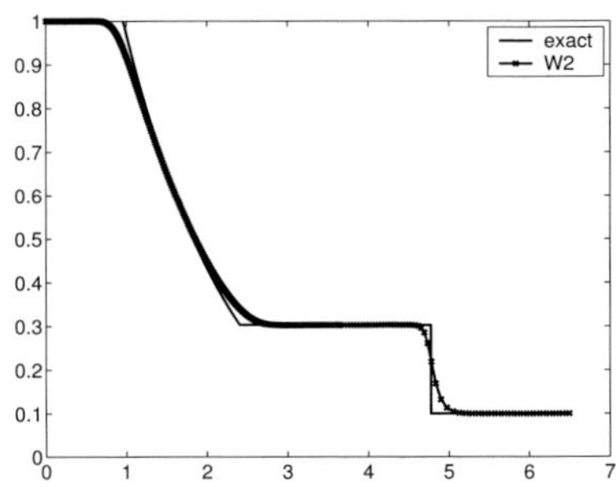

Figure 12. Sod problem - pressure. t = 1.3

5.1 Example 1. Strong Shock Wave

The first example is the strong shock wave propagation problem which consists of the following initial data

$$\begin{cases} \rho = 4, U = 1, E = 1, \; x < 0.5 \\ \rho = 1, U = 0, E = 0, \; x > 0.5. \end{cases}$$

The number of particles is equal to 110.

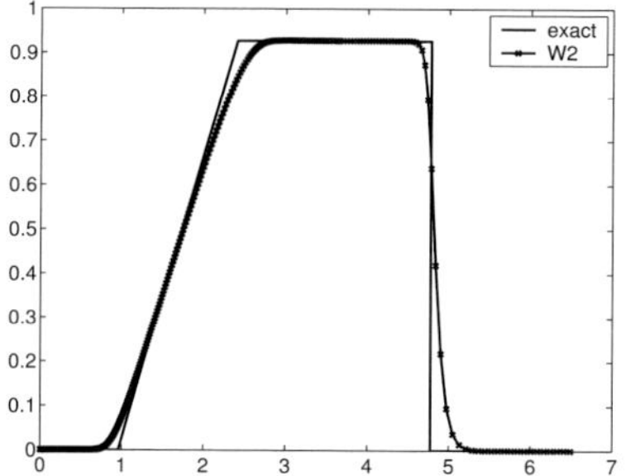

Figure 13. Sod problem - velocity. t = 1.3

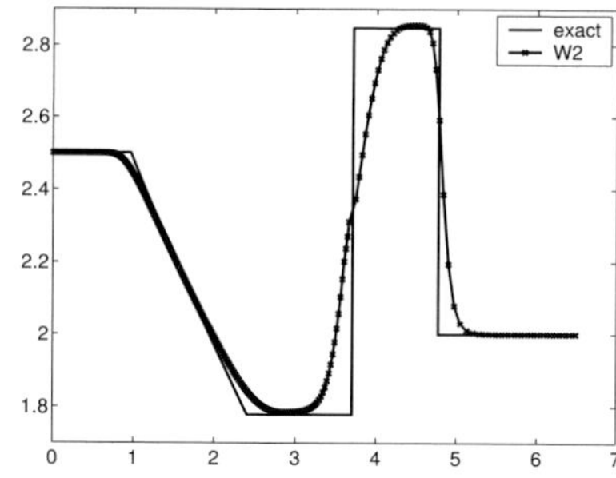

Figure 14. Sod problem - energy. t = 1.3

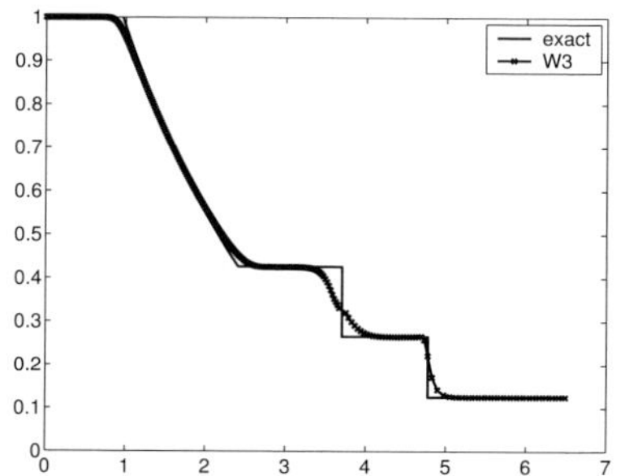

Figure 15. Sod problem - density. t = 1.3

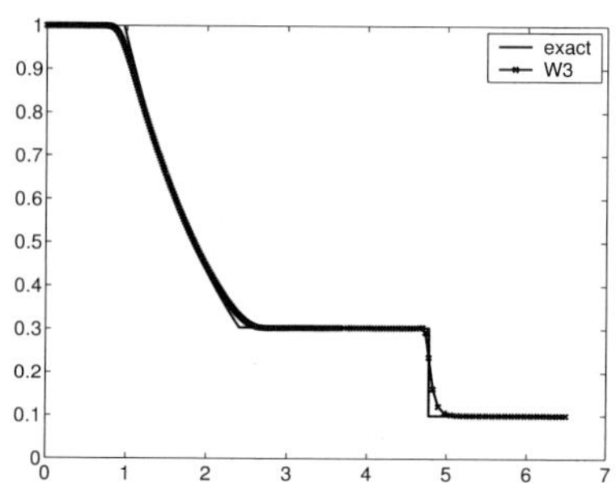

Figure 16. Sod problem - pressure. t = 1.3

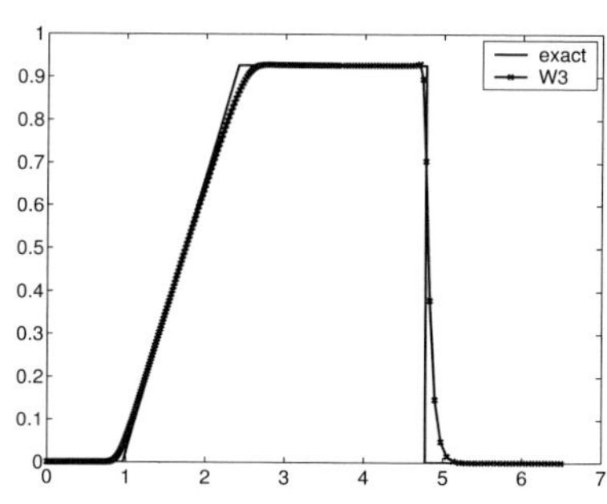

Figure 17. Sod problem - velocity. t = 1.3

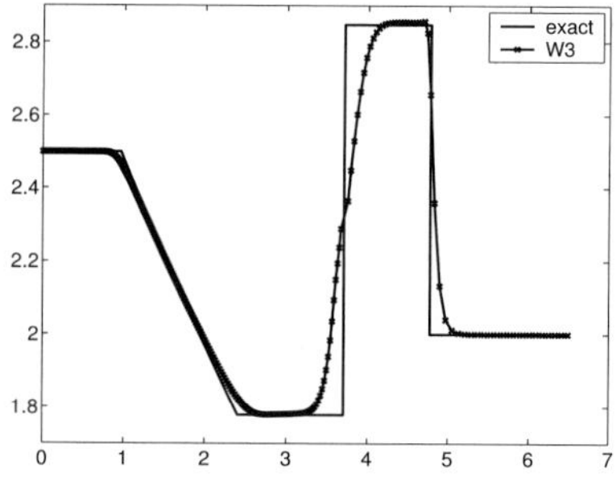

Figure 18. Sod problem - energy. t = 1.3

As we can see the best result is obtained by using **W3** method. The only disadvantage that we have small oscillation on the shock wave front.

5.2 Example 2. Sod Problem

The second example is the Riemann problem proposed by Sod [4] which consists of the following initial data

$$\begin{cases} \rho = 1, U = 0, E = 2.5, & x < 2.5 \\ \rho = 0.125, U = 0, E = 2, & x > 2.5 \end{cases}$$
$$\gamma = 1.4.$$

The number of particles is equal to 300.

6 Conclusion

We've presented a new approach to modify weight coefficients for mesh free method. The unsymmetrical modification method displays excessive dissipation properties for shock wave problem (8 points on shock wave front) and non-physical behaviour for Sod problem. The symmetrical modification method showes good spatial resolution for shock wave problem (5-6 points on shock wave front) and sufficient correspondence with exact solution for Sod problem. The only disadvantage is the slightly broad transition of the contact discontinuity. With using anti-diffusion correction procedure the discontinuity profiles are sharper and rarefaction wave is more accurate.

References

1. J. J. Monaghan, *An introduction to SPH*, Computer Physics Communications, vol. 48, 1988, pp. 89–96.
2. G. R. Liu, *Mesh Free Methods*, CRC Press, 2003.
3. A. Kurganov, E. Tadmor, *New High-Resolution Central Schemes for Nonlinear Conservation Laws and Convection-Diffusion Equations*, J. Comput. Phys., vol. 160, 2000, pp. 241–282.
4. G. A. Sod, *A Survey of Several Finite Difference Methods for Systems of Nonlinear Hyperbolic Conservation Laws*, J. Comput. Phys., vol. 27, 1978.

New Shape Functions for Arbitrary Discontinuities without Additional Unknowns

Thomas-Peter Fries and Ted Belytschko

Dep. of Mechanical Engineering, Northwestern University, 2145 Sheridan Road, 60208 Evanston, Illinois, USA
t-fries@northwestern.edu

Summary. A method is proposed for arbitrary discontinuities, without the need for a mesh that aligns with the interfaces, and without introducing additional unknowns as in the extended finite element method. The approximation space is built by special shape functions that are able to represent the discontinuity, which is described by the level-set method. The shape functions are constructed by means of the moving least-squares technique. This technique employs special mesh-based weight functions such that the resulting shape functions are discontinuous along the interface. The new shape functions are used only near the interface, and are coupled with standard finite elements, which are employed in the rest of the domain for efficiency. The coupled set of shape functions builds a linear partition of unity that represents the discontinuity. The method is illustrated for linear elastic examples involving strong and weak discontinuities.

Key words: Discontinuity, partition of unity, moving least-squares

1 Introduction

A discontinuous change of field quantities or their gradients along certain interfaces is frequently observed in the real world. For example, this may be found in structures in the presence of cracks, pores, and inclusions. In fluids, it occurs on interfaces between two different fluids. The mathematical model and the numerical methods for their approximation have to appropriately consider these interfaces. Discontinuities may be classified as strong and weak. The first involve discontinuous changes in the dependent variable of a model, i.e. of the function itself, whereas weak discontinuities describe discontinuous changes of the derivative of the function. For structural models, typical examples of strong and weak discontinuities are cracks and interfaces of different materials, respectively.

In the approximation of discontinuous fields, a suitable treatment of the interfaces is required. In standard finite element analysis [4, 22], this may be achieved by constructing a mesh whose element edges align with the interface. For moving interfaces, this requires a frequent remeshing, which often restricts this approach to problems where the interface topology does not change significantly during the simulation [18]. The extended finite element method (XFEM) [6, 16, 21] overcomes the need for aligning the elements with the discontinuity. The approximation space resulting from the standard finite element method is enriched by special functions via a partition of unity so that the discontinuity may be considered appropriately. However, additional unknowns for the enriched nodes are needed for this method. With the XFEM, arbitrary discontinuities may be treated implicitly on a fixed mesh. Often, the description of the discontinuities is realized by means of the level-set method [18, 21]. An interesting method that does not require a partition of unity is given by Hansbo and Hansbo [11], though it can be shown to have the same basis functions as the XFEM [1]. Furthermore, meshfree methods have been successfully used for arbitrary discontinuities, see e.g. the overview in [19].

In this paper a new method, which constructs the shape functions from the beginning so that they are able to represent discontinuities, is proposed. No additional unknowns are introduced. The moving least-squares (MLS) method [14] is used for the construction of the shape functions. This method is frequently used in the context of meshfree methods [3, 10], where it is often the underlying principle for the construction of meshfree shape functions. However, the functions are only meshfree if the weight functions which are involved in the MLS technique are mesh-independent. In this paper, the weight functions are defined on a standard finite element mesh, and the resulting shape functions are mesh-based. For each nodal weight function, the support consists of the neighboring and next-neighboring elements of that node. The support of the weight function is truncated along the discontinuity, consequently, a node has no influence across the interface. The visibility criterion of [17] is employed for this purpose. Thereby, the weight functions are designed so that they lead to shape functions which build a linear partition of unity and are able to represent the discontinuity.

The new shape functions have larger supports than standard finite element shape functions. The final system of equation, which results from the use of these shape functions in a weighted residual setting, is less sparse due to the increased connectivity. Therefore, it is desirable to employ the new shape functions only in the proximity of the discontinuity. In all other parts of the domain, standard finite element shape functions are used. The coupling of the two different types of shape functions is realized by a ramp function according to the approach in [7]. The set of coupled shape functions still builds a linear partition of unity with the ability to consider discontinuous changes of the sought functions or its derivatives along the interface.

An outline of the paper is as follows: In section 2, the level-set method for the description of the interface is briefly discussed. Throughout this work, the interfaces are static. Section 3 gives an outline of the moving least-squares method, which constructs shape functions based on locally defined weight functions. Shape functions, which are able to consider a discontinuity, result for specially designed weight functions. This is worked out in section 4. The coupling of the new shape functions with standard finite element shape functions is described in section 5, and enables an efficient assembly of the final system of equations. Section 6 shows numerical results with the coupled shape functions for linear elastic problems including strong and weak discontinuities. The numerical results exhibit the optimal rate of convergence. The paper ends in section 7 with a summary and conclusions.

2 Level-Set Method

The level-set method is a numerical technique for the implicit tracking of moving interfaces [18]. Throughout this work, only static interfaces Γ_{disc} in a d-dimensional domain $\Omega \in \mathbb{R}^d$ are considered. The signed distance function [18] is used for the representation of the interface position,

$$\psi(\boldsymbol{x}) = \pm \min \|\boldsymbol{x} - \boldsymbol{x}_{\Gamma_{\text{disc}}}\|, \quad \forall \boldsymbol{x}_{\Gamma_{\text{disc}}} \in \Gamma_{\text{disc}}, \forall \boldsymbol{x} \in \Omega, \tag{2.1}$$

where the sign is different on the two sides of a closed interface and $\|\cdot\|$ denotes the Euclidean norm. It follows directly from (2.1) that the zero-level of this scalar function is a representation of the discontinuity, i.e.

$$\psi(\boldsymbol{x}) = 0 \quad \forall \boldsymbol{x} \in \Gamma_{\text{disc}}. \tag{2.2}$$

If the discontinuity only partially cuts the body, it is necessary to construct another level-set function ξ such that $\xi(\boldsymbol{x}) > 0$ on the cut part, and $\xi(\boldsymbol{x}) < 0$ on the uncut part.

Consider a discretization of the domain by a mesh. The values of the level-set function are only computed at nodes $\boldsymbol{\psi} = \psi(\boldsymbol{x}_i)$, and the level-set function $\psi^h(\boldsymbol{x}) = \boldsymbol{M}^{\mathrm{T}}(\boldsymbol{x})\,\boldsymbol{\psi}$ is an approximation of $\psi(\boldsymbol{x})$ using the interpolation functions $\boldsymbol{M}(\boldsymbol{x})$. Then, also the representation of the discontinuity as the zero-level of $\psi^h(\boldsymbol{x})$ is only an approximation of the real interface position, which improves with mesh refinement. In this work, the interpolation functions $\boldsymbol{M}(\boldsymbol{x})$ are standard bilinear finite element (FE) functions [4, 22].

3 Moving Least-Squares Method

The method is discussed here following [14, 15]. For a function $u(\boldsymbol{x})$ defined on an open set $\Omega \in \mathbb{R}^d$ being sufficiently smooth, i.e. at least $u(\boldsymbol{x}) \in C^0(\Omega)$, one can define a "local" approximation around a fixed point $\overline{\boldsymbol{x}} \in \Omega$ as

$$u^h_{\text{local}}(\boldsymbol{x}, \overline{\boldsymbol{x}}) = \boldsymbol{p}^{\mathrm{T}}(\boldsymbol{x})\,\boldsymbol{a}(\overline{\boldsymbol{x}}), \tag{3.3}$$

where $\boldsymbol{p}(\boldsymbol{x})$ forms a basis of the approximation subspace, which generally consists of monomials. Throughout this paper, a linear basis is used,

$$\boldsymbol{p}^{\mathrm{T}}(\boldsymbol{x}) = [1, x, y]. \tag{3.4}$$

The coefficient vector $\boldsymbol{a}(\overline{\boldsymbol{x}})$ is obtained by minimizing the weighted least-squares discrete L_2-error norm

$$J_{\overline{\boldsymbol{x}}}(\boldsymbol{a}(\overline{\boldsymbol{x}})) = \sum_{i=1}^{r} \phi_i(\overline{\boldsymbol{x}}) \left[\boldsymbol{p}^{\mathrm{T}}(\boldsymbol{x}_i)\, \boldsymbol{a}(\overline{\boldsymbol{x}}) - \boldsymbol{u}\right]^2, \tag{3.5}$$

where $\phi_i(\overline{\boldsymbol{x}})$ are weight functions. Thereby, a relation between the unknowns $\boldsymbol{a}(\overline{\boldsymbol{x}})$ with the nodal values $\boldsymbol{u}$ is found. The vector $\boldsymbol{x}_i$ refers to the position of the r nodes within the domain. The weight function ϕ_i has small supports $\widetilde{\Omega}_i$ around each node, thereby ensuring the locality of the approximation. It plays an important role in the context of the MLS method. A mesh-independent definition of the weight functions leads to the class of meshfree methods, where the MLS is often the underlying principle for the construction of meshfree shape functions, see e.g. [3, 10]. Throughout this work, however, the weight functions are defined on a mesh and the new shape functions, to be derived in section 4, are mesh-based.

Minimization of (3.5) with respect to $\boldsymbol{a}(\overline{\boldsymbol{x}})$ results in a system of equations

$$\sum_{i=1}^{r} \phi_i(\overline{\boldsymbol{x}})\, \boldsymbol{p}(\boldsymbol{x}_i)\, \boldsymbol{p}^{\mathrm{T}}(\boldsymbol{x}_i)\, \boldsymbol{a}(\overline{\boldsymbol{x}}) = \sum_{i=1}^{r} \phi_i(\overline{\boldsymbol{x}})\, \boldsymbol{p}(\boldsymbol{x}_i)\, u_i. \tag{3.6}$$

Solving this for $\boldsymbol{a}(\overline{\boldsymbol{x}})$ and then replacing $\boldsymbol{a}(\overline{\boldsymbol{x}})$ in the local approximation (3.3) leads to

$$u^h_{\mathrm{local}}(\boldsymbol{x}, \overline{\boldsymbol{x}}) = \boldsymbol{p}^{\mathrm{T}}(\boldsymbol{x}) \left[\sum_{i=1}^{r} \phi_i(\overline{\boldsymbol{x}})\, \boldsymbol{p}(\boldsymbol{x}_i)\, \boldsymbol{p}^{\mathrm{T}}(\boldsymbol{x}_i)\right]^{-1} \sum_{i=1}^{r} \phi_i(\overline{\boldsymbol{x}})\, \boldsymbol{p}(\boldsymbol{x}_i)\, u_i. \tag{3.7}$$

Since the point $\overline{\boldsymbol{x}}$ can be chosen arbitrarily, one can let it "move" over the entire domain, $\overline{\boldsymbol{x}} \to \boldsymbol{x}$, which leads to the global approximation of $u(\boldsymbol{x})$ [15]. It should be noted that the concept of a "moving" approximation is not needed to construct the MLS functions; one can simply start with (3.5) as the definition and proceed as in [5]. Finally, the MLS approximation may be written as

$$u^h(\boldsymbol{x}) = \boldsymbol{p}^{\mathrm{T}}(\boldsymbol{x})\, [\mathbf{M}(\boldsymbol{x})]^{-1}\, \mathbf{B}(\boldsymbol{x})\, \boldsymbol{u}, \tag{3.8}$$

where

$$\mathbf{M}(\boldsymbol{x}) = \sum_{i=1}^{r} \phi_i(\boldsymbol{x})\, \boldsymbol{p}(\boldsymbol{x}_i)\, \boldsymbol{p}^{\mathrm{T}}(\boldsymbol{x}_i), \tag{3.9}$$

and

$$\mathbf{B}(\boldsymbol{x}) = \left[\phi_1(\boldsymbol{x})\, \boldsymbol{p}(\boldsymbol{x}_1) \quad \phi_2(\boldsymbol{x})\, \boldsymbol{p}(\boldsymbol{x}_2) \quad \ldots \quad \phi_r(\boldsymbol{x})\, \boldsymbol{p}(\boldsymbol{x}_r)\right]. \tag{3.10}$$

The matrix $\mathbf{M}(\boldsymbol{x})$ is of size $k \times k$, with k being the number of components in $\boldsymbol{p}(\boldsymbol{x})$. This matrix has to be inverted wherever the MLS functions are to be evaluated. Using these MLS functions as shape functions in an approximation of the form $u^h(\boldsymbol{x}) = \boldsymbol{N}^{\mathrm{T}}(\boldsymbol{x})\,\boldsymbol{u}$, one can immediately write a specific shape function N_i at a point $\boldsymbol{x}$

$$N_i(\boldsymbol{x}) = \boldsymbol{p}^{\mathrm{T}}(\boldsymbol{x})\,[\mathbf{M}(\boldsymbol{x})]^{-1}\,\phi_i(\boldsymbol{x})\,\boldsymbol{p}(\boldsymbol{x}_i)\,. \tag{3.11}$$

The set of r MLS functions $\{\boldsymbol{N}(\boldsymbol{x})\}$ builds a partition of unity (PU) of order n over the d-dimensional domain Ω [3].

4 Design of Special Shape Functions

The aim is to develop shape functions that are able to model arbitrary discontinuities without losing their interpolation properties and without additional degrees of freedom at nodes around the discontinuity. This aim is achieved by especially designed weight functions, which—through the MLS procedure—guarantee shape functions that build a linear PU in the entire domain, taking the discontinuity into account. These functions are C^0-continuous everywhere in the domain except along the interfaces, where they are constructed to be discontinuous.

4.1 Special Weight Functions

The weight functions $\phi_i(\boldsymbol{x})$ in the MLS procedure determine some important properties of the resulting shape functions. The support and the continuity of the shape functions are identical to the weight functions, that is, $\forall i = 1, \ldots, r$: $N_i = 0$ where $\phi_i = 0$, and $N_i \in C^l(\Omega)$ if $\phi_i \in C^l(\Omega)$ (assuming that $\boldsymbol{p}(\boldsymbol{x})$ is sufficiently smooth). For the new weight functions, the supports consist in the elements contiguous to a node and their neighboring elements. This is shown in Fig. 1 for quadrilateral elements.

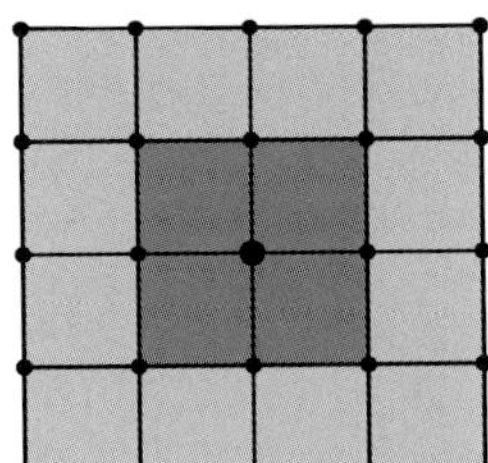

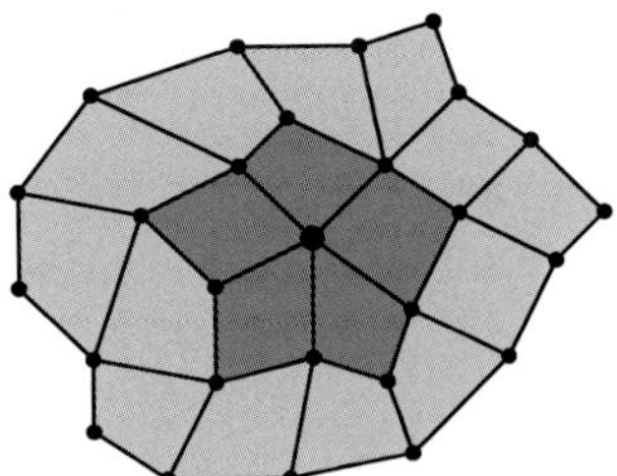

Figure 1. The weight function corresponding to the center node has a support which includes the neighboring elements of that node (dark-grey area) and the next-neighboring elements (light-grey area).

The following definition of the weight functions has been found useful: It is assumed that a domain $\Omega \in \mathbb{R}^2$ is subdivided into n^{el} elements, and each element is defined by a set $I_k^{\mathrm{el}} \in (\mathbb{N}^+)^m$, $k = 1, \ldots, n^{\mathrm{el}}$, of m element nodes. The set of neighboring nodes of a particular node ℓ is defined as

$$I^\ell = \Big(\bigcup_{i:\ell \in I_i^{\mathrm{el}}} I_i^{\mathrm{el}} \Big) \setminus \ell. \tag{4.12}$$

The weight function of node ℓ is defined as

$$\phi_\ell^\star(\boldsymbol{x}) = 2 \cdot N_\ell^{\mathrm{FEM}}(\boldsymbol{x}) + \sum_{i \in I^\ell} N_i^{\mathrm{FEM}}(\boldsymbol{x}), \tag{4.13}$$

where $N_i^{\mathrm{FEM}}(\boldsymbol{x})$ is a standard finite element shape function. This weight function is depicted in Fig. 2 for a node in a structured and unstructured quadrilateral element setting. One may use the definition of the new weight functions (4.13) for both triangular and quadrilateral elements. However, in this work, without loss of generality, only quadrilateral elements with corresponding bilinear shape functions are considered. The shape functions, used for the construction of the approximation, follow from the MLS procedure, as described in section 3, based on these weight functions. It is noted, that standard FE shape functions are employed for the definition of the special *weight* functions, which are then used to obtain C^0-continuous shape functions by the MLS technique.

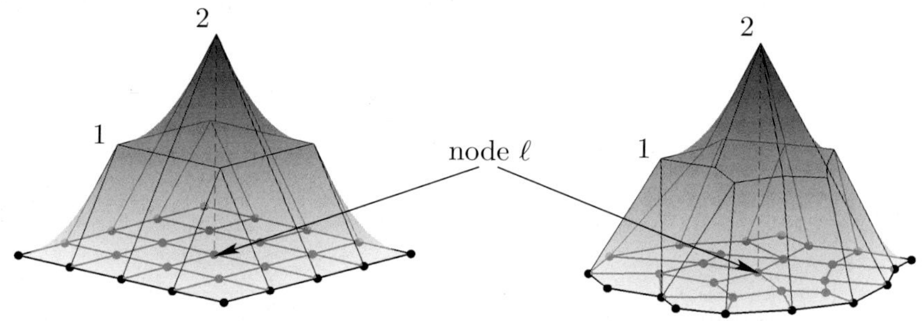

Figure 2. The proposed weight function of node ℓ in a structured and unstructured element situation when no discontinuity is present.

4.2 Modifying the Weight Function in the Presence of a Discontinuity

To introduce the discontinuity we use the visibility method [17], in which all nodes not visible from a point $\boldsymbol{x}$, when the discontinuity is considered opaque, are omitted. That is, the support of the weight function is truncated on the other side of the discontinuity, see Fig. 3. This modification of the support of the weight functions is a standard treatment of a discontinuity in the field of meshfree methods [5, 17].

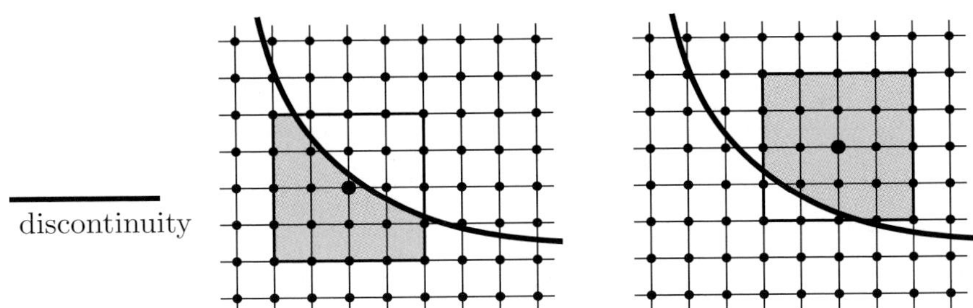

Figure 3. The modified supports of the new weight functions, as a consequence of the discontinuity, are shown for two selected nodes.

Mathematically, this is expressed as

$$\phi_\ell(\boldsymbol{x}) = \begin{cases} \phi_\ell^\star(\boldsymbol{x}) & \text{, for } \boldsymbol{x} \text{ visible from } \boldsymbol{x}_\ell \\ 0 & \text{, instead,} \end{cases} \tag{4.14}$$

where $\phi_\ell^\star(\boldsymbol{x})$ is defined in (4.13). The fact whether $\boldsymbol{x}$ is visible from $\boldsymbol{x}_\ell$ can be based on the sign of the level-set functions, see section 2. The point $\boldsymbol{x}$ is visible from $\boldsymbol{x}_\ell$ if

$$\psi(\boldsymbol{x}) \cdot \psi(\boldsymbol{x}_\ell) > 0 \quad \text{and} \quad \xi(\bar{\boldsymbol{x}}) > 0, \tag{4.15}$$

where $\bar{\boldsymbol{x}}$ is the intersection of the line going from $\boldsymbol{x}_\ell$ to $\boldsymbol{x}$ with $\psi(\boldsymbol{x}) = 0$.

It may be seen in Fig. 3 that the truncation of the weight function supports results in a *reduced* overlap of weight functions. However, due to the large supports of the new weight functions, there is still sufficient overlap of the weight functions in the cut elements so that a linear PU may be constructed through the MLS procedure. It is noted that for weight functions with supports of the same size as the standard bilinear FE shape functions, the MLS matrix $\mathbf{M}(\boldsymbol{x})$ in Eq. (3.9) would become singular near the discontinuity, and a PU of the same order as in uncut elements could not be constructed.

For *open* discontinuities, it should be noted that the visibility criterion introduces discontinuous shape functions not only along the interface. Close to the tip of the discontinuity (e.g. a crack tip), artificial discontinuities are resulting in the domain, see Fig. 4. However, these artificial discontinuities do not inhibit the convergence [13], and may be avoided by using approaches as in [17].

It may thus be found that the new shape functions resulting from the proposed definition of the weight functions share the following properties:

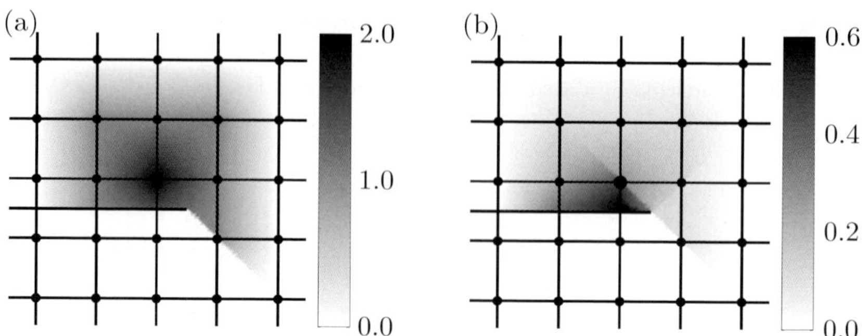

Figure 4. (a) Weight function, and (b) shape function of a node close to the tip of an open discontinuity.

- They build a linear PU in the entire domain which is able to represent the discontinuity, because no shape function has influence across the discontinuity.
- Their supports are larger than those of standard FE shape functions. This leads to an increase in the computational effort. Therefore, it is desirable to use these shape functions only near the discontinuity, i.e., where it is needed, and standard FE shape functions in all other parts of the domain. This is discussed in section 5.
- They are C^0-continuous throughout the domain except along the interface Γ_{disc}. For open discontinuities, some artificial discontinuities in the shape functions are introduced by the visibility criterion near the tip of the interface. This can be avoided by using approaches as described in [17].
- The resulting shape functions do not have Kronecker-delta property, that is $N_i(\boldsymbol{x}_j) \neq \delta_{ij}$. If the new shape functions are employed near the boundary, special treatment of the boundary conditions is necessary. This is well-known in the context of meshfree methods, see [3, 10] for an overview of different techniques to apply boundary conditions there.

The proposed truncation of the supports is directly appropriate only in case of strong discontinuities, where the function $u(\boldsymbol{x})$ has a strong discontinuity. In case of weak discontinuities, where $u(\boldsymbol{x})$ is continuous but its derivatives are not (e.g., wherever the coefficients of the underlying partial differential equation change), continuity of the function has to be enforced. One may for example enforce continuity by a penalty method or Lagrangian multipliers [4].

5 Coupling

In order to ameliorate the increased amount of computational work which results from the larger supports, it is desirable to use the new shape functions as little as possible. They are only needed near the discontinuity because there the standard FE shape functions lose their favorable approximation properties. In all other parts, standard bilinear FE shape functions may be used.

The approach of [7] is used for the coupling of the two different shape function classes. Alternatively, the method of [12] could be used. The domain is decomposed into several subdomains as shown in Fig. 5. The set of cut elements is defined as

$$Q' = \left\{ k \middle| \; i,j \in I_k^{\mathrm{el}} : \psi\left(\boldsymbol{x}_i\right) \cdot \psi\left(\boldsymbol{x}_j\right) < 0 \right\}, \tag{5.16}$$

where $I_k^{\mathrm{el}} \in \left(\mathbb{N}^+\right)^m$, $k = 1, \ldots, n^{\mathrm{el}}$, are the sets of m element nodes belonging the n^{el} elements, and i, j are any two nodes of an element. The union of the elements in Q' is called Ω^{MLS}. The set of neighboring elements is

$$Q^{\star} = \left\{ l \middle| I_k^{\mathrm{el}} \cap I_l^{\mathrm{el}} \neq \emptyset, \forall k = 1, \ldots, n^{\mathrm{el}}, \forall l \in Q' \right\} \setminus Q' \tag{5.17}$$

and the union of its elements is called transition area $\Omega^{\star}$. Finally, the domain $\Omega \setminus \Omega^{\mathrm{MLS}} \setminus \Omega^{\star}$ is called Ω^{FEM}.

The shape functions are defined as follows:

$$N_i\left(\boldsymbol{x}\right) = \left[1 - R\left(\boldsymbol{x}\right)\right] N_i^{\mathrm{FEM}}\left(\boldsymbol{x}\right) + R\left(\boldsymbol{x}\right) N_i^{\mathrm{MLS}}\left(\boldsymbol{x}\right), \tag{5.18}$$

where N_i^{FEM} are standard FE functions, and N_i^{MLS} are the shape functions of Eq. (3.11) constructed by the MLS technique. A ramp function $R\left(\boldsymbol{x}\right)$ is involved in this definition of the coupled shape functions,

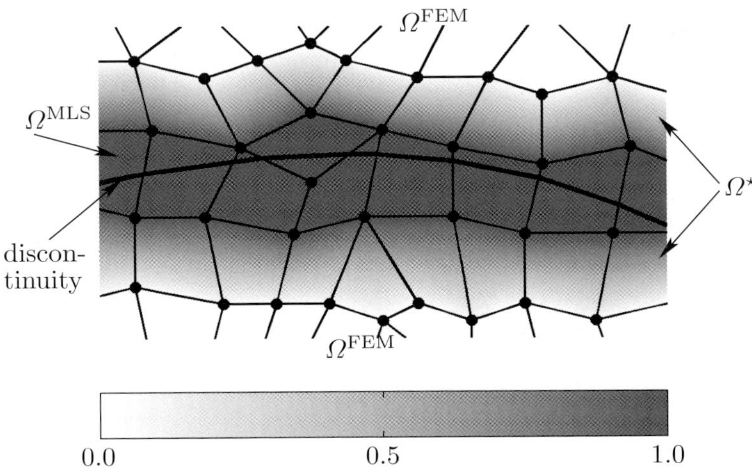

Figure 5. The decomposition of the domain into Ω^{FEM}, $\Omega^{\star}$, and Ω^{MLS} in dependence of the discontinuity is shown, together with the ramp function $R\left(\boldsymbol{x}\right)$.

$$R(\boldsymbol{x}) = \begin{cases} 0 & , \boldsymbol{x} \in \Omega^{\mathrm{FEM}} \\ 1 & , \boldsymbol{x} \in \Omega^{\mathrm{MLS}} \\ \sum_{i \in I'} N_i^{\mathrm{FEM}}(\boldsymbol{x}) & , \boldsymbol{x} \in \Omega^{\star}, I' = \{ i \,|\, \boldsymbol{x}_i \in \partial\Omega^{\mathrm{MLS}} \cap \partial\Omega^{\star} \} . \end{cases} \tag{5.19}$$

The set I' is the set of nodes that are on the boundary between Ω^{MLS} and $\Omega^{\star}$. The ramp function varies monotonically between 0 and 1 in the transition area, a graphical representation may be seen in Fig. 5.

In an element-based code, only in $\Omega^{\star}$ both shape function classes have to be evaluated. All elements in Ω^{MLS} require only the construction of the new shape functions, and all elements in Ω^{FEM} have only standard FE shape functions.

6 Numerical Results

6.1 Governing Equations of Linear Elasticity

The strong form for an elastic solid in two dimensions, undergoing small displacements and strains under static conditions, is [4, 22]

$$\nabla \cdot \boldsymbol{\sigma} = \boldsymbol{f}, \quad \text{on } \Omega \subseteq \mathbb{R}^2, \tag{6.20}$$

where $\boldsymbol{f}$ describe volume forces, and $\boldsymbol{\sigma}$ is the Cauchy stress tensor,

$$\boldsymbol{\sigma} = \mathbf{C} : \boldsymbol{\varepsilon} = \lambda \left(\mathrm{tr}\,\boldsymbol{\varepsilon}\right) \mathbf{I} + 2\mu\boldsymbol{\varepsilon}, \tag{6.21}$$

with λ and μ being the Lamé constants. The linearized strain tensor $\boldsymbol{\varepsilon}$ is

$$\boldsymbol{\varepsilon} = \frac{1}{2}\left(\nabla\boldsymbol{u} + (\nabla\boldsymbol{u})^{\mathrm{T}}\right). \tag{6.22}$$

For the approximation of the displacements $\boldsymbol{u}$, the following test and trial function spaces $\mathcal{S}_{\boldsymbol{u}}^h$ and $\mathcal{V}_{\boldsymbol{u}}^h$ are introduced as

$$\mathcal{S}_{\boldsymbol{u}}^h = \left\{ \boldsymbol{u}^h \middle| \, \boldsymbol{u}^h \in \left(\mathcal{H}^{1h}\right)^d, \, \boldsymbol{u}^h = \hat{\boldsymbol{u}}^h \text{ on } \Gamma_{\boldsymbol{u}} \right\}, \tag{6.23}$$

$$\mathcal{V}_{\boldsymbol{u}}^h = \left\{ \boldsymbol{w}^h \middle| \, \boldsymbol{w}^h \in \left(\mathcal{H}^{1h}\right)^d, \, \boldsymbol{w}^h = \mathbf{0} \text{ on } \Gamma_{\boldsymbol{u}} \right\}, \tag{6.24}$$

where $\mathcal{H}^{1h} \subseteq \mathcal{H}^1$ is a finite dimensional Hilbert space consisting of the shape functions. The space $\mathcal{H}^1$ is the set of functions which are, together with their first derivatives, square-integrable in Ω. Along the Dirichlet boundary $\Gamma_{\boldsymbol{u}}$, the displacements are prescribed as $\hat{\boldsymbol{u}}^h$. The discretized weak form may be formulated in the following Bubnov-Galerkin setting [4, 22]: Find $\boldsymbol{u}^h \in \mathcal{S}_{\boldsymbol{u}}^h$ such that

$$\int_{\Omega} \boldsymbol{\sigma}\left(\boldsymbol{u}^h\right) : \boldsymbol{\varepsilon}\left(\boldsymbol{w}^h\right) d\Omega = \int_{\Omega} \boldsymbol{w}^h \cdot \boldsymbol{f}^h d\Omega + \int_{\Gamma_t} \boldsymbol{w}^h \cdot \hat{\boldsymbol{t}}^h d\Gamma \quad \forall \boldsymbol{w}^h \in \mathcal{V}_{\boldsymbol{d}}^h, \tag{6.25}$$

where $\hat{\boldsymbol{t}}^h$ characterizes the traction along the Neumann boundary. It is noted that the numerical integration of this weak form, in elements cut by the discontinuity, has been realized by dividing these elements into suitable quadrilateral subelements, where Gauss integration is performed in the standard way.

6.2 Error Norms

Two test cases with analytical solution are considered in this work. A convergence test is realized based on the following two error norms:

$$\text{energy-norm:} \quad \left\|\boldsymbol{u}-\boldsymbol{u}^h\right\|_E = \left(\int_\Omega \left(\boldsymbol{\varepsilon}-\boldsymbol{\varepsilon}^h\right)^{\mathrm{T}} \mathbf{C} \left(\boldsymbol{\varepsilon}-\boldsymbol{\varepsilon}^h\right) d\Omega\right)^{1/2}, \tag{6.26}$$

$$L_2\text{-norm:} \quad \left\|\boldsymbol{u}-\boldsymbol{u}^h\right\|_{L_2} = \left(\int_\Omega \left(\boldsymbol{u}-\boldsymbol{u}^h\right)^{\mathrm{T}} \left(\boldsymbol{u}-\boldsymbol{u}^h\right) d\Omega\right)^{1/2}. \tag{6.27}$$

where $\boldsymbol{u}$ and $\boldsymbol{\varepsilon}$ are the analytical solutions. The norms are normalized as $\left\|\boldsymbol{u}-\boldsymbol{u}^h\right\|_E / \|\boldsymbol{u}\|_E$ and $\left\|\boldsymbol{u}-\boldsymbol{u}^h\right\|_{L_2} / \|\boldsymbol{u}\|_{L_2}$, respectively.

For standard bilinear shape functions, the optimal rate of convergence in the L_2-norm is of order 2, and in the energy-norm of order 1, as long as the discontinuities align with element edges and the exact solution is suitable for a polynomial approximation. It is found that this order of convergence is also obtained with the new shape functions if the discontinuities cut arbitrarily through the elements. However, in case of an approximation of a solution that contains a singularity, as is the case in a crack problem, a polynomial basis only leads to a reduced convergence order [20] for both, finite element and new shape functions.

6.3 Edge-Crack Problem

The first test case considers a square domain of size $L \times L$ with an edge-crack of length a, see Fig. 6 for a sketch. Along the boundary of the square domain, displacements are prescribed such that the well-known analytic solution of a near-tip crack field is the exact solution in the entire domain. The material is defined by $E = 10000$ and $\nu = 0.3$, no Neumann boundary is present. The exact solution of this problem may be found e.g. in [9]. It is given in polar coordinates as

$$\sigma_{11}(r,\theta) = \frac{k_1}{\sqrt{2\pi r}} \cos\frac{\theta}{2}\left(1-\sin\frac{\theta}{2}\sin\frac{3\theta}{2}\right), \tag{6.28}$$

$$\sigma_{22}(r,\theta) = \frac{k_1}{\sqrt{2\pi r}} \cos\frac{\theta}{2}\left(1+\sin\frac{\theta}{2}\sin\frac{3\theta}{2}\right), \tag{6.29}$$

$$\sigma_{12}(r,\theta) = \frac{k_1}{\sqrt{2\pi r}} \cos\frac{\theta}{2}\left(\sin\frac{\theta}{2}\cos\frac{3\theta}{2}\right), \tag{6.30}$$

for the stress components and

$$u_1(r,\theta) = \frac{k_1}{2\mu}\sqrt{\frac{r}{2\pi}} \cos\frac{\theta}{2}\left[\kappa-1+2\sin^2\frac{\theta}{2}\right], \tag{6.31}$$

$$u_2(r,\theta) = \frac{k_1}{2\mu}\sqrt{\frac{r}{2\pi}} \sin\frac{\theta}{2}\left[\kappa+1-2\cos^2\frac{\theta}{2}\right], \tag{6.32}$$

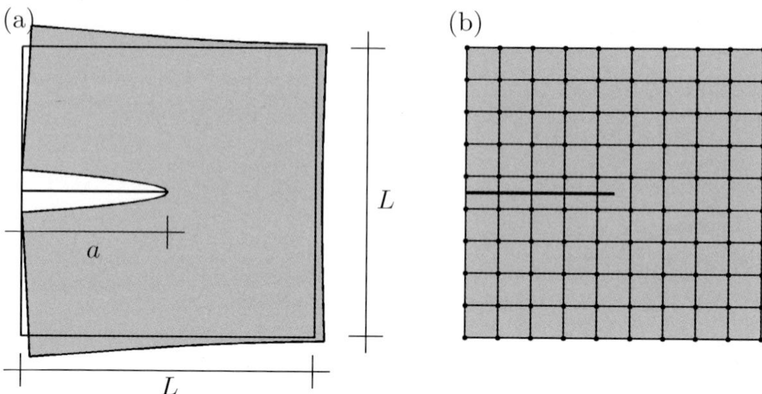

Figure 6. (a) Problem statement of the edge-crack problem and the exact displacement solution (grey) enlarged by a factor of 1000, (b) Structured mesh with 9×9 elements and the discontinuity.

for the displacements. The Kolosov constant κ is defined for

$$\text{plain strain: } \kappa = 3 - 4\nu, \qquad \text{plain stress: } \kappa = \frac{3-\nu}{1+\nu}, \tag{6.33}$$

and μ is the shear modulus. The parameter k_1 is called stress intensity factor, where the index 1 refers to the present case of a mode-1 crack [9].

For the numerical computation, we choose $L = 2$, $a = 1$, and $k_1 = 1$ is prescribed for the displacements along the Dirichlet boundary. Plane stress conditions are assumed. Only structured meshes have been used with n_d^{el} elements per dimensions, see Fig. 6b, where also the discontinuity is shown. For the convergence study, n_d^{el} is $9, 19, 29, 39, 49, 69, 99$, consequently, the discontinuity never aligns with the elements.

At the crack tip, the consideration of the visibility criterion requires special attention. It is practically impossible to divide the elements near the crack tip for integration purposes such that they align with the modified supports of the weight functions resulting from the visibility criterion. This is only relevant in the element containing the crack tip and its neighboring elements. In these elements, instead of a decomposition into subelements for integration, the trapezoidal rule is used with a large number of integration points ($n_Q = 20 \times 20$). As the element size of the affected elements decreases for higher element numbers, this does not degrade the convergence of the method [13].

Figure 7 shows the rate of convergence obtained with the new shape functions for this test case with a strong discontinuity. It is found, that due to the singularity at the crack tip, the order 2 in the L_2-norm and 1 in the energy-norm can *not* be obtained. However, comparing the results with a standard finite element computation with bilinear elements, where the crack aligns with the elements and a node is placed at the crack tip, it may be seen that the

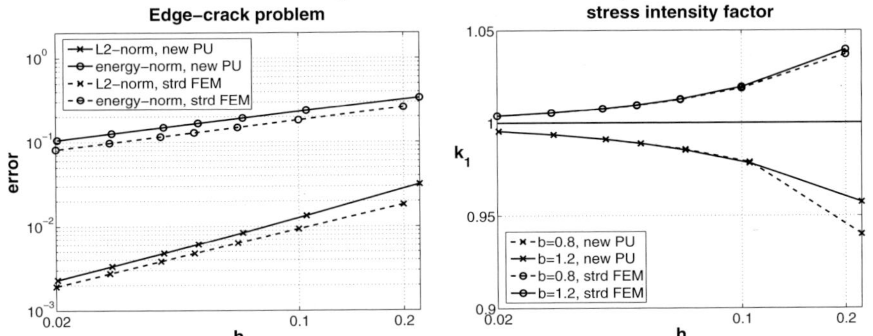

Figure 7. Convergence result in the L_2-norm and energy-norm, and convergence of the approximated stress intensity factor k_1.

same convergence order is obtained. For the present test case, the obtained convergence rate is the best possible for a linear basis. In the presence of a singularity, identical rates of convergence were found in meshfree methods in [2]. Higher-order convergence may only be found with methods that enrich this basis by appropriate terms.

The stress intensity factor k_1 has been evaluated numerically in different integration domains of length $b \times b$ around the crack tip. The interaction integral is evaluated for this purpose, see [16], and k_1 should be constant, independent of the integration domain. The results may be seen in the right part of Fig. 7. A convergence towards the exact value of $k_1 = 1$ may be seen, and the dependence form the size of the integration domain becomes virtually zero for increasing node numbers.

6.4 Bi-material Problem

This test case includes a weak discontinuity. Inside a circular plate of radius b, whose material is defined by $E_1 = 10$ and $\nu_1 = 0.3$, a circular inclusion with radius a of a different material with $E_2 = 1$ and $\nu_2 = 0.25$ is considered. The loading of the structure results from a linear displacement of the outer boundary: $u_r(b, \theta) = r$ and $u_\theta(b, \theta) = 0$. The situation is depicted in Fig. 8. The exact solution may be found in [21]. The stresses are given as

$$\sigma_{rr}(r, \theta) = 2\mu\varepsilon_{rr} + \lambda(\varepsilon_{rr} + \varepsilon_{\theta\theta}), \tag{6.34}$$

$$\sigma_{\theta\theta}(r, \theta) = 2\mu\varepsilon_{\theta\theta} + \lambda(\varepsilon_{rr} + \varepsilon_{\theta\theta}), \tag{6.35}$$

where the Lamé constants $\lambda(\boldsymbol{x})$ and $\mu(\boldsymbol{x})$ are piecewise constant functions with a discontinuity at $r = a$. The strains are

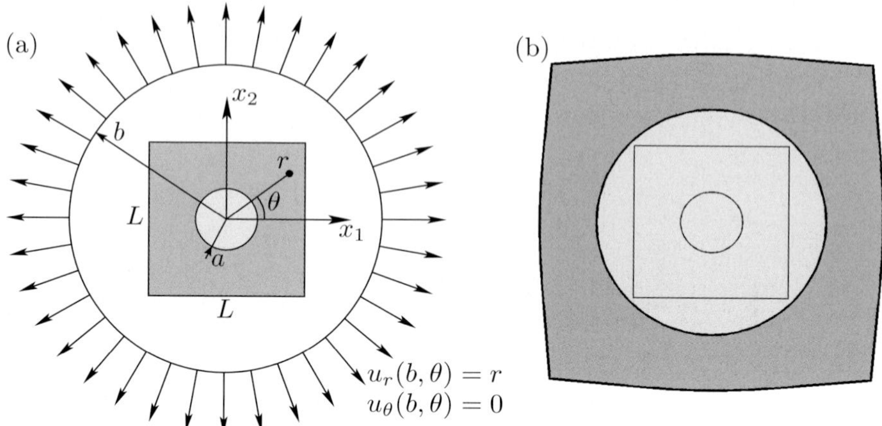

Figure 8. (a) Problem statement of the bi-material problem (the grey area is the numerical domain), (b) the exact displacement solution.

$$\varepsilon_{rr}(r,\theta) = \begin{cases} (1 - \frac{b^2}{a^2})\alpha + \frac{b^2}{a^2}, & 0 \leq r \leq a, \\ (1 + \frac{b^2}{r^2})\alpha - \frac{b^2}{r^2}, & a < r \leq b, \end{cases} \tag{6.36}$$

$$\varepsilon_{\theta\theta}(r,\theta) = \begin{cases} (1 - \frac{b^2}{a^2})\alpha + \frac{b^2}{a^2}, & 0 \leq r \leq a, \\ (1 - \frac{b^2}{r^2})\alpha + \frac{b^2}{r^2}, & a < r \leq b, \end{cases} \tag{6.37}$$

and the displacements

$$u_r(r,\theta) = \begin{cases} \left[(1 - \frac{b^2}{a^2})\alpha + \frac{b^2}{a^2}\right] r, & 0 \leq r \leq a, \\ (r - \frac{b^2}{r})\alpha + \frac{b^2}{r}, & a < r \leq b, \end{cases} \tag{6.38}$$

$$u_\theta(r,\theta) = 0. \tag{6.39}$$

The parameter α involved in these definitions is

$$\alpha = \frac{(\lambda_1 + \mu_1 + \mu_2)\, b^2}{(\lambda_2 + \mu_2)\, a^2 + (\lambda_1 + \mu_1)\,(b^2 - a^2) + \mu_2 b^2}. \tag{6.40}$$

For the numerical model, the domain is a square of size $L \times L$ with $L = 2$, the outer radius is chosen to be $b = 2$ and the inner radius $a = 0.4 + \epsilon$. The parameter ϵ is set to 10^{-3}, and avoids for the used meshes that the level-set function is exactly zero at a node (in this case, the discontinuity would directly cut through that node). The exact stresses are prescribed along the boundaries of the square domain, and displacements are prescribed as $u_1(0,\pm 1) = 0$ and $u_2(\pm 1, 0) = 0$. Plane strain conditions are assumed. In this test case, a weak discontinuity is present, and the displacement field is continuous with discontinuous strains. The continuity information is considered by introducing a penalty term in the weak form (6.25):

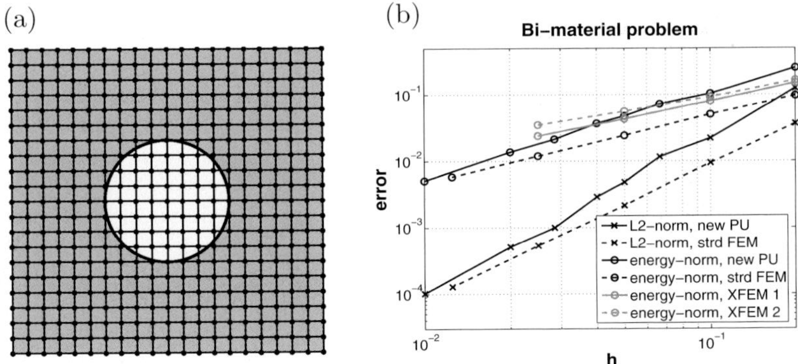

Figure 9. (a) Structured mesh with 20×20 elements and the discontinuity, (b) convergence result for the bi-material problem.

$$\gamma \cdot \int_{\Gamma_{\text{disc}}} \boldsymbol{w}^h \left[\boldsymbol{u}^h\left(\Gamma_{\text{disc}}^+\right) - \boldsymbol{u}^h\left(\Gamma_{\text{disc}}^-\right)\right] d\Gamma, \tag{6.41}$$

where Γ_{disc}^+ and Γ_{disc}^- represent each side of the discontinuity. The penalty parameter is set to $\gamma = 10^5$; this value enforces the continuity appropriately without increasing the condition number of the stiffness matrix too much.

Structured meshes have been used with n_d^{el} elements per dimensions, see Fig. 9a, where also the discontinuity is shown. For the convergence study, n_d^{el} is $10, 20, 30, 40, 50, 70, 100, 200$. Fig. 9b shows the rate of convergence of different methods employed for this test case with a weak discontinuity. The standard finite element result is obtained by a mesh which aligns with the discontinuity. The XFEM results are displayed for two different extended bases, the results are taken from [21], see this reference for details. It may be seen that these XFEM results have a convergence order of 0.75 and 0.91 in the energy norm, respectively [21]. It is noted, that the XFEM matches optimal convergence for the same test case in [8] by employing special blending elements. With the new shape function, the order is 2 for the L_2-norm and 1 for the energy-norm, which is the optimal convergence for shape functions that build a linear PU.

7 Conclusion

A method is proposed which constructs special shape functions with the ability to represent discontinuous changes of field quantities along arbitrary interfaces. The set of shape functions builds a linear partition of unity. The shape functions are C^0-continuous in the domain except along the interfaces, where they are C^{-1}-continuous (discontinuous).

Close to the interfaces, the moving least-squares technique is employed for the construction of the shape functions, and especially designed mesh-based weight functions are involved. In all other parts of the domain, standard FE shape functions are used. The coupling of the two types of shape functions is realized in transition areas by a ramp function. The transition areas depend directly on the interface position, which itself is defined by the level-set method.

The resulting method shows favorable numerical properties in a weighted residual setting for the approximation of continua with strong and weak discontinuities. In case of weak discontinuities, the continuity information of the primal variable has to be introduced into the weak form by a penalty or Lagrange multiplier method. It will be a matter of further investigation whether shape functions can be found which are also able to represent weak discontinuities from the beginning without the need of an indirect way to introduce the continuity information of the primal variable. Furthermore, it would be desirable to extend the linear partition of unity by appropriate terms such that singularities as those occurring in crack problems can be considered appropriately.

Acknowledgement. The support of the Office of Naval Research under grant N00014-98-1-0578 and the Army Research Office under grant W911NF-05-1-0049 is gratefully acknowledged.

References

1. P.M.A. Areias and T. Belytschko, *Letter to the editor*, Comp. Methods Appl. Mech. Engrg. **195** (2004), 1275–1276.
2. T. Belytschko, L. Gu, and Y. Y. Lu, *Fracture and crack growth by element free galerkin methods*, Modelling Simul. Material Science Eng. **2** (1994), 519–534.
3. T. Belytschko, Y. Krongauz, D. Organ, M. Fleming, and P. Krysl, *Meshless methods: An overview and recent developments*, Comp. Methods Appl. Mech. Engrg. **139** (1996), 3–47.
4. T. Belytschko, W.K. Liu, and B. Moran, *Nonlinear finite elements for continua and structures*, John Wiley & Sons, Chichester, 2000.
5. T. Belytschko, Y.Y. Lu, and L. Gu, *Element-free Galerkin methods*, Internat. J. Numer. Methods Engrg. **37** (1994), 229–256.
6. T. Belytschko, N. Moës, S. Usui, and C. Parimi, *Arbitrary discontinuities in finite elements*, Internat. J. Numer. Methods Engrg. **50** (2001), 993–1013.
7. T. Belytschko, D. Organ, and Y. Krongauz, *A coupled finite element–element-free Galerkin method*, Comput. Mech. **17** (1995), 186–195.
8. J. Chessa, H. Wang, and T. Belytschko, *On the construction of blending elements for local partition of unity enriched finite elements*, Internat. J. Numer. Methods Engrg. **57** (2003), 1015–1038.
9. H. Ewalds and R. Wanhill, *Fracture mechanics*, Edward Arnold, New York, 1989.
10. T.P. Fries and H.G. Matthies, *Classification and overview of meshfree methods*, Informatikbericht-Nr. 2003-03, Technical University Braunschweig, (http://opus.tu-bs.de/opus/volltexte/2003/418/), Brunswick, 2003.

11. A. Hansbo and P. Hansbo, *A finite element method for the simulation of strong and weak discontinuities in solid mechanics*, Comp. Methods Appl. Mech. Engrg. **193** (2004), 3523–3540.
12. A. Huerta and S. Fernández-Méndez, *Enrichment and coupling of the finite element and meshless methods*, Internat. J. Numer. Methods Engrg. **48** (2000), 1615–1636.
13. P. Krysl and T. Belytschko, *Element-free Galerkin method: Convergence of the continuous and discontinuous shape functions*, Comp. Methods Appl. Mech. Engrg. **148** (1997), 257–277.
14. P. Lancaster and K. Salkauskas, *Surfaces generated by moving least squares methods*, Math. Comput. **37** (1981), 141–158.
15. W.K. Liu, S. Li, and T. Belytschko, *Moving least square reproducing kernel methods (i) methodology and convergence*, Comp. Methods Appl. Mech. Engrg. **143** (1997), 113–154.
16. N. Moës, J. Dolbow, and T. Belytschko, *A finite element method for crack growth without remeshing*, Internat. J. Numer. Methods Engrg. **46** (1999), 131–150.
17. D. Organ, M. Fleming, T. Terry, and T. Belytschko, *Continous meshless approximations for nonconvex bodies by diffraction and transparency*, Comput. Mech. **18** (1996), 225–235.
18. S. Osher and R.P. Fedkiw, *Level set methods and dynamic implicit surfaces*, Springer Verlag, Berlin, 2003.
19. T. Rabczuk, T. Belytschko, S. Fernández-Méndez, and A. Huerta, *Meshfree methods*, Encyclopedia of Computational Mechanics (E. Stein, R. De Burst, T.J.R. Hughes, ed.), vol. 1, John Wiley & Sons, Chichester, 2004.
20. G. Strang and G. Fix, *An analysis of the finite element method*, Prentice-Hall, Englewood Cliffs, NJ, 1973.
21. N. Sukumar, D.L. Chopp, N. Moës, and T. Belytschko, *Modeling holes and inclusions by level sets in the extended finite-element method*, Comp. Methods Appl. Mech. Engrg. **190** (2001), 6183–6200.
22. O.C. Zienkiewicz and R.L. Taylor, *The finite element method*, vol. 1 – 3, Butterworth-Heinemann, Oxford, 2000.

A Meshless BEM for 2-D Stress Analysis in Linear Elastic FGMs

Xiaowei Gao[1], Chuanzeng Zhang[2], Jan Sladek[3], and Vladimir Sladek[3]

[1] Department of Engineering Mechanics, Southeast University, Nanjing, 210096, P.R. China
xwgao@seu.edu.cn
[2] Department of Civil Engineering, University of Siegen, D-57068 Siegen, Germany
c.zhang@uni-siegen.de
[3] Institute of Construction and Architecture, Slovak Academy of Sciences, 84503 Bratislava, Slovakia
{jan.sladek, vladimir.sladek}@savba.sk

Summary. A meshless boundary element method (BEM) for stress analysis in two-dimensional (2-D), isotropic, continuously non-homogeneous, and linear elastic functionally graded materials (FGMs) is presented in this paper. It is assumed that Young's modulus has an exponential variation, while Poisson's ratio is taken to be constant. Since no fundamental solutions are yet available for general FGMs, fundamental solutions for isotropic, homogeneous, and linear elastic solids are applied, which results in a boundary-domain integral formulation. Normalized displacements are introduced in the formulation, which avoids displacement gradients in the domain-integrals. The radial integration method (RIM) is used to transform the domain-integrals into boundary integrals along the global boundary. The normalized displacements appearing in the domain-integrals are approximated by a series of prescribed basis functions, which are taken as a combination of radial basis functions and polynomials in terms of global coordinates. Numerical examples are presented to verify the accuracy and the efficiency of the present meshless BEM.

Key words: Meshless boundary element method (BEM), Radial integration method, Partial differential equations (PDFs) with variable coefficients, Functionally graded materials (FGMs).

1 Introduction

Meshfree or meshless methods have certain advantages over many domain-type discretization methods like the Finite Element Method (FEM), the Finite-Difference Method (FDM) and the Finite Volume Method (FVM). Among many meshless methods, the global and the local weak-form formulations are often applied ([1]-[4]). The global approach uses a weak-form formulation for the global domain, which requires background meshes for the integration of the weak-form. In the local approach, the weak-form formulation is applied to local sub-domains, which doesn't need any background meshes ([2, 4, 5]). The meshless local Petrov-Galerkin (MLPG) method [2, 4] is a representative example of the local approach, where trial and test functions can be selected from different functional spaces. If the unit step function is chosen as the test function, then a local boundary-domain integral equation formulation for the sub-domains can be obtained.

An alternative way to avoid domain-type discretizations is the boundary element method (BEM) or boundary integral equation method (BIEM). In the classical BEM, the problem dimension is reduced by one, which reduces the computational effort, especially for problems with complicated geometries and moving boundary value problems, where a cumbersome mesh generation and re-meshing are needed in FEM and FDM. Unfortunately, the classical BEM with a boundary-only discretization is limited to problems where the fundamental solutions of the governing partial differential equations can be obtained in closed or simple forms. For isotropic, continuously non-homogeneous and linear elastic solids, the governing partial differential equations possess variable coefficients. Thus, the corresponding fundamental solutions in this case are either not available or they are too complicated. This fact brings significant difficulties to the extension and the application of the classical BEM to non-homogeneous linear elastic solids. This difficulty can be circumvented by using both the global and the local boundary-domain integral equation formulations. The global BEM uses fundamental solutions for homogeneous linear elastic solids, which contains a domain-integral due to the material nonhomogeneity. To transform the domain-integral into boundary integrals over the global boundary of the analyzed domain, the dual reciprocity method (DRM) [6] can be used, where radial basis functions (RBF) are applied. Another novel transform technique is the so-called radial integration method (RIM), which has been developed by Gao ([7, 8, 9]). In the local BEM, the analyzed domain is divided into sub-domains, for which local boundary-domain integral equations are formulated. Recent applications of the local BEM based on the meshless local Petrov-Galerkin (MLPG) method can be found for instance in references [10, 11, 12]. A mesh-free method based on the global weak-form formulation for elastostatic crack analysis in isotropic linear elastic FGMs has been presented by Rao and Rahman [13].

In this paper, a meshless BEM for 2-D stress analysis of isotropic, continuously non-homogeneous, and linear elastic FGMs is presented. The method uses a global boundary-domain integral equation formulation, where fundamental solutions for isotropic, homogeneous, and linear elastic solids are utilized. Normalized displacements are introduced in the formulation to avoid the appearance of displacement gradients in the domain-integrals. An exponential variation is assumed for Young's modulus, while Poisson's ratio is taken to be constant. To transform the domain-integral into boundary-integrals along the global boundary of the analyzed domain, the radial integration method

(RIM) developed by Gao [7, 8] is applied. The unknown normalized displacements are approximated by a series of prescribed basis functions, which are taken as a combination of radial basis functions and polynomials in terms of global coordinates [14]. The present meshless BEM uses interior nodes instead of domain-type meshes and it is therefore a meshfree or meshless method. To verify the accuracy and the efficiency of the present meshless BEM, numerical results are presented and discussed.

2 Formulation of Boundary-Domain Integral Equations

Let us consider an isotropic, continuously non-homogeneous and linear elastic solid with variable Young's modulus $E(x)$ and constant Poisson's ratio ν. In this case, the elasticity tensor can be written as

$$c_{ijkl}(\mathbf{x}) = \mu(\mathbf{x}) c^0_{ijkl} \,, \tag{2.1}$$

where

$$\mu(\mathbf{x}) = \frac{E(\mathbf{x})}{2(1+\nu)} \,, \qquad c^0_{ijkl} = \frac{2\nu}{1-2\nu}\delta_{ij}\delta_{kl} + \delta_{ik}\delta_{jl} + \delta_{il}\delta_{jk} \,, \tag{2.2}$$

In Eq. (2.2), ν is Poisson's ratio, $\mu(\mathbf{x})$ is the shear modulus, and δ_{ij} denotes the Kronecker delta. The stress tensor σ_{ij} and the displacement gradients $u_{i,j}$ are related by Hooke's law

$$\sigma_{ij} = \mu c^0_{ijkl} u_{k,l} \,. \tag{2.3}$$

Here and in the following analysis, a comma after a quantity represents spatial derivatives and the conventional summation rule over repeated subscripts is applied. The traction vector t_i on the boundary of the considered domain is related to the stress components by

$$t_i = \sigma_{ij} n_j \,, \tag{2.4}$$

where n_j is the unit outward normal vector to the boundary Γ of the considered domain Ω. In the absence of body forces, the equilibrium equations $\sigma_{ij,j} = 0$ can be written in the weak-form as

$$\int_\Omega U_{ij} \sigma_{jk,k} dA = 0 \,, \tag{2.5}$$

where U_{ij} is the weight-function. Substitution of Eq. (2.3) into Eq. (2.5) and application of Gauss's divergence theorem yield

$$\int_\Gamma U_{ij} t_j ds - \int_\Gamma T_{ij} \mu u_j ds + \int_\Omega U_{ir,sl} c^0_{rsjl} \mu u_j dA + \int_\Omega U_{ir,s} c^0_{rsjl} \mu_{,l} u_j dA = 0 \,, \tag{2.6}$$

where

$$T_{ij} = \Sigma_{ijl} n_l \,, \qquad \Sigma_{ijl} = c^0_{rsjl} U_{ir,s} = \frac{2\nu}{1-2\nu} U_{ik,k}\delta_{jl} + U_{ij,l} + U_{il,j} \,. \tag{2.7}$$

As weight-function the fundamental solution of the following governing equations is chosen

$$c^0_{rsjl} U_{ir,sl} = -\delta_{ij}\delta(\mathbf{x}-\mathbf{y}) \,, \tag{2.8}$$

where $\delta(\mathbf{x}-\mathbf{y})$ is the Dirac-delta function. Substitution of Eq. (2.8) into Eq. (2.6) leads to

$$\tilde{u}_i(\mathbf{y}) = \int_\Gamma U_{ij}(\mathbf{x},\mathbf{y}) t_j(\mathbf{x}) ds - \int_\Gamma T_{ij}(\mathbf{x},\mathbf{y}) \tilde{u}_j(\mathbf{x}) ds + \int_\Omega V_{ij}(\mathbf{x},\mathbf{y}) \tilde{u}_j(\mathbf{x}) dA \,, \tag{2.9}$$

where

$$\tilde{u}_i(\mathbf{x}) = \mu(\mathbf{x}) u_i(\mathbf{x}) \,, \quad V_{ij}(\mathbf{x},\mathbf{y}) = \Sigma_{ijl}(\mathbf{x},\mathbf{y}) \tilde{\mu}_{,l}(\mathbf{x}) \,, \quad \tilde{\mu}(\mathbf{x}) = \log \mu(\mathbf{x}) \,. \tag{2.10}$$

The solution $U_{ij}(\mathbf{x},\mathbf{y})$ of Eq. (2.8) is the Kelvin's displacement fundamental solution for an isotropic, homogeneous and linear elastic solid with $\mu = 1$, which, as well as the corresponding traction fundamental solution $T_{ij}(\mathbf{x},\mathbf{y})$ and the stress fundamental solution $\Sigma_{ijl}(\mathbf{x},\mathbf{y})$, can be found for instance in [15]. The fundamental solution V_{ij} in the domain-integral of Eq. (2.9) can be expressed as

$$\begin{aligned} V_{ij} = -\frac{1}{4\pi(1-\nu)r} \{ & \tilde{\mu}_{,k} r_{,k} \left[(1-2\nu)\delta_{ij} + 2r_{,i} r_{,j} \right] \\ & + (1-2\nu)(\tilde{\mu}_{,i} r_{,j} - \tilde{\mu}_{,j} r_{,i}) \} \,. \end{aligned} \tag{2.11}$$

It should be noted here that Eq. (2.9) is a representation integral for the displacement components at an arbitrary internal point. By taking the limit-process $\mathbf{y} \to \Gamma$, boundary integral equations for boundary points can be obtained (e.g., [15]).

Unlike many previous BEM formulations for isotropic, non-homogeneous and linear elastic solids (e.g., [10, 16]), Eq. (2.9) is formulated in terms of the tractions t_j and the normalized displacements $\tilde{u}_j$. The domain-integral contains only the normalized displacements instead of the displacement gradients. This feature not only facilitates the numerical implementation, but also results in highly accurate numerical results. For an exponential variation of Young's modulus or shear modulus such as that used in [10] and in this analysis, it can be seen from Eq. (2.10) that $\tilde{\mu}_{,j}$ is constant and Eq. (2.11) thus becomes much simpler for integration.

3 Transformation of Domain-Integrals into Boundary Integrals

In this analysis, the radial integration method (RIM) developed by Gao [7, 8, 9] is used to transform the domain-integral of Eq. (2.9) into boundary integrals. For this purpose, the normalized displacements in the domain-integral of Eq. (2.9) are approximated by a series of prescribed basis functions as commonly used in dual reciprocity method (DRM) [6]. As shown by many previous investigations (e.g., [14]), the combination of the radial basis functions and the polynomials in terms of global coordinates can give satisfactory results. Therefore, the normalized displacements $\tilde{u}_i(\mathbf{x})$ are approximated by

$$\tilde{u}_i(\mathbf{x}) = \sum_A \alpha_i^A \phi^A(R) + \alpha_i^k x_k + a_i^0 , \tag{3.12}$$

$$\sum_A \alpha_i^A = 0 , \tag{3.13}$$

$$\sum_A \alpha_i^A x_j^A = 0 , \tag{3.14}$$

where $R = |\mathbf{x} - \mathbf{x}^A|$ is the distance from the application point A to the field point $\mathbf{x}$, α_i^A and a_i^k are unknown coefficients to be determined, and $\mathbf{x}^A$ denotes the coordinates at the application point A which consists of all boundary nodes and some internal nodes. The commonly used radial basis functions $\phi^A(R)$ can be found in many references (e.g., [7], [14]). In this analysis, we use the following 4-th order spline-type RBF [2]

$$\phi^A(R) \begin{cases} 1 - 6\left(\frac{R}{d_A}\right)^2 + 8\left(\frac{R}{d_A}\right)^3 - 3\left(\frac{R}{d_A}\right)^4 , & 0 \leq R \leq d_A , \\ 0 & d_A \leq R , \end{cases} \tag{3.15}$$

where d_A is the support size for the application point A.

The unknown coefficients α_i^A and a_i^k in Eqs. (3.12)-(3.14) can be determined by applying the application point A in Eqs. (3.12)-(3.14) to every node. This leads to a set of linear algebraic equations, which can be written in the matrix form as

$$\tilde{\mathbf{u}} = \boldsymbol{\phi} \cdot \boldsymbol{\alpha} , \tag{3.16}$$

where $\boldsymbol{\alpha}$ is a vector consisting of the coefficients α_i^A for all application points and a_i^k. If two nodes do not coincide, i.e., share the same coordinates, the matrix $\boldsymbol{\phi}$ is invertible and thus

$$\boldsymbol{\alpha} = \boldsymbol{\phi}^{-1} \cdot \tilde{\mathbf{u}} . \tag{3.17}$$

Substitution of Eq. (3.12) into the domain-integral of Eq. (2.9) yields

$$\int_\Omega V_{ij} \tilde{u}_j dA = \alpha_j^A \int_\Omega V_{ij} \phi^A dA + a_j^k \int_\Omega V_{ij} x_k dA + a_j^0 \int_\Omega V_{ij} dA . \tag{3.18}$$

By using RIM [7, 8, 9], the domain-integrals on the right-hand side of Eq. (3.18) can be transformed into boundary integrals as

$$\int_{\Omega} V_{ij}\tilde{u}_j dA = \alpha_j^A \int_{\Gamma} \frac{1}{r}\frac{\partial r}{\partial n} F_{ij}^A ds + a_j^k \int_{\Gamma} \frac{r_{,k}}{r}\frac{\partial r}{\partial n} F_{ij}^1 ds$$

$$+ \left(a_j^k y_k + a_j^0\right) \int_{\Gamma} \frac{1}{r}\frac{\partial r}{\partial n} F_{ij}^0 ds \, , \tag{3.19}$$

where

$$F_{ij}^A = \int_0^r rV_{ij}\phi^A dr \, , \tag{3.20}$$

$$F_{ij}^1 = \int_0^r r^2 V_{ij} dr \, , \tag{3.21}$$

$$F_{ij}^0 = \int_0^r rV_{ij} dr \, . \tag{3.22}$$

Note here that in the radial integrals (3.20)-(3.22) the term $r_{,i}$ is constant [8] and the following relation is used for the transformation from $\mathbf{x}$ to r

$$x_i = y_i + r_{,i} r \, . \tag{3.23}$$

The radial integrals (3.20)-(3.22) are regular and they can be computed numerically by using standard Gaussian quadrature formula for every field point [9].

4 System of Linear Algebraic Equations

After numerical integration and substitution of Eq. (3.17) into Eq. (3.19), the domain-integral can be expressed in terms of the normalized displacement vector $\tilde{\mathbf{u}}$ at all nodes. If the BEM model consists of N_b boundary nodes and N_i internal nodes and after invoking the boundary conditions, Eq. (2.9) leads to the following system of linear algebraic equations

$$\mathbf{A}_b \mathbf{x}_b = \mathbf{y}_b + \mathbf{V}_b \tilde{\mathbf{u}} \tag{4.24}$$

for boundary nodes, and

$$\tilde{\mathbf{u}}_i = \mathbf{A}_i \mathbf{x}_b + \mathbf{y}_i + \mathbf{V}_i \tilde{\mathbf{u}} \tag{4.25}$$

for internal nodes. In Eqs. (4.24) and (4.25), the sizes of the matrices $\mathbf{A}_b$ and $\mathbf{A}_i$ are $2N_b \times 2N_b$ and $2N_i \times 2N_b$, while $\mathbf{V}_b$ and $\mathbf{V}_i$ are $2N_b \times 2N_t$ and $2N_i \times 2N_t$ with $N_t = N_b + N_i$, respectively. The vector $\mathbf{x}_b$ with a size of $2N_b \times 1$ contains the unknown normalized boundary displacements or the unknown boundary tractions. The vector $\tilde{\mathbf{u}}$ with a size of $2N_t \times 1$ consists of the unknown normalized boundary displacements and all normalized internal displacements. It should be noted here that after invoking the boundary conditions, the columns of the matrices $\mathbf{V}_b$ and $\mathbf{V}_i$ corresponding to the known boundary displacement nodes should be zero. By combining Eqs. (4.24) and (4.25) we obtain

$$\left(\begin{bmatrix} \mathbf{A}_b & \mathbf{0} \\ -\mathbf{A}_i & \mathbf{I} \end{bmatrix} - \begin{bmatrix} \mathbf{V}_b \\ \mathbf{V}_i \end{bmatrix} \right) \begin{Bmatrix} \mathbf{x}_b \\ \tilde{\mathbf{u}}_i \end{Bmatrix} = \begin{Bmatrix} \mathbf{y}_b \\ \mathbf{y}_i \end{Bmatrix}, \tag{4.26}$$

where $\mathbf{I}$ is the identity matrix. By solving Eq. (4.26) numerically the boundary unknowns $\mathbf{x}_b$ and the normalized internal displacements $\tilde{\mathbf{u}}_i$ can be obtained. Subsequently, the true displacements can be computed by using the first equation of Eq. (2.10).

5 Computation of Stresses

Once the unknown boundary data and the internal normalized displacements are obtained from Eq. (4.26), stress components at boundary and internal nodes can be computed by using these quantities. By taking Eq. (2.10) into account, the generalized Hooke's law (2.3) can be rewritten as

$$\sigma_{ij} = \mu c^0_{ijkl} \frac{\partial u_k}{\partial x_l} = \mu c^0_{ijkl} \frac{\partial}{\partial x_l} \left(\frac{\tilde{u}_k}{\mu} \right) = c^0_{ijkl} \frac{\partial \tilde{u}_k}{\partial x_l} - c^0_{ijkl} \tilde{u}_k \tilde{\mu}_{,l} \,. \tag{5.27}$$

From Eq. (2.9) we obtain

$$\frac{\partial \tilde{u}_k}{\partial y_l} = \int_\Gamma \frac{\partial U_{kj}(\mathbf{x},\mathbf{y})}{\partial y_l} t_j(\mathbf{x}) ds - \int_\Gamma \frac{\partial T_{kj}(\mathbf{x},\mathbf{y})}{\partial y_l} \tilde{u}_j(\mathbf{x}) ds$$
$$+ \int_\Omega \frac{\partial V_{kj}(\mathbf{x},\mathbf{y})}{\partial y_l} \tilde{u}_j(\mathbf{x}) dA \,, \tag{5.28}$$

where all integrals exist in the sense of Cauchy principal values except the last one which will be discussed later. Substituting Eqs. (2.7) and (2.11) into Eq. (5.28) and using the relation $\partial(\cdot)/\partial y_l = -\partial(\cdot)/\partial x_l$, the following integral representation for the stress components can be obtained from Eq. (5.27)

$$\sigma_{ij}(\mathbf{y}) = \int_\Gamma U_{ijk}(\mathbf{x},\mathbf{y}) t_k(\mathbf{x}) ds - \int_\Gamma T_{ijk}(\mathbf{x},\mathbf{y}) \tilde{u}_k(\mathbf{x}) ds$$
$$+ \int_\Omega V_{ijk}(\mathbf{x},\mathbf{y}) \tilde{u}_k(\mathbf{x}) dA - c^0_{ijkl} \tilde{u}_k \tilde{\mu}_{,l} \,, \tag{5.29}$$

where the kernels U_{ijk} and T_{ijk} have the same expressions as in the conventional BEM formulation with $\mu = 1$ (e.g. in [15]), and

$$V_{ijk}(\mathbf{x},\mathbf{y}) = -c^0_{ijmn} \frac{\partial V_{mk}(\mathbf{x},\mathbf{y})}{\partial x_n} \,. \tag{5.30}$$

By using the second equation of Eq. (2.2) the last term in Eq. (5.27) or (5.29) can be written as

$$c^0_{ijkl}\tilde{u}_k\tilde{\mu}_{,l} = \left(\frac{2\nu}{1-2\nu}\delta_{ij}\tilde{\mu}_{,k} + \delta_{ik}\tilde{\mu}_{,j} + \delta_{jk}\tilde{\mu}_{,i}\right)\tilde{u}_k\,. \tag{5.31}$$

Since the differentiation of $V_{mk}(\mathbf{x},\mathbf{y})$ causes a strong singularity in the domain integral of Eq. (5.29), a jump term exists. Following the procedure as described in [15], the jump term can be obtained by cutting-out an infinitesimal circle around the source point $\mathbf{y}$ and finally, after putting the jump term and the term given in Eq. (5.31) together, equation (5.29) can be rewritten as

$$\begin{aligned}\sigma_{ij}(\mathbf{y}) &= \int_\Gamma U_{ijk}(\mathbf{x},\mathbf{y})t_k(\mathbf{x})ds - \int_\Gamma T_{ijk}(\mathbf{x},\mathbf{y})\tilde{u}_k(\mathbf{x})ds \\ &+ \int_\Omega V_{ijk}(\mathbf{x},\mathbf{y})\tilde{u}_k(\mathbf{x})dA + F_{ijk}(\mathbf{y})\tilde{u}_k(\mathbf{y})\,,\end{aligned} \tag{5.32}$$

where

$$\begin{aligned}V_{ijk} = &\frac{1}{2\pi(1-\nu)r^2}\{2\tilde{\mu}_{,m}r_{,m}\left[(1-2\nu)\delta_{ij}r_{,k} + \nu(\delta_{ik}r_{,j} + \delta_{jk}r_{,i})\right.\\ &- \left.4r_{,i}r_{,j}r_{,k}\right] + 2\nu(\tilde{\mu}_{,i}r_{,j} + \tilde{\mu}_{,j}r_{,i})r_{,k} - (1-4\nu)\tilde{\mu}_{,k}\delta_{ij}\\ &+ (1-2\nu)(2\tilde{\mu}_{,k}r_{,i}r_{,j} + \tilde{\mu}_{,j}\delta_{ik} + \tilde{\mu}_{,i}\delta_{jk})\}\,,\end{aligned} \tag{5.33}$$

$$F_{ijk} = -\frac{1}{4(1-\nu)}\left(\delta_{ij}\tilde{\mu}_{,k} + \delta_{ik}\tilde{\mu}_{,j} + \delta_{jk}\tilde{\mu}_{,i}\right)\,. \tag{5.34}$$

Now the domain-integral in Eq. (5.32) exists in the sense of Cauchy principal values. This means that cutting-out an infinitesimal circle around the source point $\mathbf{y}$ does not change the integration result. Based on this property, the conventional singularity-separation technique [15] can be applied to regularize this strongly singular domain-integral. The domain-integral in Eq. (5.32) can be computed by using the radial integration method as described in section 3. Note here that the procedure is exactly the same as in the treatment of domain-integrals with initial stresses arising in plasticity problems [8].

It should be remarked here that the stress integral representation (5.32) is only applicable to compute stresses at internal points. For boundary points when the source point $\mathbf{y}$ approaches the field point $\mathbf{x}$, hypersingularities arise. Although the hypersingular integrals can be evaluated directly by using the methods suggested in [9] and [17], the Fortran subroutines as presented in [15] based on the 'traction-recovery' technique is adopted in this analysis to compute the stresses at the boundary points.

6 Numerical Example

We consider an isotropic, continuously non-homogeneous and linear elastic rectangular plate with the dimensions $L \times W$ as depicted in Fig. 6.1. The FGM plate is subjected to a uniform tensile stress loading σ=1. The plate is discretized into 48 equally-spaced linear boundary elements: 20 along the longitudinal and 4 along the transversal directions, with a total of 48 boundary nodes. An exponential variation of Young's modulus in the transversal direction is considered, which is described by

$$E(\mathbf{x}) = E_0 e^{\beta y} , \qquad \beta = \frac{1}{W} \log\left(\frac{E_w}{E_0}\right) , \tag{6.35}$$

where $E_0 = 10000$ and $E_w = 20000$. Plane stress condition is assumed and Poisson's ratio is taken as $\nu = 0.25$.

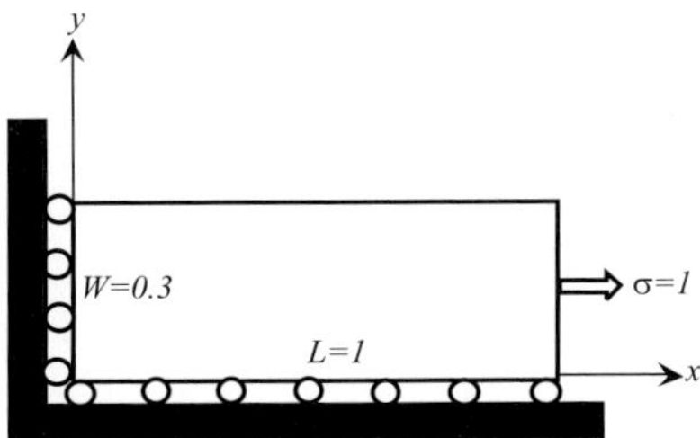

Figure 1. A rectangular FGM plate subjected to a tensile loading.

Table 6.1 and Fig. 6.3 show a comparison of the computed displacement component u_x at $y=0.3$ by the present meshless BEM with the results obtained by Sladek et al. [10], who used a meshless local Petrov-Galerkin method. To investigate the influences of the number and the distribution of internal nodes on the numerical results, 48 boundary nodes are used, while the internal nodes are varied from 57 to 0 as shown in Fig. 6.2. The comparison shows that our numerical results agree very well with that of Sladek et al. [10]. In addition, it is seen here that the present meshless BEM is pretty insensitive to the selected node number and distribution, at least for the case considered here. Even with very few or no internal nodes, the present meshless BEM can still yield very accurate numerical results.

In Tabs. 6.2 and 6.3 as well as Figs. 6.4 and 6.5, numerical results for the stress component σ_{11} at $x = 0.0$ and $x = 0.75$ obtained by the present meshless BEM are presented and compared with that of Sladek et al. [10]. Numerical calculations are carried out by using 48 boundary nodes and 57 internal nodes (see Fig. 6.2a). The present numerical results for the stress component σ_{11} show a very good agreement with that obtained by Sladek et al. [10].

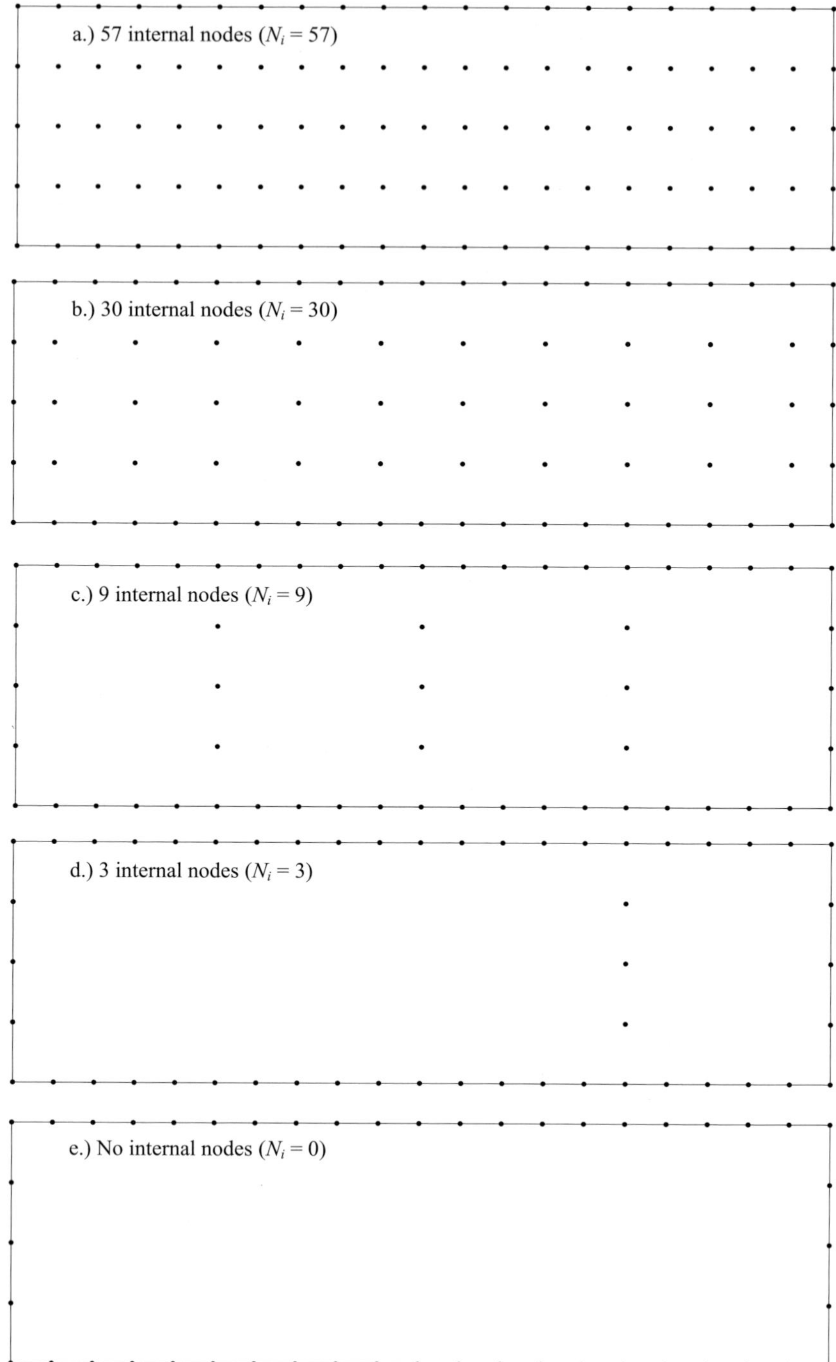

Figure 2. Boundary and internal nodes.

Table 1. Comparison of computed displacement component u_x at $y = 0.3$.

x	homog.	Ref. [10]	$N_i = 57$	$N_i = 30$	$N_i = 9$	$N_i = 3$	$N_i = 0$
0.00	0	0	0	0	0	0	0
0.05	5.0_{-6}	3.44353_{-6}	3.46979_{-6}	3.46983_{-6}	3.46897_{-6}	3.47958_{-6}	3.47949_{-6}
0.10	1.0_{-5}	6.88585_{-6}	6.94299_{-6}	6.94292_{-6}	6.94249_{-6}	6.96360_{-6}	6.96383_{-6}
0.15	1.5_{-5}	1.03293_{-5}	1.04200_{-5}	1.04196_{-5}	1.04200_{-5}	1.04513_{-5}	1.04530_{-5}
0.20	2.0_{-5}	1.37755_{-5}	1.38985_{-5}	1.38983_{-5}	1.38987_{-5}	1.39398_{-5}	1.39447_{-5}
0.25	2.5_{-5}	1.72253_{-5}	1.73781_{-5}	1.73783_{-5}	1.73779_{-5}	1.74287_{-5}	1.74386_{-5}
0.30	3.0_{-5}	2.06784_{-5}	2.08579_{-5}	2.08585_{-5}	2.08579_{-5}	2.09175_{-5}	2.09338_{-5}
0.35	3.5_{-5}	2.41341_{-5}	2.43360_{-5}	2.43367_{-5}	2.43372_{-5}	2.44036_{-5}	2.44280_{-5}
0.40	4.0_{-5}	2.75896_{-5}	2.78096_{-5}	2.78103_{-5}	2.78121_{-5}	2.78825_{-5}	2.79179_{-5}
0.45	4.5_{-5}	3.10426_{-5}	3.12738_{-5}	3.12747_{-5}	3.12766_{-5}	3.13487_{-5}	3.13981_{-5}
0.50	5.0_{-5}	3.44827_{-5}	3.47214_{-5}	3.47222_{-5}	3.47222_{-5}	3.47947_{-5}	3.48605_{-5}
0.55	5.5_{-5}	3.78967_{-5}	3.81418_{-5}	3.81422_{-5}	3.81386_{-5}	3.82104_{-5}	3.82942_{-5}
0.60	6.0_{-5}	4.12715_{-5}	4.15206_{-5}	4.15210_{-5}	4.15125_{-5}	4.15815_{-5}	4.16840_{-5}
0.65	6.5_{-5}	4.45876_{-5}	4.48380_{-5}	4.48389_{-5}	4.48249_{-5}	4.48888_{-5}	4.50109_{-5}
0.70	7.0_{-5}	4.78189_{-5}	4.80700_{-5}	4.80710_{-5}	4.80511_{-5}	4.81082_{-5}	4.82512_{-5}
0.75	7.5_{-5}	5.09342_{-5}	5.11884_{-5}	5.11897_{-5}	5.11634_{-5}	5.12123_{-5}	5.13781_{-5}
0.80	8.0_{-5}	5.39030_{-5}	5.41644_{-5}	5.41663_{-5}	5.41338_{-5}	5.41747_{-5}	5.43644_{-5}
0.85	8.5_{-5}	5.67056_{-5}	5.69748_{-5}	5.69774_{-5}	5.69430_{-5}	5.69776_{-5}	5.71886_{-5}
0.90	9.0_{-5}	5.93457_{-5}	5.96176_{-5}	5.96197_{-5}	5.95901_{-5}	5.96209_{-5}	5.98480_{-5}
0.95	9.5_{-5}	6.18648_{-5}	6.21251_{-5}	6.21267_{-5}	6.21042_{-5}	6.21333_{-5}	6.23695_{-5}
1.00	1.0_{-4}	6.43231_{-5}	6.45682_{-5}	6.45699_{-5}	6.45503_{-5}	6.45787_{-5}	6.48155_{-5}

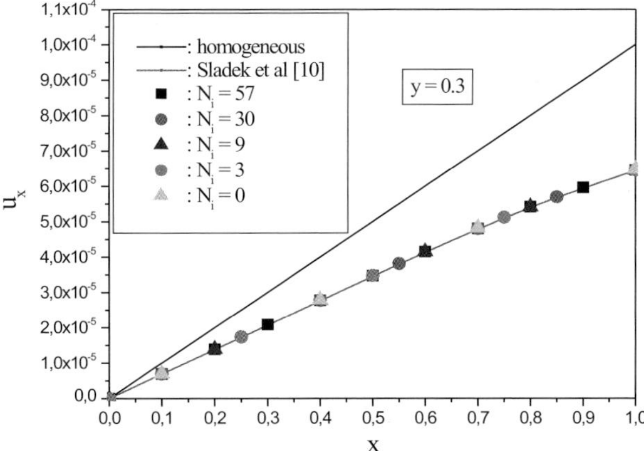

Figure 3. Comparison of computed displacement component u_x at $y = 0.3$.

Table 2. Comparison of numerical results ($x = 0.0$, 57 internal nodes).

y	Sladek et al. [10]	This work
0.000	0.685000	0.693971
0.075	0.814000	0.821044
0.150	0.969000	0.977980
0.225	1.155000	1.167289
0.300	1.375000	1.383252

Table 3. Comparison of numerical results ($x = 0.75$, 57 internal nodes).

y	Sladek et al. [10]	This work
0.000	0.758000	0.772665
0.075	0.884000	0.897074
0.150	0.995000	1.004405
0.225	1.102000	1.105088
0.300	1.220000	1.218766

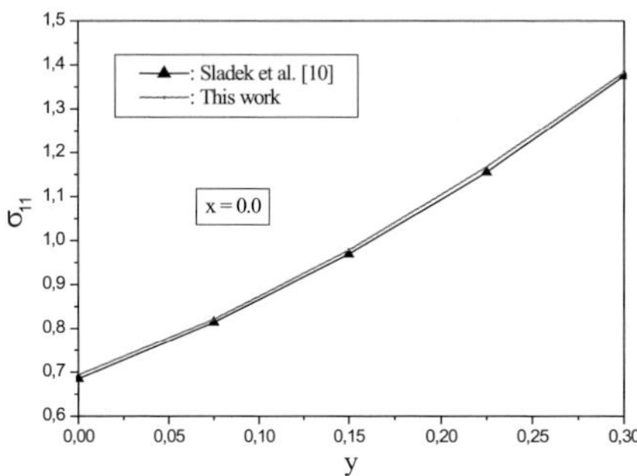

Figure 4. Comparison of numerical results ($x = 0.0$, 57 internal nodes).

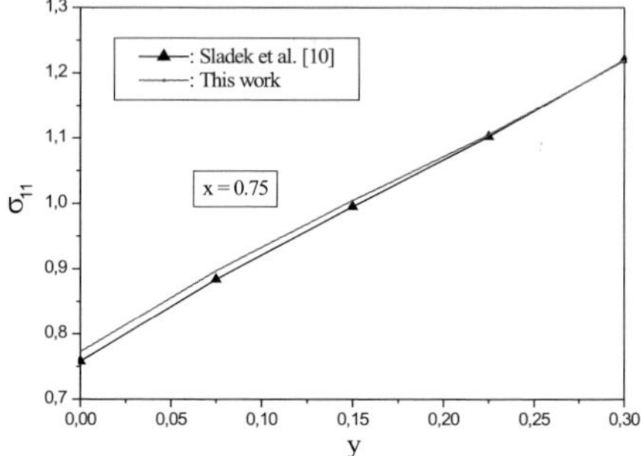

Figure 5. Comparison of numerical results ($x = 0.75$, 57 internal nodes).

The effects of the number and the distribution of used internal nodes on the computed stress component σ_{11} are shown in Tabs. 6.4 and 6.5 as well as Figs. 6.6 and 6.7. Numerical calculations are performed for 48 fixed boundary nodes and different internal nodes as shown in Fig. 6.2. Here again, it can be concluded that the present meshless BEM for the stress computation is quite insensitive to the selected node number and distribution. This confirms the accuracy, the efficiency, and the robustness of the present meshless BEM.

Table 4. Numerical results for different internal nodes ($x = 0.0$).

y	N_i=57	N_i= 30	N_i=9	N_i=3	N_i=0
0.000	0.693971	0.693672	0.692579	0.686051	0.686164
0.075	0.821044	0.821500	0.821216	0.816922	0.817061
0.150	0.977980	0.978044	0.978779	0.979832	0.979900
0.225	1.167289	1.166979	1.167411	1.171966	1.171949
0.300	1.383252	1.383287	1.382814	1.387026	1.386973

Table 5. Numerical results for different internal nodes ($x = 0.75$).

y	N_i=57	N_i= 30	N_i=9	N_i=3
0.000	0.772665	0.772592	0.773472	0.776418
0.075	0.897074	0.897191	0.898036	0.900220
0.150	1.004405	1.004311	1.004399	1.004362
0.225	1.105088	1.105005	1.104611	1.102561
0.300	1.218766	1.219057	1.216538	1.213336

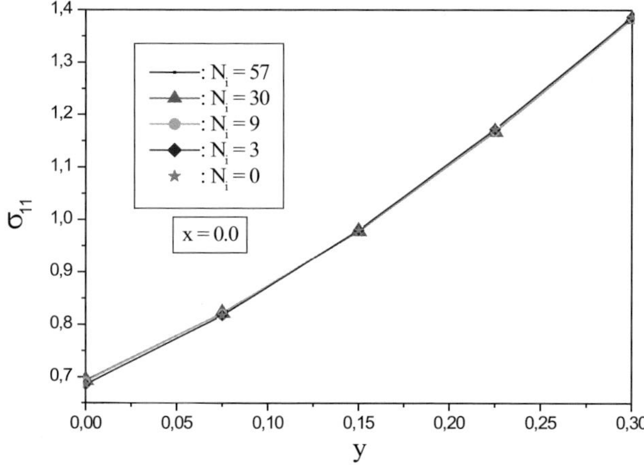

Figure 6. Numerical results for different internal nodes ($x = 0.0$).

7 Conclusions

In this paper, a meshless BEM for stress analysis in 2-D, isotropic, non-homogeneous and linear elastic solids is presented. Normalized displacements are introduced in the boundary-domain integral formulation, which avoids displacement gradients in the domain-integrals. To transform domain-integrals into boundary integrals along the global boundary of the analyzed domain, the radial integration method of Gao [7, 8, 9] is applied, which results in a meshless scheme. The normalized displacements in the domain-integrals are approximated by a series of prescribed basis functions, which are taken as a combination of radial basis functions and polynomials in terms of global coordinates. A 4-th order spline-type radial basis function is chosen in the present analysis. Numerical results are presented to show the accuracy and efficiency of the present meshless BEM. The present meshless BEM is easy to implement, very accurate and quite insensitive to the selected node number and distribution. For simple boundary value problems as shown in this analysis, very few internal or even no internal nodes are required to achieve a sufficient accuracy for the stress computation.

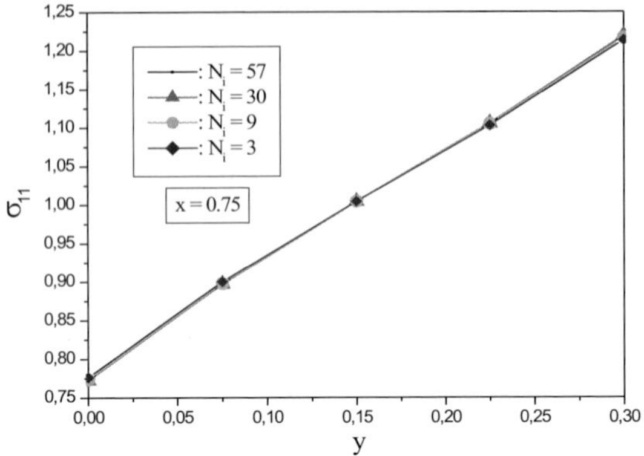

Figure 7. Numerical results for different internal nodes ($x = 0.75$).

Acknowledgement. The support by the German Academic Exchange Service (DAAD) and the Ministry of Education of Slovak Republic under the project number D/04/25722 is gratefully acknowledged.

References

1. T. Belytschko, Y. Krogauz, D. Organ, M. Fleming and P. Krysl, *Meshless Methods: An Overview and Recent Developments*, Comp. Meth. Appl. Mech. Eng. **139** (1996), 3–47.
2. S. N. Atluri and S. Shen, *The Meshless Local Petrov-Galerkin (MLPG) Method*, Tech Science Press, 2002.
3. G. R. Liu, *Mesh Free Methods: Moving Beyond the Finite Element Method*, CRC Press, 2003.
4. S. N. Atluri, *The Meshless Method (MLPG) for Domain & BIE Discretizations*, Tech Science Press, 2004.
5. S. E. Mikhailov, *Localized Boundary-Domain Integral Formulations for Problems With Variable Coefficients*, Eng. Anal. Bound. Elem. **26** (2002), 681–690.
6. N. Nardini and C. A. Brebbia, *A New Approach for Free Vibration Analysis Using Boundary Elements*, Boundary Element Methods in Engineering (C. A. Brebbia, ed.), Springer, Berlin, 1982, pp. 312–326.
7. X. W. Gao, *The Radial Integration Method for Evaluation of Domain Integrals With Boundary-Only Discretization*, Eng. Anal. Bound. Elem. **26** (2002), 905–916.
8. X. W. Gao, *A Boundary Element Method Without Internal Cells for Two-Dimensional and Three-Dimensional Elastoplastic Problems*, ASME Journal of Applied Mechanics **69** (2002), 154–160.
9. X. W. Gao, *Evaluation of Regular and Singular Domain Integrals With Boundary-Only Discetization - Theory and Fortran Code*, Journal of Computational and Applied Mathematics **175** (2005), 265–290.

10. J. Sladek, V. Sladek and S. N. Atluri, *Local Boundary Integral Equation (LBIE) Method for Solving Problems of Elasticity With Nonhomogeneous Material Properties*, Computational Mechanics **24** (2000), 456–462.
11. J. Sladek, V. Sladek and Ch. Zhang, *Application of Meshless Local Petrov-Galerkin (MLPG) Method to Elastodynamic Problems in Continuously Nonhomogeneous Solids*, Computer Modeling in Eng. & Sciences **4** (2003), 637–648.
12. V. Sladek, J. Sladek and Ch. Zhang, *Local Integro-Differential Equations With Domain Elements for Numerical Solution of PDE With Variable Coefficients*, J. Eng. Math. **51** (2005), 261–282.
13. B. N. Rao and S. Rahman, *Mesh-Free Analysis of Cracks in Isotropic Functionally Graded Materials*, Eng. Fract. Mech. **70** (2003), 1–27.
14. M. A. Golberg, C. S. Chen and H. Bowman, *Some Recent Results and Proposals for the Use of Radial Basis Functions in the BEM*, Eng. Anal. Bound. Elem. **23** (1999), 285–296.
15. X. W. Gao and T. G. Davies, *Boundary Element Programming in Mechanics*, Cambridge University Press, Cambridge, 2002.
16. V. Sladek, J. Sladek and I. Markechova, *An Advanced Boundary Element Method for Elasticity Problems in Nonhomogeneous Media*, Acta Mechanica **97** (1993), 71–90.
17. X. W. Gao, *Numerical Evaluation of Two-Dimensional Singular Boundary Integrals - Theory and Fortran Code*, Journal of Computational and Applied Mathematics, **188** (2006), 44–64.

A Particle-Partition of Unity Method Part VII: Adaptivity

Michael Griebel and Marc Alexander Schweitzer

Institut für Numerische Simulation, Universität Bonn, Wegelerstr. 6, D-53115 Bonn, Germany
{griebel, schweitzer}@ins.uni-bonn.de

Summary. This paper is concerned with the adaptive multilevel solution of elliptic partial differential equations using the partition of unity method. While much of the work on meshfree methods is concerned with convergence-studies, the issues of fast solution techniques for the discrete system of equations and the construction of optimal order algorithms are rarely addressed. However, the treatment of large scale real-world problems by meshfree techniques will become feasible only with the availability of fast adaptive solvers.

The adaptive multilevel solver proposed in this paper is a main step toward this goal. In particular, we present an h-adaptive multilevel solver for the partition of unity method which employs a subdomain-type error indicator to control the refinement and an efficient multilevel solver within a nested iteration approach. The results of our numerical experiments in two and three space dimensions clearly show the efficiency of the proposed scheme.

Key words: Meshfree method, partition of unity method, adaptive refinement, multilevel method.

1 Introduction

One main purpose of this paper is to investigate adaptive h-type refinement strategies for meshfree methods and their interplay with multilevel solution techniques; in particular we address these issues for the partition of unity method [1, 15]. To this end, we employ a classical a posteriori error estimation technique due to Babuška and Rheinboldt [2]. The resulting error indicator is used to steer the local refinement procedure in our tree-based cover construction procedure. To obtain an adaptive solver with optimal complexity we combine the multilevel techniques developed in [7, 9, 15] for the PUM with the nested iteration approach [12]. The results of our numerical experiments in two and three space dimensions demonstrate the effectiveness of the proposed approach and its overall efficiency.

In this paper we restrict ourselves to the study of a scalar elliptic partial differential equation, namely we consider the diffusion problem

$$\begin{aligned} -\Delta u &= f \quad \text{in } \Omega \subset \mathbb{R}^d, \\ u &= g_D \quad \text{on } \Gamma_D \subset \partial\Omega, \\ \frac{\partial u}{\partial n} &= g_N \quad \text{on } \Gamma_N = \partial\Omega \setminus \Gamma_D. \end{aligned} \tag{1.1}$$

The remainder of this paper is organized as follows. In section 2 we give a short overview of the PUM and its convergence properties. Furthermore, we outline the implementation of essential boundary conditions using Nitsche's method and the Galerkin discretization of the arising variational problem. The main theme of this paper, the adaptive meshfree multilevel solution of an elliptic PDE, is presented in section 3. There, we introduce our refinement algorithm and show that it leads to point sets and covers that are consistent with the multilevel construction of [9]. Moreover, we present the construction of our error indicator and discuss how we obtain an adaptive multilevel solver with optimal complexity using the nested iteration approach. Then, we present the results of our numerical experiments in two and three space dimensions in section 4. These results clearly demonstrate the efficiency of the proposed scheme. Finally, we conclude with some remarks in section 5.

2 Partition of Unity Method

In the following, we shortly review the construction of a partition of unity space V^{PU} and the Galerkin discretization of an elliptic partial differential equation using V^{PU} as trial and test space, see [15] for details.

2.1 Construction of a Partition of Unity Space

In a PUM, we define a global approximation u^{PU} simply as a weighted sum of local approximations u_i,

$$u^{\text{PU}}(x) := \sum_{i=1}^{N} \varphi_i(x) u_i(x). \tag{2.2}$$

These local approximations u_i are completely independent of each other, i.e., the local supports $\omega_i := \text{supp}(u_i)$, the local basis $\{\psi_i^n\}$ and the order of approximation p_i for every single $u_i := \sum u_i^n \psi_i^n \in V_i^{p_i}$ can be chosen independently of all other u_j. Here, the functions φ_i form a partition of unity (PU). They are used to splice the local approximations u_i together in such a way that the global approximation u^{PU} benefits from the local approximation orders p_i yet it still fulfills global regularity conditions. Hence, the global approximation space on Ω is defined as

$$V^{\mathrm{PU}} := \sum_i \varphi_i V_i^{p_i} = \sum_i \varphi_i \operatorname{span}\langle\{\psi_i^n\}\rangle = \operatorname{span}\langle\{\varphi_i\psi_i^n\}\rangle. \qquad (2.3)$$

The starting point in the implementation of a PUM approximation space V^{PU} is the construction of an appropriate PU, see Definition 1 and Definition 2.

Definition 1 (Partition of Unity). *Let $\Omega \subset \mathbb{R}^d$ be an open set. Let $\{\varphi_i\}$ be a collection of Lipschitz functions with*

$$\begin{aligned} 0 \leq \varphi_i(x) \leq 1, &\qquad \textstyle\sum_i \varphi_i \equiv 1 \text{ on } \overline{\Omega}, \\ \|\varphi_i\|_{L^\infty(\mathbb{R}^d)} \leq C_\infty, &\quad \|\nabla\varphi_i\|_{L^\infty(\mathbb{R}^d)} \leq \frac{C_\nabla}{\operatorname{diam}(\omega_i)}, \end{aligned}$$

where $\omega_i := \operatorname{supp}(\varphi_i)$, C_∞ and C_∇ are two positive constants. The sets ω_i are called patches and their collection is referred to as a cover $C_\Omega := \{\omega_i\}$ of the domain Ω.

For PUM spaces (2.3) which employ such a PU $\{\varphi_i\}$ there hold the following error estimates due to [1].

Theorem 1. *Let $\Omega \subset \mathbb{R}^d$ be given. Let $\{\varphi_i\}$ be a partition of unity according to Definition 1. Let us further introduce the covering index $\lambda_{C_\Omega} : \Omega \to \mathbb{N}$ such that*

$$\lambda_{C_\Omega}(x) = \operatorname{card}(\{i \,|\, x \in \omega_i\}) \qquad (2.4)$$

and let us assume that $\lambda_{C_\Omega}(x) \leq M \in \mathbb{N}$ for all $x \in \Omega$. Let a collection of local approximation spaces $V_i^{p_i} = \operatorname{span}\langle\{\psi_i^n\}\rangle \subset H^1(\Omega \cap \omega_i)$ be given. Let $u \in H^1(\Omega)$ be the function to be approximated. Assume that the local approximation spaces $V_i^{p_i}$ have the following approximation properties: On each patch $\Omega \cap \omega_i$, the function u can be approximated by a function $u_i \in V_i^{p_i}$ such that

$$\|u - u_i\|_{L^2(\Omega\cap\omega_i)} \leq \hat{\epsilon}_i, \quad \text{and} \quad \|\nabla(u - u_i)\|_{L^2(\Omega\cap\omega_i)} \leq \tilde{\epsilon}_i \qquad (2.5)$$

hold for all i. Then the function

$$u^{\mathrm{PU}} := \sum_{\omega_i \in C_\Omega} \varphi_i u_i \in V^{\mathrm{PU}} \subset H^1(\Omega)$$

satisfies the global estimates

$$\|u - u^{\mathrm{PU}}\|_{L^2(\Omega)} \leq \sqrt{M} C_\infty \Big(\sum_{\omega_i \in C_\Omega} \hat{\epsilon}_i^2 \Big)^{\frac{1}{2}}, \qquad (2.6)$$

$$\|\nabla(u - u^{\mathrm{PU}})\|_{L^2(\Omega)} \leq \sqrt{2M} \Big(\sum_{\omega_i \in C_\Omega} \big(\frac{C_\nabla}{\operatorname{diam}(\omega_i)}\big)^2 \hat{\epsilon}_i^2 + C_\infty^2 \tilde{\epsilon}_i^2 \Big)^{\frac{1}{2}}. \qquad (2.7)$$

The estimates (2.6) and (2.7) show that the global error is of the same order as the local errors provided that the covering index is bounded independent of the size of the cover, i.e. $M = O(1)$. Note that we need to assume a slightly stronger condition to obtain a *sparse* linear system by the Galerkin approach. To this end, we introduce the notion of a local neighborhood or local cover $C_{\Omega,i} \subset C_\Omega$ of a particular cover patch $\omega_i \in C_\Omega$ by

$$C_{\Omega,i} := \{\omega_j \in C_\Omega \,|\, \omega_j \cap \omega_i \neq \emptyset\} \tag{2.8}$$

and require $\max_{\omega_i \in C_\Omega} \operatorname{card}(C_{\Omega,i}) = O(1)$.

Note furthermore that the conditions imposed on the PU in Definition 1 do not ensure that the product functions $\varphi_i \psi_i^n$ of (2.3) are linearly independent. However, to obtain the linear independence of the product functions $\varphi_i \psi_i^n$ it is sufficient to require that the PU has the following property.

Definition 2 (Flat top property). *Let $\{\varphi_i\}$ be a partition of unity according to Definition 1. Let us define the sub-patches $\omega_{\mathrm{FT,i}} := \{x \,|\, \lambda_{C_\Omega}(x) = 1\}$ such that $\varphi_i|_{\omega_{\mathrm{FT,i}}} \equiv 1$. Then, the PU is said to have the flat top property, if there exists a constant C_{FT} such that for all patches ω_i*

$$\mu(\omega_i) \leq C_{\mathrm{FT}}\, \mu(\omega_{\mathrm{FT,i}}) \tag{2.9}$$

where $\mu(A)$ denotes the Lebesgue measure of $A \subset \mathbb{R}^d$. We have $C_\infty = 1$ for a PU with the flat top property.

Obviously the product functions $\varphi_i \psi_i^n$ are linearly independent if we assume that the PU has the flat top property and that each of the local bases $\{\psi_i^n\}$ is locally linearly independent on the sub-patches $\omega_{\mathrm{FT,i}} \subset \omega_i$.[1]

The PU concept is employed in many meshfree methods. However, in most cases very smooth PU functions $\varphi_i \in C^k(\Omega)$ with $k \geq 2$ are used and the functions φ_i have rather large supports ω_i which overlap substantially. Hence in most meshfree methods $\operatorname{card}(C_{\Omega,i})$ is large and the employed PU does *not* have the flat top property. This makes it easier to control $\|\nabla \varphi_i\|_{L^\infty}$, compare Definition 1 and (2.7), but it can lead to ill-conditioned and even singular stiffness matrices. For a flat top PU we obviously have $\nabla \varphi_i|_{\omega_{\mathrm{FT,i}}} \equiv 0$ so that it is sufficient to bound $\nabla \varphi_i$ on the complement $\omega_i \setminus \omega_{\mathrm{FT,i}}$ which requires some additional properties, compare (2.13) and (2.14). Hence, the cover construction for a flat top PU is somewhat more challenging.

A PU can for instance be constructed by simple averaging, often referred to as Shepard's method. Let us assume that we have a cover $C_\Omega = \{\omega_i\}$ of the domain Ω such that $1 \leq \lambda_{C_\Omega}(x) \leq M$ for all $x \in \Omega$. With the help of non-negative weight functions W_k defined on these cover patches ω_k, i.e. $W_k(x) > 0$ for all $x \in \omega_k \setminus \partial\omega_k$, we can easily generate a partition of unity by

$$\varphi_i(x) := \frac{W_i(x)}{S_i(x)} \quad \text{where } S_i(x) := \sum_{\omega_j \in C_{\Omega,i}} W_j(x). \tag{2.10}$$

[1] Note that the flat top property is a sufficient condition only. It is not a necessary requirement. In practice we already obtain a linearly independent set of shape functions if the flat top property is satisfied by most but not necessarily all patches ω_i of the cover C_Ω.

Obviously, the smoothness of the resulting PU functions φ_i is determined entirely by the smoothness of the employed weight functions. Hence, on a cover with tensor product patches ω_i we can easily construct partitions of unity of any regularity for instance by using tensor products of splines with the desired regularity as weight functions.[2] Hence, let us assume that the weight functions W_i are all given as linear transformations of a generating normalized spline weight function $\mathcal{W} : \mathbb{R}^d \to \mathbb{R}$ with $\text{supp}(\mathcal{W}) = [0,1]^d$, i.e.,

$$W_i(x) = \mathcal{W} \circ T_i(x), \quad T_i : \omega_i \to [0,1]^d, \quad \|DT_i\|_\infty \leq \frac{C_T}{\text{diam}(\omega_i)} \tag{2.11}$$

and $\|\mathcal{W}\|_\infty = 1$, $\|\nabla\mathcal{W}\|_\infty \leq C_{\mathcal{W},U}$. To show that the PU arising from (2.10) is valid according to Definition 1 it is sufficient to make the following additional assumptions:

- Comparability of neighboring patches: There exist constants C_L and C_U such that for all local neighborhoods $C_{\Omega,i}$ there holds the implication

$$\omega_j \in C_{\Omega,i} \quad \Rightarrow \quad C_L \,\text{diam}(\omega_i) \leq \text{diam}(\omega_j) \leq C_U \,\text{diam}(\omega_i) \tag{2.12}$$

  with absolute constants C_L and C_U.
- Sufficient overlap: There exists a constant $K > 0$ such that for any $x \in \Omega$ there is at least one cover patch ω_i with the property

$$x \in \omega_i, \quad \text{dist}(x, \partial\omega_i) \geq K \,\text{diam}(\omega_i). \tag{2.13}$$

- Weight function and cover are compatible: There exists a constant $C_{\mathcal{W},L}$ such that for all cover patches ω_i

$$|\nabla W_i(x)| > \frac{C_{\mathcal{W},L}}{\text{diam}(\omega_i)} \quad \text{holds for all } x \in \Omega \text{ with } \lambda_{C_\Omega}(x) > 1, \tag{2.14}$$

  compare Figure 1.

Lemma 1. *The PU defined by* (2.10) *with weights* (2.11) *is valid according to Definition 1 under the assumptions* (2.12), (2.13), *and* (2.14).

Proof. For $x \in \Omega$ with $\lambda_{C_\Omega}(x) = 1$ we have $\nabla\varphi_i(x) = 0$. Note that we have

$$|S_i(x)| \geq |W_l(x)| = |W_l(x) - W_l(y)|$$

where ω_l denotes the cover patch with property (2.13) for $x \in \Omega$ and $y \in \partial\omega_l$ is arbitrary. For any $x \in \Omega$ with $\lambda_{C_\Omega} > 1$ we therefore obtain with the mean value theorem, (2.13) and (2.14)

[2] Other shapes of the cover patches $\omega_i \in C_\Omega$ are of course possible, e.g. balls or ellipsoids, but the resulting partition of unity functions φ_i are more challenging to integrate numerically. For instance a subdivision scheme based on the piecewise constant covering index λ_{C_Ω} leads to integration cells with very complicated geometry.

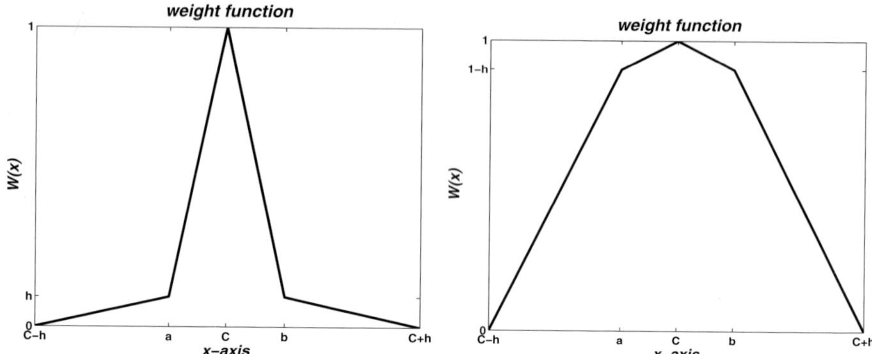

Figure 1. A one-dimensional weight function W_i on a patch $\omega_i = (C-h, C+h)$ with $\omega_{\mathrm{FT},i} = (a,b)$ that does not satisfy (left) the compatibility condition (2.14), and one that does (right).

$$|S_i(x)| \geq C_{\mathcal{W},L} K.$$

Together with (2.11) and (2.12) this yields the point-wise estimate

$$\begin{aligned}
|\nabla \varphi_i(x)| &= \Big| \frac{W_i(x)\nabla S_i(x) - \nabla W_i(x) S_i(x)}{S_i^2(x)} \Big| \\
&\leq \frac{\Big(|\nabla \mathcal{W} \circ T_i(x) DT_i(x) S_i(x)| + |W_i(x) \sum_k \nabla \mathcal{W} \circ T_k(x) DT_k(x)| \Big)}{|S_i^2(x)|} \\
&\leq (C_{\mathcal{W},L} K)^{-2} \frac{2MC_T C_{\mathcal{W},U}}{\operatorname{diam}(\omega_i)},
\end{aligned}$$

which gives the asserted bound $\|\nabla \varphi_i\|_{L^\infty(\mathbb{R}^d)} \leq \frac{C_\nabla}{\operatorname{diam}(\omega_i)}$ with $C_\nabla \geq \frac{2MC_T C_{\mathcal{W},U}}{(C_{\mathcal{W},L}K)^{-2}}$. □

In general any local space which provides some approximation property such as (2.5) can be used in a PUM. Furthermore, the local approximation spaces are independent of each other. Hence, if there is a priory knowledge about the (local) behavior of the solution u available, it can be utilized to choose operator-dependent approximation spaces. For instance, the a priori information can be used to enrich a space of polynomials by certain singularities or it maybe used to choose systems of eigenfunction of (parts) of the considered differential operator as local approximation spaces. Such specialized local function spaces may be given analytically or numerically.

If no a priori knowledge about the solution is available classical multipurpose expansion systems like polynomials are used. In this paper we employ products of univariate Legendre polynomials throughout, i.e., we use

$$V_i^{p_i}(\omega_i) := \mathcal{P}^{p_i} \circ \widetilde{T}_i, \quad \widetilde{T}_i : \omega_i \to (-1,1)^d$$

$$\mathcal{P}^{p_i}((-1,1)^d) = \operatorname{span}\langle \{ \psi^n \mid \psi^n = \prod_{l=1}^d \mathcal{L}^{\hat{n}_l}, \|\hat{n}\|_1 = \sum_{l=1}^d \hat{n}_l \leq p_i \} \rangle,$$

where $\mathcal{L}^k$ denotes the Legendre polynomial of degree k.

2.2 Essential Boundary Conditions and Galerkin Discretization

The treatment of essential boundary conditions in meshfree methods is not straightforward and a number of different approaches have been suggested. In [10] we have presented how Nitsche's method [13] can be applied successfully in the meshfree context.

In order to formulate the resulting weak formulation of (1.1) arising from Nitsche's approach, we introduce some additional notation. Let $\partial_n u := \frac{\partial u}{\partial n}$ denote the normal derivative, $\Gamma_{D,i} := \omega_i \cap \Gamma_D$, and

$$C_{\Gamma_D} := \{\omega_i \in C_\Omega \,|\, \Gamma_{D,i} \neq \emptyset\}$$

denote the cover of the Dirichlet boundary. Furthermore, we define the cover-dependent norm

$$\|\partial_n u\|^2_{-\frac{1}{2},C_\Omega} := \sum_{\omega_i \in C_{\Gamma_D}} \operatorname{diam}(\Gamma_{D,i}) \|\partial_n u\|^2_{L^2(\Gamma_{D,i})}.$$

With these conventions we obtain the weak formulation

$$a_\beta(u,v) = l_\beta(v) \quad \text{for all } v \in V^{\text{PU}} \tag{2.15}$$

with the cover-dependent bilinear form

$$a_\beta(u,v) := \int_\Omega \nabla u \nabla v - \int_{\Gamma_D} (\partial_n u v + u \partial_n v) + \beta \sum_{\omega_i \in C_{\Gamma_D}} \operatorname{diam}(\Gamma_{D,i})^{-1} \int_{\Gamma_{D,i}} uv \tag{2.16}$$

and the corresponding linear form

$$l_\beta(v) := \int_\Omega fv - \int_{\Gamma_D} g_D \partial_n v + \int_{\Gamma_N} g_N v + \beta \sum_{\omega_i \in C_{\Gamma_D}} \operatorname{diam}(\Gamma_{D,i})^{-1} \int_{\Gamma_{D,i}} g_D v \tag{2.17}$$

from the minimization of the functional

$$J_\beta(w) := \int_\Omega |\nabla w|^2 - 2\int_{\Gamma_D} \partial_n w w + \beta \sum_{\omega_i \in C_{\Gamma_D}} \operatorname{diam}(\Gamma_{D,i})^{-1} \int_{\Gamma_{D,i}} |w|^2. \tag{2.18}$$

Note that this minimization is completed for the error in V^{PU}, i.e., we are looking for $\min_{u^{\text{PU}} \in V^{\text{PU}}} J_{N,\beta}(u - u^{\text{PU}})$. There is a unique solution u^{PU} if the regularization parameter β is chosen large enough; i.e., the regularization parameter is dependent on the discretization space V^{PU}. The solution u^{PU} of (2.15) satisfies an optimal error estimate if the space V^{PU} admits the following inverse estimate

$$\|\partial_n u\|_{-\frac{1}{2},C_\Omega} \leq C_{\text{inv}} \|\nabla u\|_{L^2(\Omega)} \quad \text{for all } v \in V^{\text{PU}} \tag{2.19}$$

with an absolute constant C_{inv} depending on the cover C_Ω, the generating weight function $\mathcal{W}$ and the employed local bases $\{\psi_i^n\}$. If C_{inv} is known, the regularization parameter β can be chosen as $\beta > 2C_{\mathrm{inv}}^2$. Hence, the main task is the automatic computation of the constant C_{inv}. Fortunately, C_{inv}^2 can be approximated very efficiently, see [10]. To this end, we consider the inverse assumption (2.19) as a generalized eigenvalue problem locally on each patch $\omega_i \in C_{\Gamma_D}$ which intersects the Dirichlet boundary and solve for the largest eigenvalue to obtain an approximation to C_{inv}^2.

Note that this (overlapping) variant of Nitsche's approach is slightly different from the one employed in [10, 15], e.g., there the inverse assumption (2.19) was formulated using a cover-independent norm. To attain a convergent scheme from (2.18) it is essential that the covering index $\lambda_{C_{\Gamma_D}}(x) < M$ is bounded. The implementation of (2.16) and (2.17) is somewhat more involved since the numerical integration scheme must be capable of handling the overlap region correctly. The main advantage of this overlapping variant is that the regularization parameter β is dependent on the employed local approximation spaces, i.e., on the employed polynomial degrees, and on the maximal level difference L close to the boundary only. It is not dependent on $\mathrm{diam}(\omega_i)$. Hence, it is sufficient to pre-compute β for the maximal allowable value of L and the maximal polynomial degree. This value can then be used for all patches on all levels.

For the Galerkin discretization of (2.15), which yields the linear system

$$A\tilde{u} = \hat{f}, \quad \text{with } A_{(i,k),(j,n)} = a_\beta(\varphi_j \psi_j^n, \varphi_i \psi_i^k), \text{ and } \hat{f}_{(i,k)} = l_\beta(\varphi_i \psi_i^k),$$

we need to employ numerical integration since the PU functions are in general piecewise rational functions. Note that the flat top property is also beneficial to the numerical integration since all PU functions are constant on each $\omega_{\mathrm{FT,i}}$ so that the integration on a large part of the domain $\bigcup_i \omega_{\mathrm{FT,i}} \subset \Omega$ involves only the local basis functions ψ_i^n. Therefore, a subdivision scheme based on the covering index λ_{C_Ω} which employs sparse grid numerical integration rules of higher order on the cover-dependent integration cells seems to be the best approach, see [8] for details. Note that the use of an automatic construction procedure for the numerical integration scheme is a must for adaptive computations since an a priori prescribed background integration scheme can hardly account for the (possibly) huge variation in the support sizes $\mathrm{diam}(\omega_i)$ and may lead to stability problems.

With respect to the assembly of the system matrix A for a refined PUM space it is important to note that we do not need to compute all its entries $A_{(i,k),(j,n)}$. We can re-use the entries $A_{(i,k),(j,n)}$ which stem from a patch ω_i with the property that none of its neighbors $\omega_j \in C_{\Omega,i}$ have been refined. Hence, there are a number of complete block-rows $A_{(i,\cdot),(\cdot,\cdot)}$ that do not need to be computed for the refined space and we need to compute only a minimal number of matrix entries $A_{(i,k),(j,n)}$ from level to level.

3 Adaptive Multilevel Solution

In [8, 9] we have developed a tree-based cover construction scheme that gives a sequence of covers $\{C_\Omega^k\}$ based on a given point set $P = \{x_i\}$. The fundamental construction principle employed in [8] is a d-binary tree. Based on the given point data P, we sub-divide a bounding-box $\mathcal{C}_\Omega \supset \Omega$ until each of the tree cells

$$\mathcal{C}_i = \prod_{l=1}^{d} (c_i^l - h_i^l, c_i^l + h_i^l)$$

contains at most a single point $x_i \in P$, see Figure 2 (left). We obtain a valid cover from this tree by choosing

$$\omega_i = \prod_{l=1}^{d} (c_i^l - \alpha h_i^l, c_i^l + \alpha h_i^l), \quad \text{with } \alpha > 1. \tag{3.20}$$

Note that we define a cover patch ω_i and a corresponding PU function φ_i for cells that contain a point $x_i \in P$ as well as for *empty* cells that do not contain any point from P.[3] This procedure increases the dimension of the resulting PUM space, yet (under some assumptions) only by a constant factor [5, 8]. The main benefit of using a larger number of cover patches is that the resulting neighborhoods $C_{\Omega,i}$ are smaller and that we therefore obtain a smaller number of entries in the stiffness matrix. The coarser covers C_Ω^k are defined considering coarser versions of the constructed tree, i.e., by removing the complete set of leaves of the tree. For details of this construction see [8].

The constructed covers C_Ω^k all satisfy the conditions of the previous section.

Figure 2. Subdivision corresponding to an initial cover (left). Subdivision with cells of neighborhood $C_{\Omega,i}$ (light gray) and cell corresponding to patch ω_i (dark gray) (center). Subdivision with cells of subset $R_{\Omega,i}$ (gray) (right).

[3] This approach can be interpreted as a saturation technique for our d-binary tree. To this end, we can define an additional point set $\tilde{P} = \{\xi_i\}$ such that each cell of the tree now contains *exactly* one point of the union $\tilde{P} \cup P$, compare [8, 9].

Lemma 1. *A cover $C_\Omega = \{\omega_i\}$ arising from the stretching of a d-binary tree cell decomposition $\{C_i\}$ according to* (3.20) *with $\alpha > 1$ satisfies conditions* (2.11), (2.12) *and* (2.13).

Proof. For the ease of notation let us assume $h_i = h_i^k$ for $k = 1, \dots, d$. Then, we have $h_i \asymp 2^{-l_i} \operatorname{diam}(\Omega)$ where l_i refers to the tree-level of the cell C_i. Obviously, we have $C_T = 1$,

$$C_U = \max_{\omega_i \in C_\Omega} \max_{\omega_j \in C_{\Omega,i}} 2^{|l_i - l_j|}, \quad \text{and} \quad C_L = C_U^{-1}.$$

Due to the stretching of the tree cells we can find for any $x \in \Omega$ at least one cover patch ω_i such that $x \in \omega_i$ and that the inequality

$$\operatorname{dist}(x, \partial\omega_i) \geq \frac{\alpha - 1}{2} \min_{\omega_j \in C_{\Omega,i}} 2^{-l_j}$$

holds. Hence, with the maximal difference in tree levels of two overlapping cover patches $L := \max_{\omega_i \in C_\Omega} \max_{\omega_j \in C_{\Omega,i}} |l_i - l_j|$ we obtain $C_U = 2^L$ and $K = (\alpha - 1)2^{-L-1}$. □

Therefore, the resulting PU defined by (2.10) using a tensor product B-spline as generating weight function satisfies the assumptions of Definition 1 and the error estimates (2.6) and (2.7) hold for our multilevel PUM on each level k. Furthermore, we can easily enforce that each PU of the resulting sequence $\{\varphi_{i,k}\}_k$ has the flat top property.

Corollary 1. *The PU resulting from* (2.10) *based on a cover $C_\Omega = \{\omega_i\}$ arising from the stretching of a d-binary tree cell decomposition $\{C_i\}$ with $\alpha > 1$ according to* (3.20) *has the flat top property if $\alpha \in (1, 1 + 2^{-L})$.*

3.1 Particle Refinement

In the following we consider the refinement of the given point set P and the respective sequence of covers C_Ω^k obtained from the tree construction reviewed above. One of the properties this refinement procedure should have is that we are able to bound the maximal level difference L of the resulting tree. Only then will we obtain a sequence of covers $\{C_\Omega^k\}$ and a sequence of PUs $\{\varphi_{i,k}\}$ which satisfy the conditions given above with *uniform* constants.

In general any local refinement procedure employs a Boolean refinement indicator function $r : C_\Omega \to \{\texttt{true}, \texttt{false}\}$ which identifies or marks regions which should be refined. Often the refinement indicator is based on thresholding of local error estimates $\eta_i \approx \|u - u^{\mathrm{PU}}\|_{H^1(\Omega \cap \omega_i)}$, e.g.

$$r(\omega_i) = \begin{cases} \texttt{true} & \text{if} \quad 2\eta_i \geq \eta_{\max} \\ \texttt{false} & \text{else} \end{cases}, \quad \text{with } \eta_{\max} := \max_{\omega_i \in C_\Omega} \eta_i. \tag{3.21}$$

Based on this refinement indicator we then employ certain refinement rules to improve the resolution on a patch ω_i. Since we are interested in the refinement of a particle set, these refinement rules must essentially create new particles in the regions $\{\omega_i \in C_\Omega \,|\, r(\omega_i) = \texttt{true}\}$. Furthermore, this refinement process should be consistent with our tree construction.

Before we consider our refinement rules for particles, let us first consider if a simple refinement indicator function like (3.21) is suitable for the PUM. To this end let us assume that we have $\eta_i \approx \|u - u^{\mathrm{PU}}\|_{H^1(\Omega\cap\omega_i)}$ an estimate of the error $u - u^{\mathrm{PU}}$ locally on each patch ω_i, see section 3.3 for the construction of such an error estimator. Note that the η_i are (overlapping) subdomain estimators which are fundamentally different from the more common (disjoint) element estimators in the FEM.

Recall that the local error is given by

$$\|u - u^{\mathrm{PU}}\|_{H^1(\Omega\cap\omega_i)} = \|\sum_{\omega_j \in C_{\Omega,i}} \varphi_j(u - u_j)\|_{H^1(\Omega\cap\omega_i)} \tag{3.22}$$

since $\sum_{\omega_j \in C_{\Omega,i}} \varphi_j(x) \equiv 1$ for all $x \in \omega_i$. Thus, using a simple indicator like (3.21) which would just mark the patch ω_i for refinement may not be sufficient to reduce the error on ω_i. It seems necessary that at least some of the neighbors are also marked for refinement to achieve a sufficient error reduction. Another reason why the highly local indicator (3.21) is not suitable for our PUM is the fact that we need to bound the maximal level difference L of neighboring patches. This can hardly be achieved using (3.21). A simple solution to this issue could be to refine all patches $\omega_j \in C_{\Omega,i}$ since they all contribute to the error on ω_i. This will certainly ensure that the error on ω_i is reduced, however, it may lead to a substantial yet unnecessary increase in computational work and storage. Taking a closer look at (3.22) we find that the contribution of a patch ω_j to the error on ω_i lives on the intersection $\omega_j \cap \omega_i$ only. Hence, it is a promising approach to select an appropriate subset of neighbors $\omega_j \in C_{\Omega,i}$ via the (relative) size of these intersections. This approach is further in perfect agreement with our constraint of bounding the maximal level difference since the intersection $\omega_j \cap \omega_i$ will be (relatively) large if $l_j < l_i$, i.e., ω_j is a coarser patch than ω_i. Hence, we introduce the sets

$$R_{\Omega,i} := \{\omega_i\} \cup \{\omega_j \in C_{\Omega,i} \,|\, l_j < l_i\}$$

of patches with a large contribution to the local error on ω_i. With the help of these sets we define our refinement indicator function as

$$r(\omega_j) = \begin{cases} \texttt{true} & \text{if} \quad \omega_j \in R_{\Omega,i} \text{ and } 2\eta_i \geq \eta_{\max} \\ \texttt{false} & \text{else} \end{cases}, \tag{3.23}$$

see Figure 2. Note however that this does not guarantee that the maximal level difference L stays constant. But it ensures that L increases somewhat slower and only in regions where the error is already small. Hence, the adverse effect of larger constants in the error bound (2.7) is almost negligible. Also note that the presented procedure can be interpreted as a refinement with implicit smoothing of the cover due to the selection of a subset of neighbors. This strategy gave very favorable results in all our numerical studies and it is employed in the numerical experiments presented in this paper.

Now that we have identified the refinement region, we need to consider the refinement rules for our particle set P based on the patch-wise refinement indicator function (3.23). Our goal is to define a set of refinement rules which create new points x_n and a respective cover $\tilde{C}_\Omega$, so that our original cover construction algorithm with the input $P \cup \{x_n\}$ will give the refined cover $\tilde{C}_\Omega$. Hence, the locations of the created points are constrained to certain cells of our tree. In this paper we employ a very simple and numerically cheap positioning scheme[4] for the new points based on a local center of gravity, which is defined as

$$g_i = \frac{1}{\mathrm{card}(G_{\Omega,i})} \sum_{x_k \in G_{\Omega,i}} x_k, \quad G_{\Omega,i} = \{x_k \in P \,|\, x_k \in \omega_k \in C_{\Omega,i}\}.$$

Note that g_i is well-defined for all patches. Due to our tree-based cover construction we can always find at least one given point x_i in the local neighborhoods $C_{\Omega,e}$ even for empty patches ω_e, i.e. $\omega_e \cap P = \emptyset$. Besides the local centers of gravity g_i we furthermore use the geometric centers c_i of the tree-cell C_i associated with the considered patch ω_i and the centers $c_{i,q}$ of the refined tree-cells $C_{i,q}$ with $q = 1, \ldots, 2^d$ for our positioning scheme. The overall refinement scheme for a patch ω_i reads as follows.

Algorithm 1 (Particle Refinement).

1. Set counter $w = 0$.
2. If there is $x_i \in P$ with $x_i \in C_i \subset \omega_i$, then determine sub-cell $C_{i,\tilde{q}} \subset \omega_{i,\tilde{q}}$ with $x_i \in C_{i,\tilde{q}}$.
3. If $g_i \in C_i \subset \omega_i$, then determine sub-cell $C_{i,\hat{q}} \subset \omega_{i,\hat{q}}$ with $g_i \in C_{i,\hat{q}}$. Set $P = P \cup \{g_i\}$ and $w = w + 1$. If $w \geq 2^{d-1}$, then stop.
4. If $\hat{q} \neq \tilde{q}$, then
   For $q = 1, \ldots, 2^d$ compute projection $p_{i,q}$ of sub-cell center $c_{i,q}$ on line $\overline{x_i g_i}$ and the projection $\tilde{p}_{i,q}$ of $c_{i,q}$ on line $\overline{x_i c_i}$.
   If $p_{i,q} \in C_{i,q}$, then set $P = P \cup \{p_{i,q}\}$ and $w = w + 1$. If $w \geq 2^{d-1}$, then stop.
   If $p_{i,q} \notin C_{i,q}$ and $\tilde{p}_{i,q} \in C_{i,q}$, then set $P = P \cup \{\tilde{p}_{i,q}\}$ and $w = w + 1$. If $w \geq 2^{d-1}$, then stop.
5. If $\hat{q} = \tilde{q}$ and $g_i = c_i$, then assume that data is gridded and set $P = P \cup \{c_{i,q}\}$ with $q = 1, \ldots, 2^d$.

[4] Note that many other approaches to the construction of new points are possible. For instance we can minimize the local fill distance or the separation radius under the constraint of positioning the new points within the sub-cells of the tree construction. Such approaches, however, involve the solution of a linear system and hence are computationally more expensive.

Now that we have our refined point set P, let us consider the question of how to define the respective cover patches ω_i. The refinement of a cover is straightforward since a d-binary tree is an adaptive data-structure. Here, it is sufficient to use a single subdivision step to split the tree-cell $\mathcal{C}_i$ into 2^d sub-cells $\mathcal{C}_{i,q}$ if the refinement indicator function $r(\omega_i) = \texttt{true}$ for the associated cover patch ω_i. Then, we insert the created particles in the respective cells and set the patches $\omega_{i,q}$ on all refined (sub-)cells $\mathcal{C}_{i,q}$ with $q = 1, \ldots, 2^d$ using (3.20) with the same overlap parameter α.

Due to our careful selection of the positions of the new points $\{x_n\}$ in our refinement scheme we ensure that the refined cover $\tilde{C}_\Omega$ is identical to the cover obtained by our original cover construction algorithm using the refined point set $P \cup \{x_n\}$ as input. Hence, a refined cover is guaranteed to have the same properties as the covers obtained from the original algorithm.

Recall that our tree-based cover construction algorithm provides a complete sequence of covers C_Ω^k. Hence, we must deal with the question how to introduce a refined cover into an existing sequence of covers C_Ω^k such that the resulting refined sequence is consistent with our multilevel construction [9, 15].

3.2 Iterative Solution

In [9] we have constructed a sequence of covers $\{C_\Omega^k = \{\omega_{i,k}\}\}$ with $k = 0, \ldots, J$ where J denotes the maximal subdivision level of the tree, that is all covers C_Ω^k have the property

$$k = \max_{\omega_i \in C_\Omega^k} l_i. \tag{3.24}$$

The respective PUM spaces V_k^{PU} are defined as

$$V_k^{\mathrm{PU}} := \sum_{\omega_{i,k} \in \Omega_k} \varphi_{i,k} V_{i,k}^{p_{i,k}}$$

with the PU functions (2.10) based on the cover $C_{\Omega,k}$ and local approximation spaces $V_{i,k}^{p_{i,k}}$ of degree $p_{i,k}$.

Note that property (3.24) ensures that we have a minimal number of levels $J+1$ and thus minimal work and storage in the iterative multilevel solver. Hence, the covers of a refined sequence must also satisfy (3.24).

Let us assume that we have a sequence of covers $\{C_\Omega^k\}$ with $k = 0, \ldots, J$ satisfying the level property (3.24) and that we refine the cover C_Ω^J. To obtain a sequence $\{C_\Omega^k\}$ with this property by the refinement scheme presented above we need to distinguish two cases.

First we consider the simple case where we refine the cover C_Ω^J in such a way that at least one patch ω_i with $l_i = J$ is marked for refinement. Then, the resulting cover $R(C_\Omega^J)$ has at least one element ω_j with $l_j = J+1$ and we can extend our sequence of covers $\{C_\Omega^k\}$ with $k = 0, \ldots, J+1$ where

$C_\Omega^{J+1} = R(C_\Omega^J)$. In the case where we refine only patches ω_i with $l_i < J$, we obtain a refined cover $R(C_\Omega^J)$ for which $\max_{\omega_i \in R(C_\Omega^J)} l_i = J$ holds and we cannot extend the existing sequence of covers by $R(C_\Omega^J)$. We rather need to replace the cover C_Ω^J by its refined version $R(C_\Omega^J)$ to obtain a consistent sequence of covers. Thus, we end up with the modified sequence $\{C_\Omega^k\}$ with $k = 0, \ldots, J$ where we assign $C_\Omega^J = R(C_\Omega^J)$.

With these conventions it is clear that our refinement scheme leads to a sequence of covers $\{C_\Omega^k\}$ that satisfies all assumptions of our multilevel construction and our iterative multilevel solver is applicable also in adaptive computations.

To reduce the computational work even further, we couple our multilevel solver with the nested iteration technique [12], see Algorithm 2.

Algorithm 2 (Nested Iteration).

1. If $l > 0$, then set initial guess
   $\tilde{u}_l^0 = P_{l-1}^l \tilde{u}_{l-1}^{k_{l-1}}$.
   Else, set initial guess
   $\tilde{u}_l^0 = 0$.
2. Set $\tilde{u}_l^{k_l} \leftarrow \mathcal{IS}_l^{k_l}(\tilde{u}_l^0, \hat{f}_l, A_l)$.

The ingredients of a nested iteration are the basic iterative solution procedure $\mathcal{IS}_l$ (in our case $\mathcal{IS}_l$ will be a multilevel iteration $\mathcal{MG}_l(0, \hat{f}_J)$) defined on each level l and prolongation operators P_{l-1}^l. One key observation which lead to the development of Algorithm 2 is that the approximate solution $\tilde{u}_{l-1}^{k_{l-1}}$ obtained on level $l-1$ is a good initial guess $\tilde{u}_l^0$ for the iterative solution on level l. To this end, we need the prolongation operator P_{l-1}^l to transfer a coarse solution on level $l-1$ to the next finer level l, see step 1 of Algorithm 2. Another property that is exploited in our nested iteration solver is that there is nothing to gain from solving the linear system of equations (almost) exactly since its solution describes an approximate solution of the considered PDE only. The iterative solution process on level l can be stopped once the error of the iteration is of the same order as the discretization error on level l. Thus, if the employed iterative solver $\mathcal{IS}_l$ has a constant error reduction rate, as it is the case for an optimal multilevel iteration $\mathcal{MG}_l$, then a (very small) constant number of iterations k_l that is independent of l in step 2 is sufficient to obtain an approximate solution $\tilde{u}_l$ on each level l within discretization accuracy. The overall iterative process is also referred to as full multigrid [6, 11].

In all our numerical studies no more than 2 applications of a $V(1,1)$-cycle with block-Gauss-Seidel smoothing (compare [9]) were necessary to obtain an approximate solution within discretization accuracy.

3.3 Error Estimation

The final ingredient of our adaptive multilevel PUM is the local error estimator $\eta_i \approx \|u - u^{\mathrm{PU}}\|_{H^1(\Omega \cap \omega_i)}$ which steers our particle refinement process and can be used to assess the quality of the computed global approximation [2, 3, 16, 17]. In this section we now construct an error estimator η_i for the PUM based on the subdomain approach due to [2].

We employ an a posteriori error estimation technique based on the solution of local Dirichlet problems defined on (overlapping) subdomains introduced in [2] which is very natural to the PUM. To this end let us consider the additional local problems

$$\begin{aligned} -\Delta w_i &= f && \text{in } \Omega \cap \omega_i, \\ w_i &= u^{\mathrm{PU}} && \text{on } \partial(\Omega \cap \omega_i) \setminus \Gamma_N, \\ \frac{\partial w_i}{\partial n} &= g_N && \text{on } \Gamma_N \cap \partial(\Omega \cap \omega_i) \end{aligned} \tag{3.25}$$

to approximate the error $u - u^{\mathrm{PU}}$ on $\Omega \cap \omega_i$ by $w_i - u^{\mathrm{PU}} \in H^1(\Omega \cap \omega_i)$, see [2]. This leads to the local error estimator $\eta_i := \|w_i - u^{\mathrm{PU}}\|_{H^1(\Omega \cap \omega_i)}$.

Note that the local problems (3.25) employ inhomogeneous Dirichlet boundary values. As discussed in section 2.2, the implementation of essential boundary conditions is somewhat more involved in the PUM. There we have presented a non-conforming approach due to Nitsche to realize the global Dirichlet conditions of our model problem (1.1). Of course this technique can also be pursued here, however, since we consider (3.25) on very special subdomains, i.e., on the support of a PU function φ_i, there is a much simpler and conforming approach.

Enforcing homogeneous boundary conditions on the boundary $\partial\omega_i$ is trivial since $\varphi_i|_{\partial\omega_i} \equiv 0$. For patches close to the boundary we can easily enforce homogeneous boundary values on $\partial(\Omega \cap \omega_i) \setminus \Gamma_N \setminus \Gamma_D$. Hence, if we reformulate (3.25) in such a way that we have to deal with vanishing boundary data on $\partial(\Omega \cap \omega_i) \setminus \Gamma_N \setminus \Gamma_D$ only, we can realize the (artificial) boundary conditions in a conforming way. Only for the global boundary data on Γ_D we need to employ the non-conforming Nitsche technique. Therefore, we employ a discrete version of the following equivalent formulation of (3.25)

$$\begin{aligned} -\Delta \tilde{w}_i &= f - f^{\mathrm{PU}} && \text{in } \Omega \cap \omega_i, \\ \tilde{w}_i &= 0 && \text{on } \partial(\Omega \cap \omega_i) \setminus \Gamma_N \setminus \Gamma_D, \\ \tilde{w}_i &= g_D - u^{\mathrm{PU}} && \text{on } \partial(\Omega \cap \omega_i) \cap \Gamma_D, \\ \frac{\partial \tilde{w}_i}{\partial n} &= g_N - \frac{\partial u^{\mathrm{PU}}}{\partial n} && \text{on } \Gamma_N \cap \partial(\Omega \cap \omega_i), \end{aligned} \tag{3.26}$$

where f^{PU} denotes the best approximation of f in V^{PU}, with mostly homogeneous boundary conditions within our implementation.

We approximate (3.26) using the trial and test spaces $V_{i,*}(\Omega \cap \omega_i) := \varphi_i V_i^{p_i+q_i}$ with $q_i > 0$. Obviously, the functions $w_i \in V_{i,*}(\Omega \cap \omega_i)$ satisfy the homogeneous boundary conditions on $\partial(\Omega \cap \omega_i) \setminus \Gamma_N \setminus \Gamma_D$ due to the multiplication with the partition of unity function φ_i. Note that these local problems fit very well with the *global* Nitsche formulation (2.18) since the solution of (3.26) coincides with the minimizer of

$$J_{\gamma_i}(u - u^{\mathrm{PU}} - \tilde{w}_i) \rightarrow \min\{\tilde{w}_i \in V_{i,*}(\Omega \cap \omega_i)\}$$

where the parameter γ_i now depends on the local discretization space $V_{i,*}(\Omega \cap \omega_i) \subset H^1(\Omega \cap \omega_i)$ and not on $V^{\mathrm{PU}} \subset H^1(\Omega)$.[5] Note that the utilization of the global Nitsche functional is possible due to the use of a conforming approach for the additional boundary $\partial(\Omega \cap \omega_i) \setminus \Gamma_N \setminus \Gamma_D$ only.

We obtain our local approximate error estimator

$$\eta_i := \|\tilde{w}_i\|_{H^1(\Omega \cap \omega_i)} \tag{3.27}$$

from the approximate solution $\tilde{w}_i \in V_{i,*}(\Omega \cap \omega_i)$ of the local problem (3.26). The global error is then estimated by

$$\eta := \Big(\sum_{\omega_i \in C_\Omega} (\eta_i)^2 \Big)^{\frac{1}{2}} = \Big(\sum_{\omega_i \in C_\Omega} \|\tilde{w}_i\|^2_{H^1(\Omega \cap \omega_i)} \Big)^{\frac{1}{2}}. \tag{3.28}$$

Note that we solve (3.26) in the complete space $V_{i,*}(\Omega \cap \omega_i) = \varphi_i V_i^{p_i+q_i}$ and not just the space $\varphi_i V_i^{p_i+q_i \setminus p_i}$ where $V_i^{p_i+q_i \setminus p_i}$ denotes the hierarchical complement of $V_i^{p_i}$ in $V_i^{p_i+q_i}$.

This subdomain error estimation approach was already analyzed in the PUM context in [1]. There it was shown that the subdomain estimator is efficient and reliable, i.e., there holds the equivalence

$$C^{-1} \sum_{\omega_i \in C_\Omega} \|w_i\|^2_{H^1(\Omega \cap \omega_i)} \leq \|u - u^{\mathrm{PU}}\|^2_{H^1(\Omega)} \leq C \sum_{\omega_i \in C_\Omega} \|w_i\|^2_{H^1(\Omega \cap \omega_i)}. \tag{3.29}$$

Yet, it was assumed that the variational problem is globally positive definite and that a globally conforming implementation of essential boundary conditions is employed. However, both these assumptions are not satisfied in our PUM due to the Nitsche approach. The analysis of the presented estimator is an open issue.

Also note that there are other a posteriori error estimation techniques based on the strong residual in Mortar finite elements based Nitsche's approach e.g. [4] which can be used in the PUM context. Finally, let us point out an interesting property of the PUM which might be beneficial in the construction of error estimators based on the strong residual. Recall from Theorem 1 that the global error is essentially given as an overlapping sum of the local errors with respect to the local approximation spaces. The properties of the PU required by Definition 1 enter in the constants of the estimates (2.6) and (2.7) only. They do not affect the attained approximation order. Hence, the global approximation error in a PUM is essentially invariant of the employed PU — if the PU is based on the same cover C_Ω.

Corollary 2. *Let $\Omega \subset \mathbb{R}^d$ be given. Let $\{\varphi_i^1\}$ and $\{\varphi_i^2\}$ be partitions of unity according to Definition 1 employing the same cover $C_\Omega = \{\omega_i\}$, i.e. for all i assume that $\omega_i^1 = \omega_i^2 = \omega_i$. Let us assume that $\lambda_{C_\Omega}(x) \leq M \in \mathbb{N}$ for all $x \in \Omega$.*

[5] We may also pre-compute the Nitsche regularization parameter $\beta_{\max}$ for maximal *total* degree $\max p_i + q_i$ and employ $\beta_{\max}$ on all levels and for all local problems.

Let a collection of local approximation spaces $V_i^{p_i} = \mathrm{span}\langle\{\psi_i^n\}\rangle \subset H^1(\Omega\cap\omega_i)$ *be given as in Theorem 1. Let* $u \in H^1(\Omega)$ *be the function to be approximated. Then there hold the global equivalencies*

$$C_{1,2}^{-1}\|u^{\mathrm{PU},1} - u\|_{L^2(\Omega)} \le \|u^{\mathrm{PU},2} - u\|_{L^2(\Omega)} \le C_{2,1}\|u^{\mathrm{PU},1} - u\|_{L^2(\Omega)} \quad (3.30)$$

and

$$C_{1,2,\nabla}^{-1}\|\nabla(u^{\mathrm{PU},1}-u)\|_{L^2(\Omega)} \le \|\nabla(u^{\mathrm{PU},2}-u)\|_{L^2(\Omega)} \le C_{2,1,\nabla}\|\nabla(u^{\mathrm{PU},1}-u)\|_{L^2(\Omega)} \quad (3.31)$$

for the functions

$$u^{\mathrm{PU},1} := \sum_{\omega_i\in C_\Omega} \varphi_i^1 u_i \in V^{\mathrm{PU},1} \quad \text{and} \quad u^{\mathrm{PU},2} := \sum_{\omega_i\in C_\Omega} \varphi_i^2 u_i \in V^{\mathrm{PU},2}$$

with constants $C_{1,2}$, $C_{2,1}$, $C_{1,2,\nabla}$, $C_{2,1,\nabla}$ *depending on the partitions of unity only.*

Due to this equivalence it is easily possible to obtain an approximation $\tilde{u}^{\mathrm{PU}}$ with higher regularity $k > 0$ from a $\mathcal{C}^0$ approximation u^{PU} in our PUM simply be changing the employed generating weight function $\mathcal{W}$. The smoother approximation $\tilde{u}^{\mathrm{PU}}$ can for instance be used to evaluate/approximate higher order derivatives without the need to consider jumps or other discontinuities explicitly.

3.4 Overall Algorithm

Let us shortly summarize our overall adaptive multilevel algorithm which employs three user-defined parameters: $\epsilon > 0$ a global error tolerance, $q > 0$ the increment in the polynomial degree for the estimation of the error and $k > 0$ the number of multilevel iterations employed in the nested iteration.

Algorithm 3 (Adaptive Multilevel PUM).

1. Let us assume that we are given an initial point set P and that we have a sequence of PUM spaces $V_k^{\mathrm{PU}} = \sum_{\omega_{i,k}\in C_{\Omega,k}} \varphi_{i,k} V_{i,k}^{p_{i,k}}$ with $k = 0,\dots,J$ based on a respective sequence of covers $C_{\Omega,k} = \{\omega_{i,k}\}$ arising from a d-binary tree construction using the point set P, see [8, 15] for details. Let $P_{l-1}^l : V_{l-1}^{\mathrm{PU}} \to V_l^{\mathrm{PU}}$ and $R_l^{l-1} : V_l^{\mathrm{PU}} \to V_{l-1}^{\mathrm{PU}}$ denote transfer operators and $\mathcal{S}_l : V_l^{\mathrm{PU}} \times V_l^{\mathrm{PU}} \to V_l^{\mathrm{PU}}$ appropriate smoothing schemes so that we can define a multilevel iteration $\mathcal{MG}_J : V_J^{\mathrm{PU}} \times V_J^{\mathrm{PU}} \to V_J^{\mathrm{PU}}$, see [7, 9, 15] for details. Set $\tilde{u}_J = \mathcal{MG}_J^{k_{\mathrm{init}}}(0, \hat{f}_J)$ where the number of iterations k_{init} is assumed to be large enough.
2. Compute the local error estimates η_i from (3.27) using the local spaces $\varphi_{J,i} V_{J,i}^{p_{J,k}+q}$. Estimate the global error by (3.28).
3. If the global estimate satisfies $\eta < \epsilon$: STOP.

4. Define the refinement indicator function (3.23) on the cover $C_{\Omega,J}$ based on the local estimates η_i.
5. Using the refinement rules of section 3.1 define a refined point set P, a refined cover $R(C_{\Omega,J})$ and its associated PUM space $R(V_J^{\mathrm{PU}})$.
6. If $R(C_{\Omega,J})$ satisfies the level property (3.24) with J:
   a) Delete the transfer operators P_{J-1}^J and R_J^{J-1}.
   b) Compute an intermediate transfer operator $\mathcal{P}_J : V_J^{\mathrm{PU}} \to R(V_J^{\mathrm{PU}})$.
   c) Set $\tilde{v}_J = \mathcal{P}_J \tilde{u}_J$.
   d) Delete the intermediate transfer operator $\mathcal{P}_J$.
   e) Remove the cover $C_{\Omega,J}$ and its associated PUM space V_J^{PU} from the respective sequences.
   f) Set $C_{\Omega,J} := R(C_{\Omega,J})$ and $V_J^{\mathrm{PU}} := R(V_J^{\mathrm{PU}})$.
7. If $R(C_{\Omega,J})$ satisfies the level property (3.24) with $J+1$:
   a) Extend the sequence of covers by $C_{\Omega,J+1} := R(C_{\Omega,J})$ and the sequence of PUM spaces by $V_{J+1}^{\mathrm{PU}} := R(V_J^{\mathrm{PU}})$.
   b) Set $\tilde{v}_J = 0$.
   c) Set $J = J + 1$.
8. Set up the stiffness matrix A_J and right-hand side $\hat{f}_J$ using an appropriate numerical integration scheme.
9. Compute transfer operators $P_{J-1}^J : V_{J-1}^{\mathrm{PU}} \to V_J^{\mathrm{PU}}$ and $R_J^{J-1} : V_J^{\mathrm{PU}} \to V_{J-1}^{\mathrm{PU}}$ and define appropriate smoother $\mathcal{S}_J : V_J^{\mathrm{PU}} \times V_J^{\mathrm{PU}} \to V_J^{\mathrm{PU}}$ on level J.
10. If $\tilde{v}_J = 0$, set $\tilde{v}_J = P_{J-1}^J \tilde{u}_{J-1}$.
11. Apply $k > 0$ iterations and set $\tilde{u}_J = \mathcal{MG}_J^k(\tilde{v}_J^{\mathrm{PU}}, \hat{f}_J)$.
12. GOTO 2.

4 Numerical Results

In this section we present some results of our numerical experiments using the adaptive PUM discussed above. To this end, we introduce some shorthand notation for various error norms, i.e., we define

$$e_{L^\infty} := \frac{\|u - u^{\mathrm{PU}}\|_{L^\infty}}{\|u\|_{L^\infty}}, \quad e_{L^2} := \frac{\|u - u^{\mathrm{PU}}\|_{L^2}}{\|u\|_{L^2}}, \quad \text{and} \quad e_{H^1} := \frac{\|(u - u^{\mathrm{PU}})\|_{H^1}}{\|u\|_{H^1}}. \tag{4.32}$$

Analogously, we introduce the notion

$$e_{H^1}^* := \frac{\eta}{\|u\|_{H^1}} = \frac{\left(\sum_{\omega_i \in C_\Omega} \eta_i^2\right)^{\frac{1}{2}}}{\|u\|_{H^1}}$$

for the estimated (relative) error using (3.27) and (3.28). These norms are approximated using a numerical integration scheme with very fine resolution, see [15]. For each of these error norms we can compute the respective convergence rate ρ by considering the error norms of two consecutive levels $l-1$ and l

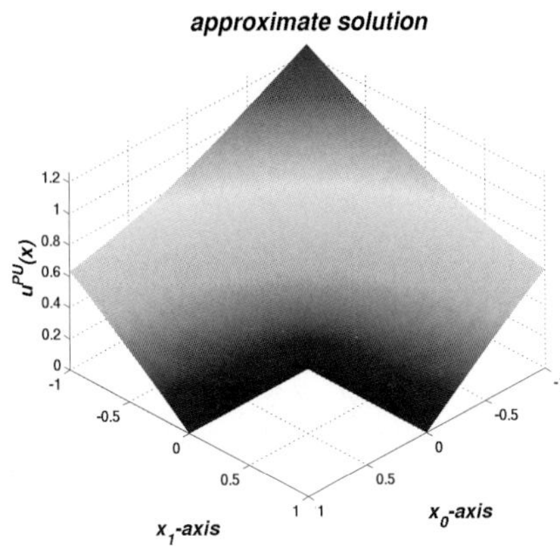

Figure 3. Surface plot of approximate solution u^{PU} on level $J = 11$.

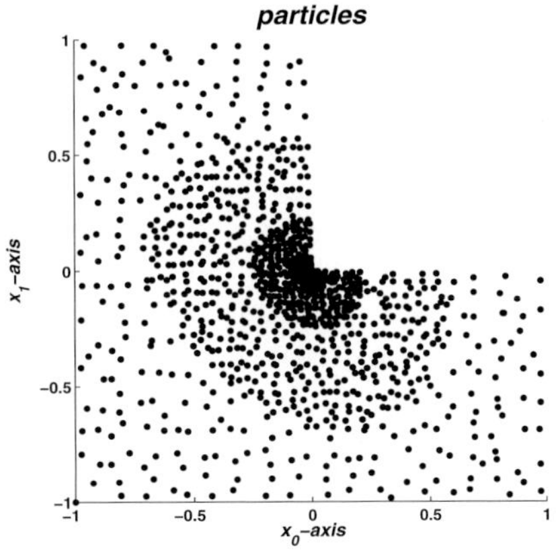

Figure 4. Refined point set P on level $J = 11$ for Example 1 using quadratic polynomials for error estimation.

$$\rho := -\frac{\log\left(\frac{\|u-u_l^{\mathrm{PU}}\|}{\|u-u_{l-1}^{\mathrm{PU}}\|}\right)}{\log\left(\frac{\mathsf{dof}_l}{\mathsf{dof}_{l-1}}\right)}, \qquad \mathsf{dof}_k := \sum_{\omega_{i,k}\in C_\Omega^k} \dim(V_{i,k}^{p_{i,k}}). \tag{4.33}$$

To assess the quality of our error estimator we give its effectivity index with respect to the H^1-norm

$$\epsilon_{H^1}^* := \frac{e_{H^1}^*}{e_{H^1}} = \frac{\eta}{\|(u-u^{\mathrm{PU}})\|_{H^1}}$$

in the tables. We also give the maximal subdivision level J of our tree for the cover construction, and the total number of degrees of freedom dof of the constructed PUM space V^{PU} on level J.

Example 1. In our first example we consider the standard test case of an L-shaped domain in two space dimensions with homogeneous boundary conditions at the re-entrant corner. That is we discretize the problem

$$\begin{aligned} -\Delta u &= f \quad \text{in } \Omega = (-1,1)^2 \setminus [0,1)^2, \\ u &= g_D \quad \text{on } \partial\Omega \end{aligned}$$

with our adaptive PUM where we choose f and g_D such that the solution $u \in H^{\frac{3}{2}}(\Omega)$ in polar coordinates is given by $u(r,\theta) = r^{\frac{2}{3}}\sin(\frac{2\theta-\pi}{3})$, see Figure 3. We employ linear Legendre polynomials as local approximation spaces V_i^1 and estimate the local errors once with quartic (Table 1 and Figure 5) and once with quadratic (Table 2 and Figure 6) Legendre polynomials, i.e. $V_{i,*} = \varphi_i V_i^4$ and $V_{i,*} = \varphi_i V_i^2$ respectively.

It is well-known that in this two-dimensional example uniform refinement will yield a convergence rate of $\rho_{H^1} = \frac{1}{3}$ only instead of the optimal $\rho_{H^1} =$

Table 1. Relative errors e (4.32) and convergence rates ρ (4.33) for Example 1 using quartic Legendre polynomials for error estimation.

J	dof	e_{L^∞}	ρ_{L^∞}	e_{L^2}	ρ_{L^2}	e_{H^1}	ρ_{H^1}	$e^*_{H^1}$	$\rho^*_{H^1}$	$\epsilon^*_{H^1}$
0	3	3.303_{-1}	1.01	2.469_{-1}	1.27	5.272_{-1}	0.58	4.150_{-1}	0.80	0.79
1	9	1.162_{-1}	0.95	5.301_{-2}	1.40	2.153_{-1}	0.82	2.327_{-1}	0.53	1.08
2	36	7.431_{-2}	0.32	2.084_{-2}	0.67	1.572_{-1}	0.23	1.344_{-1}	0.40	0.85
3	144	4.628_{-2}	0.34	8.947_{-3}	0.61	1.034_{-1}	0.30	8.702_{-2}	0.31	0.84
4	252	2.908_{-2}	0.83	4.207_{-3}	1.35	7.291_{-2}	0.62	6.030_{-2}	0.66	0.83
5	360	1.830_{-2}	1.30	2.774_{-3}	1.17	5.632_{-2}	0.72	4.566_{-2}	0.78	0.81
6	468	1.152_{-2}	1.76	2.389_{-3}	0.57	4.821_{-2}	0.59	3.836_{-2}	0.66	0.80
7	576	7.252_{-3}	2.23	2.277_{-3}	0.23	4.457_{-2}	0.38	3.505_{-2}	0.43	0.79
8	1008	4.564_{-3}	0.83	1.164_{-3}	1.20	3.244_{-2}	0.57	2.554_{-2}	0.57	0.79
9	1566	2.874_{-3}	1.05	6.580_{-4}	1.29	2.524_{-2}	0.57	1.981_{-2}	0.58	0.78
10	2124	1.811_{-3}	1.52	5.282_{-4}	0.72	2.174_{-2}	0.49	1.701_{-2}	0.50	0.78
11	3636	1.140_{-3}	0.86	2.669_{-4}	1.27	1.635_{-2}	0.53	1.280_{-2}	0.53	0.78
12	5418	7.183_{-4}	1.16	1.834_{-4}	0.94	1.306_{-2}	0.56	1.023_{-2}	0.56	0.78
13	8226	4.525_{-4}	1.11	1.230_{-4}	0.96	1.063_{-2}	0.49	8.324_{-3}	0.49	0.78
14	13491	2.850_{-4}	0.93	6.996_{-5}	1.14	8.298_{-3}	0.50	6.493_{-3}	0.50	0.78
15	20412	1.796_{-4}	1.12	4.731_{-5}	0.94	6.618_{-3}	0.55	5.191_{-3}	0.54	0.78
16	30438	1.131_{-4}	1.16	3.305_{-5}	0.90	5.455_{-3}	0.48	4.277_{-3}	0.48	0.78
17	49842	7.125_{-5}	0.94	1.926_{-5}	1.09	4.288_{-3}	0.49	3.359_{-3}	0.49	0.78
18	77256	4.489_{-5}	1.05	1.225_{-5}	1.03	3.385_{-3}	0.54	2.657_{-3}	0.53	0.79
19	115326	2.828_{-5}	1.15	8.611_{-6}	0.88	2.786_{-3}	0.49	2.187_{-3}	0.49	0.78
20	189585	1.781_{-5}	0.93	5.119_{-6}	1.05	2.193_{-3}	0.48	1.720_{-3}	0.48	0.78
21	298440	1.122_{-5}	1.02	3.129_{-6}	1.09	1.719_{-3}	0.54	1.350_{-3}	0.53	0.79
22	446850	7.069_{-6}	1.14	2.201_{-6}	0.87	1.411_{-3}	0.49	1.109_{-3}	0.49	0.79
23	737478	4.453_{-6}	0.92	1.321_{-6}	1.02	1.110_{-3}	0.48	8.715_{-4}	0.48	0.78
24	1171548	2.805_{-6}	1.00	7.915_{-7}	1.11	8.672_{-4}	0.53	6.814_{-4}	0.53	0.79
25	1756818	1.767_{-6}	1.14	5.570_{-7}	0.87	7.109_{-4}	0.49	5.585_{-4}	0.49	0.79

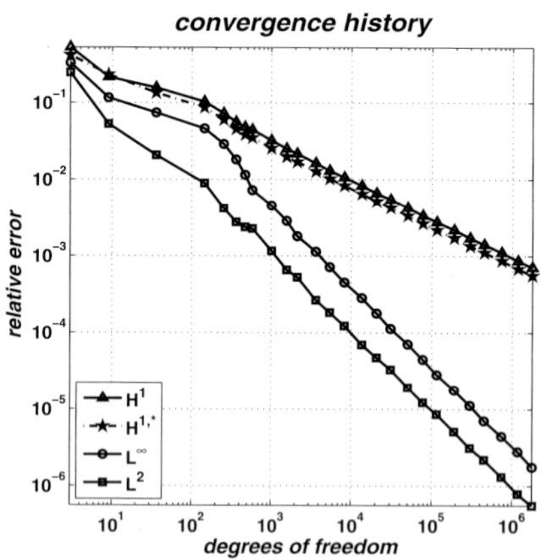

Figure 5. Convergence history for Example 1 using quartic polynomials for error estimation.

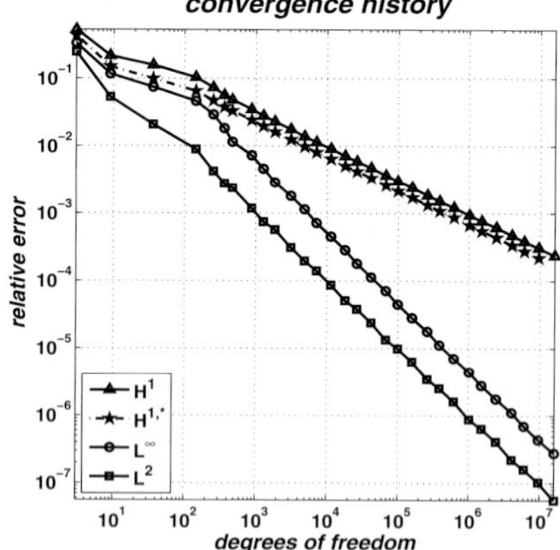

Figure 6. Convergence history for Example 1 using quadratic polynomials for error estimation.

Table 2. Relative errors e (4.32) and convergence rates ρ (4.33) for Example 1 using quadratic Legendre polynomials for error estimation.

J	dof	e_{L^∞}	ρ_{L^∞}	e_{L^2}	ρ_{L^2}	e_{H^1}	ρ_{H^1}	$e^*_{H^1}$	$\rho^*_{H^1}$	$\epsilon^*_{H^1}$
0	3	3.303_{-1}	1.01	2.469_{-1}	1.27	5.272_{-1}	0.58	4.287_{-1}	0.77	0.81
1	9	1.162_{-1}	0.95	5.301_{-2}	1.40	2.153_{-1}	0.82	1.496_{-1}	0.96	0.69
2	36	7.431_{-2}	0.32	2.084_{-2}	0.67	1.572_{-1}	0.23	9.882_{-2}	0.30	0.63
3	144	4.628_{-2}	0.34	8.947_{-3}	0.61	1.034_{-1}	0.30	6.502_{-2}	0.30	0.63
4	252	2.908_{-2}	0.83	4.207_{-3}	1.35	7.291_{-2}	0.62	4.696_{-2}	0.58	0.64
5	360	1.830_{-2}	1.30	2.774_{-3}	1.17	5.632_{-2}	0.72	3.753_{-2}	0.63	0.67
6	468	1.152_{-2}	1.76	2.389_{-3}	0.57	4.821_{-2}	0.59	3.306_{-2}	0.48	0.69
7	873	7.250_{-3}	0.74	1.181_{-3}	1.13	3.484_{-2}	0.52	2.418_{-2}	0.50	0.69
8	1278	4.566_{-3}	1.21	7.443_{-4}	1.21	2.778_{-2}	0.59	1.950_{-2}	0.56	0.70
9	1854	2.875_{-3}	1.24	5.693_{-4}	0.72	2.298_{-2}	0.51	1.615_{-2}	0.51	0.70
10	3123	1.811_{-3}	0.89	3.101_{-4}	1.16	1.765_{-2}	0.51	1.243_{-2}	0.50	0.70
11	4842	1.140_{-3}	1.05	1.963_{-4}	1.04	1.380_{-2}	0.56	9.784_{-3}	0.55	0.71
12	7146	7.184_{-4}	1.19	1.408_{-4}	0.85	1.136_{-2}	0.50	8.054_{-3}	0.50	0.71
13	11385	4.525_{-4}	0.99	8.684_{-5}	1.04	9.076_{-3}	0.48	6.422_{-3}	0.49	0.71
14	17937	2.851_{-4}	1.02	5.109_{-5}	1.17	7.102_{-3}	0.54	5.047_{-3}	0.53	0.71
15	26235	1.796_{-4}	1.22	3.806_{-5}	0.77	5.863_{-3}	0.50	4.173_{-3}	0.50	0.71
16	41598	1.131_{-4}	1.00	2.404_{-5}	1.00	4.710_{-3}	0.47	3.347_{-3}	0.48	0.71
17	67266	7.126_{-5}	0.96	1.344_{-5}	1.21	3.654_{-3}	0.53	2.602_{-3}	0.52	0.71
18	99162	4.489_{-5}	1.19	9.895_{-6}	0.79	2.999_{-3}	0.51	2.138_{-3}	0.51	0.71
19	157779	2.828_{-5}	0.99	6.324_{-6}	0.96	2.410_{-3}	0.47	1.714_{-3}	0.48	0.71
20	259047	1.781_{-5}	0.93	3.465_{-6}	1.21	1.861_{-3}	0.52	1.325_{-3}	0.52	0.71
21	383805	1.122_{-5}	1.18	2.532_{-6}	0.80	1.521_{-3}	0.51	1.085_{-3}	0.51	0.71
22	612792	7.070_{-6}	0.99	1.621_{-6}	0.95	1.220_{-3}	0.47	8.686_{-4}	0.47	0.71
23	1014804	4.454_{-6}	0.92	8.828_{-7}	1.20	9.396_{-4}	0.52	6.695_{-4}	0.52	0.71
24	1509102	2.806_{-6}	1.16	6.403_{-7}	0.81	7.659_{-4}	0.51	5.465_{-4}	0.51	0.71
25	2412603	1.767_{-6}	0.98	4.109_{-7}	0.95	6.143_{-4}	0.47	4.375_{-4}	0.47	0.71
26	4014459	1.113_{-6}	0.91	2.230_{-7}	1.20	4.723_{-4}	0.52	3.366_{-4}	0.51	0.71
27	5983155	7.014_{-7}	1.16	1.610_{-7}	0.82	3.844_{-4}	0.52	2.743_{-4}	0.51	0.71
28	9575469	4.419_{-7}	0.98	1.034_{-7}	0.94	3.082_{-4}	0.47	2.195_{-4}	0.47	0.71
29	15969915	2.784_{-7}	0.90	5.600_{-8}	1.20	2.368_{-4}	0.52			

$\frac{1}{2}$. An efficient self-adaptive method however must achieve the optimal rate $\rho_{H^1} = \frac{1}{2}$ and show a very sharp local refinement near the re-entrant corner, compare Figure 4. From the numbers displayed in Table 1 and the graphs depicted in Figure 5 we can clearly observe that our adaptive PUM achieves this optimal value of $\rho_{H^1} \approx \frac{1}{2}$. The corresponding L^2-convergence rate ρ_{L^2} and L^∞-convergence rate ρ_{L^∞} are also optimal with a value close to 1. We can also observe the convergence of the effectivity index $\epsilon^*_{H^1}$ to 0.79 from Table 1. Hence, we see that our approximation to the local error using quartic Legendre polynomials is rather accurate. The convergence of $\epsilon^*_{H^1}$ is clear numerical evidence that the subdomain estimator described in section 3.3 satisfies an equivalence such as (3.29) also for the non-conforming Nitsche approach and solutions with less than full elliptic regularity.

Of course, an approximation of the error estimator using quartic polynomials is rather expensive. We have to solve a local problem of dimension 15 on each patch. If we are interested in steering the refinement only, then this amount of computational work might be too expensive. Hence, we carried out the same experiment using quadratic Legendre polynomials for the approximation of (3.26) only. Here, we need to solve local problems of dimension 6. The measured errors and convergence rates are displayed in Table 2 and in Figure 6. From these numbers we can clearly observe that we retain the optimal rates of $\rho_{H^1} \approx \frac{1}{2}$ and $\rho_{L^2} \approx 1$ also with this coarser approximation. However, we also see that the quality of our approximate estimate is slightly reduced, i.e., the effectivity index converges to the smaller value 0.71. Furthermore, we find that our refinement scheme based on the quadratic approximation selects more patches for refinement than in the quartic case; e.g., on level 9 we have $\mathsf{dof} = 1854$ for the quadratic polynomials and only $\mathsf{dof} = 1566$ for the quartic polynomials. However, we obtain covers with a maximal level difference of $L = 1$ for both approximation to the local errors. Obviously, the use of the quadratic approximation for the error estimation leads to an unnecessary increase in the total number of degrees of freedom, however, since we attain optimal convergence rates with both approximations this cheaper approximation for the error may well pay off with respect to the total compute time.

The solution of the arising linear systems using our nested iteration multilevel solver required an almost negligible amount of compute time. In this experiment it was sufficient to employ a single $V(1,1)$-cycle with block-Gauss-Seidel smoothing (compare [9]) within the nested iteration solver to obtain an approximate solution within discretization accuracy.

Finally, let us point out that the obtained numerical approximations u^{PU} to the considered singular solution are highly accurate with $e_{L^\infty} \approx 10^{-7}$. Such quality requires an accurate and stable numerical integration scheme which can account for the sharp localization in the adaptive refinement and the singular character of the solution u automatically. Our subdivision sparse grid integration scheme meets these requirements.

In summary we can say that the results of this numerical experiment indicate that our adaptive PUM can handle problems with singular solutions with optimal complexity. We obtain a stable approximation with optimal convergence rates already with a relatively cheap approximation to the local errors. The application of a nested iteration with our multilevel method as inner iteration yields approximate solution of very high quality with a minimal amount of computational work.

Example 2. In our second example we consider our model-problem (1.1) with Dirichlet boundary conditions on the cube $(0,1)^3$ in three space dimensions. We choose f and g_D such that the solution is given by $u(x) = |x|^{\frac{1}{3}}$. Again, we use linear Legendre polynomials for the approximation and estimate the local errors using quadratic Legendre polynomials. The optimal convergence

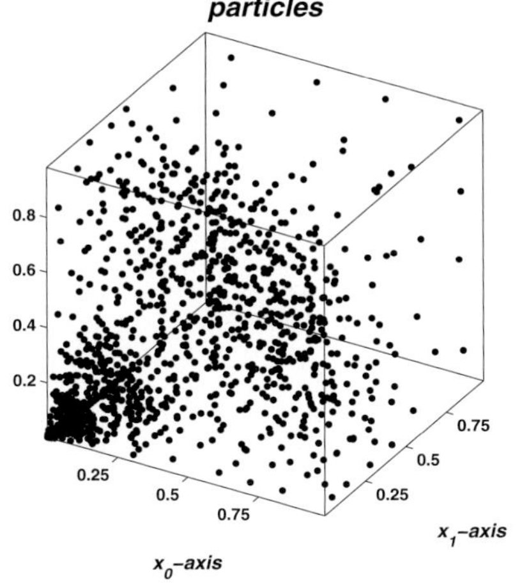

Figure 7. Refined point set P on level $J = 9$ for Example 2.

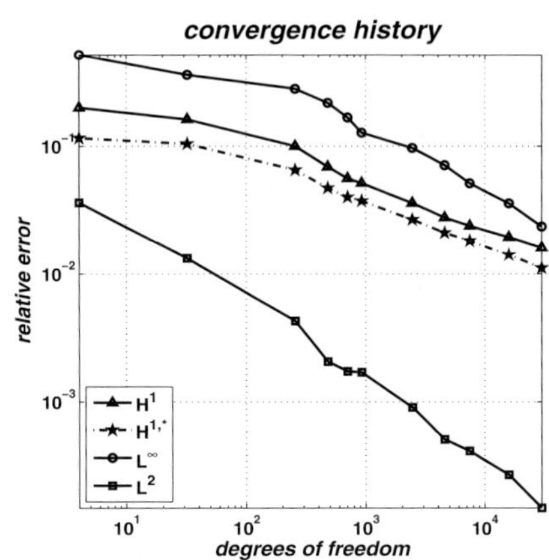

Figure 8. Convergence history for Example 2.

Table 4.3. Relative errors e (4.32) and convergence rates ρ (4.33) for Example 2.

J	dof	e_{L^∞}	ρ_{L^∞}	e_{L^2}	ρ_{L^2}	e_{H^1}	ρ_{H^1}	$e^*_{H^1}$	$\rho^*_{H^1}$	$\epsilon^*_{H^1}$
0	4	5.204_{-1}	0.47	3.635_{-2}	2.39	2.001_{-1}	1.16	1.157_{-1}	1.56	0.58
1	32	3.605_{-1}	0.18	1.323_{-2}	0.49	1.619_{-1}	0.10	1.043_{-1}	0.05	0.64
2	256	2.804_{-1}	0.12	4.285_{-3}	0.54	9.998_{-2}	0.23	6.530_{-2}	0.23	0.65
3	480	2.173_{-1}	0.41	2.061_{-3}	1.16	6.908_{-2}	0.59	4.728_{-2}	0.51	0.68
4	704	1.673_{-1}	0.68	1.733_{-3}	0.45	5.615_{-2}	0.54	4.003_{-2}	0.44	0.71
5	928	1.276_{-1}	0.98	1.703_{-3}	0.06	5.159_{-2}	0.31	3.745_{-2}	0.24	0.73
6	2440	9.678_{-2}	0.29	9.047_{-4}	0.65	3.600_{-2}	0.37	2.673_{-2}	0.35	0.74
7	4540	7.116_{-2}	0.50	5.046_{-4}	0.94	2.774_{-2}	0.42	2.108_{-2}	0.38	0.76
8	7424	5.140_{-2}	0.66	4.099_{-4}	0.42	2.386_{-2}	0.31	1.822_{-2}	0.30	0.76
9	15964	3.580_{-2}	0.47	2.673_{-4}	0.56	1.941_{-2}	0.27	1.412_{-2}	0.33	0.73
10	30076	2.355_{-2}	0.66	1.472_{-4}	0.94	1.613_{-2}	0.29	1.112_{-2}	0.38	0.69

rate with respect to the H^1-norm in three dimensions is $\rho_{H^1} = \frac{1}{3}$. From the numbers given in Table 4.3 we can clearly observe this optimal convergence behavior of our adaptive PUM. The rates ρ_{L^2} and ρ_{L^∞} obtained for the L^2-norm and L^∞-norm respectively are comparable to the optimal value of $\frac{2}{3}$. The effectivity index of our error estimator converges to a value of $\epsilon^*_{H^1} \approx 0.64$. Hence, the quality of the quadratic approximation to the error estimator in three dimensions is of comparable quality to that in two dimensions.

The maximal level difference in this example was $L = 1$ as in the previous example, see also Figure 7. Again, it was sufficient to use a single $V(1,1)$-cycle within the nested iteration to obtain an approximate solution within discretization accuracy.

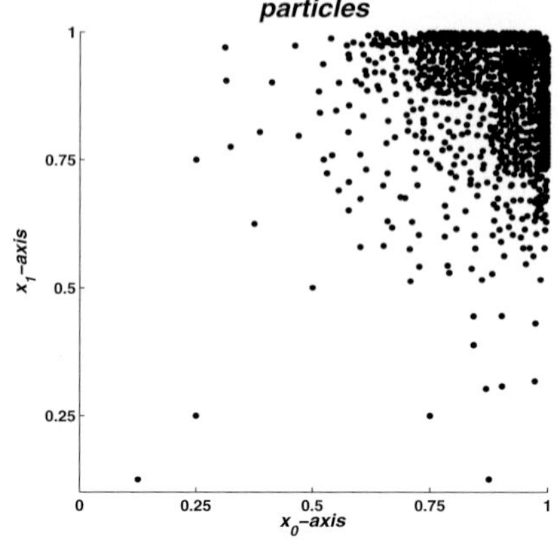

Figure 9. Refined point set P on level $J = 8$ for for Example 1 with $p = 1$.

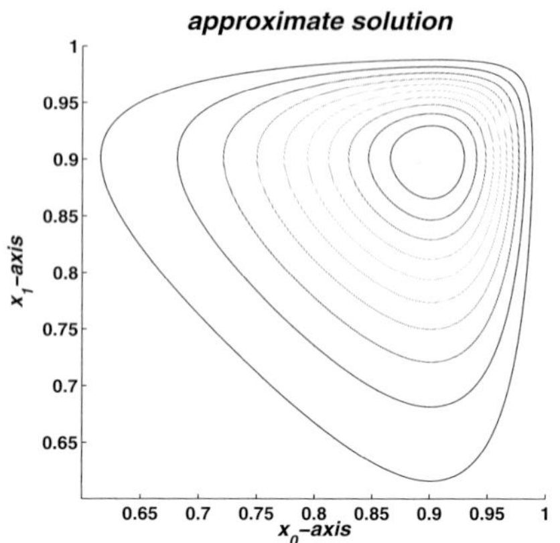

Figure 10. Zoom of isolines of the approximate solution computed on level $l = 10$ with $p = 1$.

Table 4.4. Relative errors e (4.32) and convergence rates ρ (4.33) for Example 3 using linear Legendre polynomials.

J	dof	e_{L^∞}	ρ_{L^∞}	e_{L^2}	ρ_{L^2}	e_{H^1}	ρ_{H^1}	$e^*_{H^1}$	$\rho^*_{H^1}$	$\epsilon^*_{H^1}$
3	84	5.853_{-1}	0.92	4.360_{-1}	1.24	6.941_{-1}	0.52	5.591_{-1}	0.45	0.81
4	156	2.940_{-1}	1.11	1.371_{-1}	1.87	4.764_{-1}	0.61	3.117_{-1}	0.94	0.65
5	291	1.041_{-1}	1.67	5.489_{-2}	1.47	3.281_{-1}	0.60	2.332_{-1}	0.47	0.71
6	822	3.772_{-2}	0.98	1.944_{-2}	1.00	2.042_{-1}	0.46	1.566_{-1}	0.38	0.77
7	1524	2.102_{-2}	0.95	1.428_{-2}	0.50	1.508_{-1}	0.49	1.178_{-1}	0.46	0.78
8	5619	4.824_{-3}	1.13	4.716_{-3}	0.85	7.710_{-2}	0.51	5.963_{-2}	0.52	0.77
9	13332	2.648_{-3}	0.69	1.805_{-3}	1.11	4.947_{-2}	0.51	3.841_{-2}	0.51	0.78
10	74838	3.061_{-4}	1.25	2.875_{-4}	1.06	2.106_{-2}	0.50	1.649_{-2}	0.49	0.78
11	275997	1.168_{-4}	0.74	8.154_{-5}	0.97	1.090_{-2}	0.50	8.520_{-3}	0.51	0.78
12	899823	2.819_{-5}	1.20	2.872_{-5}	0.88	6.002_{-3}	0.51	4.667_{-3}	0.51	0.78
13	2885646	9.056_{-6}	0.97	1.025_{-5}	0.88	3.358_{-3}	0.50	2.612_{-3}	0.50	0.78
14	13579752	1.841_{-6}	1.03	2.062_{-6}	1.04	1.546_{-3}	0.50	1.201_{-3}	0.50	0.78

Example 3. In our last example we consider the Poisson problem (1.1) with Dirichlet boundary conditions where we choose f and g_D such that the solution [14] is given by

$$u(x) = \frac{1}{2000} \prod_{l=1}^{2} (x^l)^2 (1 - x^l)^2 (\exp(10(x^l)^2) - 1),$$

see also Figure 10.

Here, we now consider not only a linear approximation but also a higher order approach with quadratic polynomials since the solution is smooth enough. First, we approximate the solution using linear Legendre polynomials and estimate the error with quadratic Legendre polynomials as before. Then, we

Table 4.5. Relative errors e (4.32) and convergence rates ρ (4.33) for Example 3 using quadratic Legendre polynomials.

J	dof	e_{L^∞}	ρ_{L^∞}	e_{L^2}	ρ_{L^2}	e_{H^1}	ρ_{H^1}	$e^*_{H^1}$	$\rho^*_{H^1}$	$\epsilon^*_{H^1}$
3	168	2.073_{-1}	1.49	9.459_{-2}	2.48	4.184_{-1}	0.58	2.375_{-1}	1.36	0.57
4	240	4.879_{-2}	4.06	3.147_{-2}	3.09	2.504_{-1}	1.44	1.527_{-1}	1.24	0.61
5	600	1.608_{-2}	1.21	1.100_{-2}	1.15	9.248_{-2}	1.09	6.380_{-2}	0.95	0.69
6	1824	2.492_{-3}	1.68	1.742_{-3}	1.66	2.523_{-2}	1.17	1.628_{-2}	1.23	0.65
7	4974	4.917_{-4}	1.62	2.885_{-4}	1.79	7.416_{-3}	1.22	4.774_{-3}	1.22	0.64
8	18492	7.648_{-5}	1.42	6.033_{-5}	1.19	1.854_{-3}	1.06	1.158_{-3}	1.08	0.62
9	61134	9.078_{-6}	1.78	6.262_{-6}	1.89	5.518_{-4}	1.01	3.351_{-4}	1.04	0.61
10	222414	1.392_{-6}	1.45	1.216_{-6}	1.27	1.512_{-4}	1.00	9.025_{-5}	1.02	0.60
11	959100	1.359_{-7}	1.59	1.163_{-7}	1.61	3.444_{-5}	1.01	2.028_{-5}	1.02	0.59
12	3580440	1.952_{-8}	1.47	1.608_{-8}	1.50	9.302_{-6}	0.99	5.451_{-6}	1.00	0.59
13	13422120	2.766_{-9}	1.48	2.321_{-9}	1.46	2.507_{-6}	0.99	1.467_{-6}	0.99	0.59

consider the case when we approximate the solution with quadratic polynomials and use cubic Legendre polynomials to estimate the errors locally. With respect to the measured convergence rates we expect to find the optimal rates $\rho_{H^1} \approx \frac{1}{2}$ for the linear approximation, see Table 4.4 and Figure 11, and $\rho_{H^1} \approx 1$ for the quadratic approximation, see Table 4.5 and Figure 12.[6] Our adaptive PUM achieves this anticipated optimal convergence behavior with respect to the H^1-norm as well as in the L^2-norm for which we find the optimal rates $\rho_{L^2} \approx 1$ and $\rho_{L^2} \approx \frac{3}{2}$ respectively. The cover refinement carried out for a linear approximation (compare Figure 9) is in fact different from the one attained for the quadratic approximation. For instance we find a maximal level difference of $L = 3$ for the linear approximation and $L = 1$ for the quadratic approximation, i.e., the point sets and covers obtain for the higher order approximation is somewhat smoother.

The quality of the quadratic approximation of the error estimator is again similar to those obtain in the previous examples, i.e., we observe $\epsilon^*_{H^1} \approx 0.78$. Since the relative increase in the number of degrees of freedom going from quadratic to cubic polynomials is smaller than when we use quadratic polynomials to estimate the error of a linear approximation we can expect to find a smaller value of $\epsilon^*_{H^1}$ in Table 4.5. In fact the effectivity index converges to 0.59 only. Note that the results of further numerical experiments confirm that the quality of the estimator is essentially influenced by the relative increase of the polynomial degree. For instance we found a value of $\epsilon^*_{H^1} \approx 0.8$ again, when we use polynomials of order 6 to approximate the error of a quadratic approximation. Hence, this example demonstrates that using only a polynomial of degree $p+1$ to estimate the error of an approximation of order p may

[6] Note that not all refinement steps are given in the tables and graphs for Example 3. Due to the smoothness of the solution our refinement scheme constructs several refined covers $R(C^J_\Omega)$ with $J = \max_{\omega_i \in R(C^J_\Omega)} l_i$. For better readability we only give the final results on the respective level J.

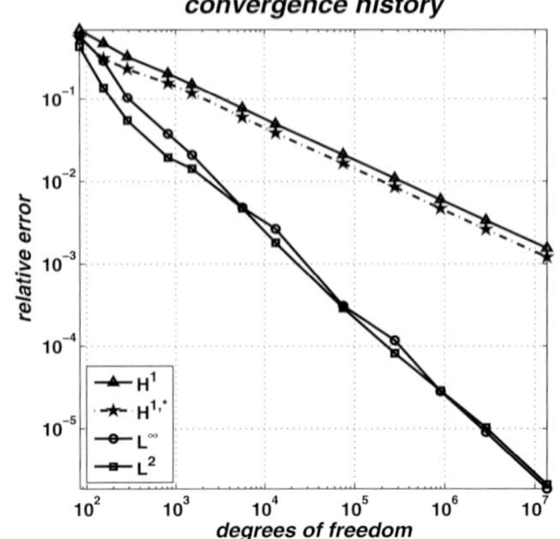

Figure 11. Convergence history for Example 3 using linear Legendre polynomials.

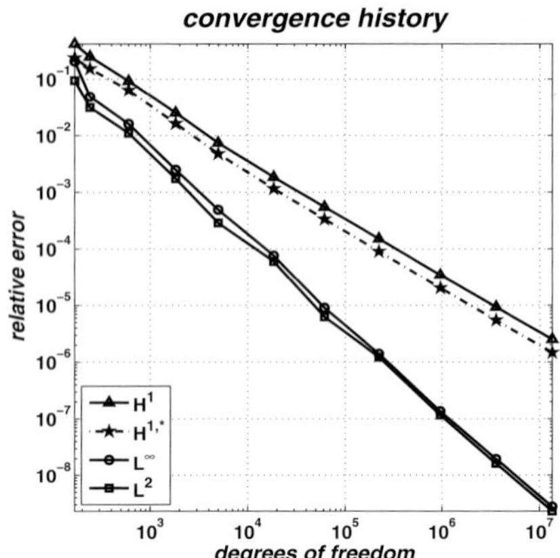

Figure 12. Convergence history for Example 3 using quadratic Legendre polynomials.

not yield very accurate estimates for large p. Nonetheless, our experiments also indicate that the actual refinement is only very slightly affected by this issue.

5 Concluding Remarks

In this paper we have considered the adaptive multilevel solution of a scalar elliptic PDE by the PUM. We have presented a particle refinement scheme for h-type adaptivity in the PUM which is steered by a local subdomain-type error estimator which is in turn approximated by local p-type enrichment. The results of our numerical experiments in two and three space dimensions are strong numerical evidence that the estimator is efficient and reliable.

Note that the adaptively constructed point sets may provide a novel way to approximate density distributions which can be related to solutions of PDEs. Also note that a local hp-type refinement is (in principle) straightforward in the PUM due to the independence of the local approximation spaces. The extension of the presented scheme to hp-type adaptivity however is subject of current research.

The nested multilevel iteration developed in this paper provides a highly efficient solver with optimal computational complexity. In summary, the presented meshfree scheme is a main step toward the availability of efficient adaptive meshfree methods which will enable us to tackle large scale complicated problems.

Acknowledgement. This work was supported in part by the Sonderforschungsbereich 611 *Singular phenomena and scaling in mathematical models* funded by the *Deutsche Forschungsgemeinschaft.*

References

1. I. BABUŠKA AND J. M. MELENK, *The Partition of Unity Method*, Int. J. Numer. Meth. Engrg., 40 (1997), pp. 727–758.
2. I. BABUŠKA AND W. C. RHEINBOLDT, *Error Estimates for Adaptive Finite Element Computations*, SIAM J. Numer. Anal., 15 (1978), pp. 736–754.
3. R. E. BANK AND A. WEISER, *Some A Posteriori Error Estimators for Elliptic Partial Differential Equations*, Math. Comp., 44 (1985).
4. R. BECKER, P. HANSBO, AND R. STENBERG, *A Finite Element Method for Domain Decomposition with Non-Matching Grids*, Math. Modell. Numer. Anal., 37 (2003), pp. 209–225.
5. M. BERN, D. EPPSTEIN, AND J. GILBERT, *Provably Good Mesh Generation*, J. Comput. Sys. Sci., 48 (1994), pp. 384–409.
6. A. BRANDT, *Multigrid Techniques: 1984 Guide with Applications to Fluid Dynamics*, tech. rep., GMD, 1984.
7. M. GRIEBEL, P. OSWALD, AND M. A. SCHWEITZER, *A Particle-Partition of Unity Method—Part VI: A p-robust Multilevel Preconditioner*, in Meshfree Methods for Partial Differential Equations II, M. Griebel and M. A. Schweitzer, eds., vol. 43 of Lecture Notes in Computational Science and Engineering, Springer, 2005, pp. 71–92.
8. M. GRIEBEL AND M. A. SCHWEITZER, *A Particle-Partition of Unity Method—Part II: Efficient Cover Construction and Reliable Integration*, SIAM J. Sci. Comput., 23 (2002), pp. 1655–1682.
9. ——, *A Particle-Partition of Unity Method—Part III: A Multilevel Solver*, SIAM J. Sci. Comput., 24 (2002), pp. 377–409.
10. ——, *A Particle-Partition of Unity Method—Part V: Boundary Conditions*, in Geometric Analysis and Nonlinear Partial Differential Equations, S. Hildebrandt and H. Karcher, eds., Springer, 2002, pp. 517–540.
11. W. HACKBUSCH, *Multi-Grid Methods and Applications*, vol. 4 of Springer Series in Computational Mathematics, Springer, 1985.
12. L. KRONSJØ AND G. DAHLQUIST, *On the Design of Nested Iterations for Elliptic Difference Equations*, BIT, 12 (1972), pp. 63–71.
13. J. NITSCHE, *Über ein Variationsprinzip zur Lösung von Dirichlet-Problemen bei Verwendung von Teilräumen, die keinen Randbedingungen unterworfen sind*, Abh. Math. Sem. Univ. Hamburg, 36 (1970–1971), pp. 9–15.
14. N. PARÉS, P. DÍEZ, AND A. HUERTA, *Subdomain-based flux-free a posteriori error estimators*, Comput. Meth. Appl. Mech. Engrg., 195 (2006), pp. 297–323.
15. M. A. SCHWEITZER, *A Parallel Multilevel Partition of Unity Method for Elliptic Partial Differential Equations*, vol. 29 of Lecture Notes in Computational Science and Engineering, Springer, 2003.
16. R. VERFÜRTH, *A Review of A Posteriori Error Estimation and Adaptive Mesh-Refinement Techniques*, Wiley Teubner, 1996.
17. O. C. ZIENKIEWICZ AND J. Z. ZHU, *A Simple Error Estimator and Adapative Procedure for Practical Engineering Analysis*, Int. J. Numer. Meth. Engrg., (1987).

Enriched Reproducing Kernel Particle Approximation for Simulating Problems Involving Moving Interfaces

Pierre Joyot[1], Jean Trunzler[1], and Fransisco Chinesta[2]

[1] LIPSI-ESTIA, Technopole Izarbel, 64210 Bidart, France
{j.trunzler,p.joyot}@estia.fr
[2] LMSP, 151 Bd. de l'Hôpital, 75013 Paris, France
francisco.chinesta@paris.ensam.fr

Summary. In this paper we propose a new approximation technique within the context of meshless methods able to reproduce functions with discontinuous derivatives. This approach involves some concepts of the reproducing kernel particle method (RKPM), which have been extended in order to reproduce functions with discontinuous derivatives. This strategy will be referred as Enriched Reproducing Kernel Particle Approximation (E-RKPA). The accuracy of the proposed technique will be compared with standard RKP approximations (which only reproduces polynomials).

Key words: Meshless methods, discontinuous derivatives, enriched approximation, reproducing kernel particle method

1 Introduction

Meshless methods are an appealing choice for constructing functional approximations (with different degrees of consistency and continuity) without a mesh support. Thus, this kind of techniques seem to be specially appropriated for treating 3D problems involving large displacements, due to the fact that the approximation is constructed only from the cloud of nodes whose positions evolve during the material deformation. In this manner neither remeshing nor fields projections are a priori required.

Other important point lies in the easy introduction of some known information related to the problem solution within the approximation functional basis. For this purpose, different reproduction conditions are enforced in the construction of the approximation functions. This approach has been widely used in the context of the moving least squares approximations involved in the diffuse meshless techniques [1] as well as in the element free Galerkin method [3]. Very accurate results were obtained for example in fracture mechanics by introducing the crack tip behavior into the approximation basis [4].

In this work we propose a numerical strategy, based on the reproducing kernel particle techniques, able to construct approximation functions with discontinuous derivatives on fixed or moving interfaces. This problem was treated in the context of the partition of unity by Kronggauz et al. [7]. In our approach the size of the discrete system of equations remains unchanged because no additional degrees of freedom are introduced related to the enrichment. However, the fact of enriching the approximation implies a bigger moment matrix that can result ill conditioned when the enrichment is applied in the whole domain. To circumvent this difficulty local enrichments seem more appropriate. This paper focuses on local enrichments with particular reproduction conditions.

The starting point of our development is the reproducing kernel particle approximation (RKPA). The RKP approximation was introduced by Liu et al. [10] for enforcing some degree of consistency to standard smooth particle approximations, i.e. they proved that starting from a SPH (smooth particle hydrodynamics) approximation [5] it is possible to enhance the kernel function for reproducing a certain degree of polynomials. We have extended or generalized this procedure in order to reproduce any function, and more concretely, functions involving discontinuous derivatives. The question of the local enrichment will be then addressed.

2 Enriched Functional Approximations

2.1 Reproducing Conditions for Enriched Shape Function

The approximation of a function $u(\boldsymbol{x})$ is defined by

$$u(\boldsymbol{x}) = \sum_{I=1}^{NP} \psi_I(\boldsymbol{x}) u(\boldsymbol{x}_I) \tag{2.1}$$

where $u(\boldsymbol{x}_I)$ are the nodal values of the approximated function and NP the number of nodes used to discretize the domain Ω. $\psi_I(\boldsymbol{x})$ is the shape function which can be written in the general form:

$$\psi_I(\boldsymbol{x}) = C\phi(\boldsymbol{x} - \boldsymbol{x}_I) \tag{2.2}$$

where ϕ is the kernel function which has a compact support. Consequently $\psi_I(\boldsymbol{x})$ will be non zero only for a small set of nodes.

We represent by $\Lambda(\boldsymbol{x})$ the set of nodes whose supports include the point $\boldsymbol{x}$. Thus, we can write Eq. (2.1) as

$$u(\boldsymbol{x}) = \sum_{\lambda \in \Lambda(\boldsymbol{x})} C\phi(\boldsymbol{x} - \boldsymbol{x}_\lambda) u(\boldsymbol{x}_\lambda) \tag{2.3}$$

In the following, we use the simplified notation:

$$u(\boldsymbol{x}) = \sum_{\lambda} C\phi_{\lambda} u(\boldsymbol{x}_{\lambda}) \tag{2.4}$$

In the RKPM context Liu and al. [10] define C as the correction function used to ensure the reproduction conditions. Thus, any linear combination of the functions used to define the reproduction conditions can be approximated exactly. Usually the reproduction conditions are imposed to ensure that the approximation can reproduce polynomials up to a specified degree that represents the approximation order of consistency. In the present work, we want to reproduce a function consisting of a polynomial part and others additional functions used to include known information about the approximated field, as for example discontinuous normal derivatives across fixed or moving interfaces.

Let $u(\boldsymbol{x})$ the function to be reproduced:

$$u(\boldsymbol{x}) = \sum_{\alpha} a_{\alpha} \boldsymbol{x}^{\alpha} + \sum_{j=1}^{ne} e_j \chi^j(\boldsymbol{x}). \tag{2.5}$$

where α is a multi-index used to represent the polynomial part of u, χ is the enrichment function and j a simple index that refers to the power of χ (we want to reproduce the enrichment function and its power up to the degree ne). Multi index notation is deeply described in [9].

First we consider the reproducing conditions for the polynomial part of u. When $|\alpha| = 0$, we obtain the partition of unity

$$\sum_{\lambda} C\phi_{\lambda} 1 = 1 \tag{2.6}$$

and for $|\alpha| \leq m, |\alpha| \neq 0$ (where m is the degree of the polynomial part)

$$\sum_{\lambda} C\phi_{\lambda} \boldsymbol{x}_{\lambda}^{\alpha} = \boldsymbol{x}^{\alpha} \tag{2.7}$$

Now, we consider the non-polynomial reproducing conditions:

$$\sum_{\lambda} C\phi_{\lambda} \chi^i(\boldsymbol{x}_{\lambda}) = \chi^i(\boldsymbol{x}) \quad 1 \leq i \leq ne \tag{2.8}$$

All these reproducing conditions can be written in the matrix form:

$$\sum_{\lambda} C\phi_{\lambda} \boldsymbol{R}(\boldsymbol{x}_{\lambda}) = \boldsymbol{R}(\boldsymbol{x}) \tag{2.9}$$

where $\boldsymbol{R}$ denotes the reproducing vector that consists of the polynomial part $\boldsymbol{R}_p$ and the non-polynomial one $\boldsymbol{R}_e$ ($\boldsymbol{R}^{\top}(\boldsymbol{x}) = \left[\boldsymbol{R}_p{}^{\top}(\boldsymbol{x})\ \boldsymbol{R}_e{}^{\top}(\boldsymbol{x})\right]$)

2.2 Direct Formulation of the Shape Functions

To construct the approximation the correction function must be defined. The choice of C lead to different formulations.

In the direct formulation, the correction function is defined by

$$C_{\mathsf{d}} = \boldsymbol{H}_{\mathsf{d}}^{\top}(\boldsymbol{x}, \boldsymbol{x}_{\lambda}, \boldsymbol{x}_{\lambda} - \boldsymbol{x})\boldsymbol{b}_{\mathsf{d}}(\boldsymbol{x}) \tag{2.10}$$

where $\boldsymbol{H}_{\mathsf{d}}$ consists of a polynomial part of degree m: $P_m(\boldsymbol{x}_{\lambda} - \boldsymbol{x})$, and the non-polynomial one $\boldsymbol{H}_{e_{\mathsf{d}}} = \left[\chi(\boldsymbol{x}_{\lambda}) - \chi(\boldsymbol{x}) \cdots (\chi(\boldsymbol{x}_{\lambda}) - \chi(\boldsymbol{x}))^{ne}\right]$:

$$\boldsymbol{H}_{\mathsf{d}}^{\top}(\boldsymbol{x}, \boldsymbol{x}_{\lambda}, \boldsymbol{x}_{\lambda} - \boldsymbol{x}) = \left[P_m(\boldsymbol{x}_{\lambda} - \boldsymbol{x})\ \chi(\boldsymbol{x}_{\lambda}) - \chi(\boldsymbol{x}) \cdots (\chi(\boldsymbol{x}_{\lambda}) - \chi(\boldsymbol{x}))^{ne}\right] \tag{2.11}$$

From the definition of the correction function (2.10) the reproducing conditions (2.9) becomes

$$\left(\sum_{\lambda} \boldsymbol{R}(\boldsymbol{x}_{\lambda})\boldsymbol{H}_{d}^{\top}(\boldsymbol{x}, \boldsymbol{x}_{\lambda}, \boldsymbol{x}_{\lambda} - \boldsymbol{x})\phi_{\lambda}\right)\boldsymbol{b}_{\mathsf{d}}(\boldsymbol{x}) = \boldsymbol{R}(\boldsymbol{x}) \tag{2.12}$$

that can by written in the matrix form:

$$\boldsymbol{M}_{\mathsf{d}}(\boldsymbol{x})\boldsymbol{b}_{\mathsf{d}}(\boldsymbol{x}) = \boldsymbol{R}(\boldsymbol{x}) \tag{2.13}$$

where the moment matrix $\boldsymbol{M}_{\mathsf{d}}(\boldsymbol{x})$ is defined by

$$\boldsymbol{M}_{\mathsf{d}}(\boldsymbol{x}) = \sum_{\lambda} \boldsymbol{R}(\boldsymbol{x}_{\lambda})\boldsymbol{H}_{\mathsf{d}}^{\top}(\boldsymbol{x}, \boldsymbol{x}_{\lambda}, \boldsymbol{x}_{\lambda} - \boldsymbol{x})\phi_{\lambda} \tag{2.14}$$

Finally, from Eq. (2.2) the shape function in the direct formulation will be defined by

$$\psi_{\mathsf{d}\lambda}(\boldsymbol{x}) = \boldsymbol{H}_{\mathsf{d}}^{\top}(\boldsymbol{x}, \boldsymbol{x}_{\lambda}, \boldsymbol{x}_{\lambda} - \boldsymbol{x})\boldsymbol{M}_{\mathsf{d}}^{-1}(\boldsymbol{x})\boldsymbol{R}(\boldsymbol{x})\phi_{\lambda} \tag{2.15}$$

2.3 Direct and RKPM Shape Function Equivalence

In [9] Liu et al. show that the RKPM formulation satisfies the polynomial reproducing conditions. Here, we prove that this result can be extended to non polynomial reproducing conditions.

For this purpose, firstly, we derive the enriched RKPM shape functions from the reproducing conditions.

The polynomial part can be written as:

$$\sum_{\lambda} C\phi_{\lambda}1 = 1 \quad |\alpha| = 0 \tag{2.16}$$

$$\sum_{\lambda} C\phi_{\lambda}(\boldsymbol{x}_{\lambda} - \boldsymbol{x})^{\alpha} = 0 \quad |\alpha| \leq m, |\alpha| \neq 0 \tag{2.17}$$

the proof can be found for example in [9].

Using the same procedure, we can also write

$$\sum_{\lambda} C\phi_{\lambda}(\chi(\boldsymbol{x}_{\lambda}) - \chi(\boldsymbol{x}))^{i} = 0 \quad 1 \leq i \leq ne \tag{2.18}$$

According to Eqs. (2.17) and (2.18) the reproducing conditions can be written as

$$\sum_{\lambda} C\phi_{\lambda}\boldsymbol{R}(\boldsymbol{x}, \boldsymbol{x}_{\lambda}, \boldsymbol{x}_{\lambda} - \boldsymbol{x}) = \boldsymbol{R}(0) \tag{2.19}$$

where $\boldsymbol{R}^{\top}(\boldsymbol{x}, \boldsymbol{x}_{\lambda}, \boldsymbol{x}_{\lambda} - \boldsymbol{x}) = \left[\, P_m(\boldsymbol{x}_{\lambda} - \boldsymbol{x}) \; \chi(\boldsymbol{x}_{\lambda}) - \chi(\boldsymbol{x}) \cdots \; (\chi(\boldsymbol{x}_{\lambda}) - \chi(\boldsymbol{x}))^{ne} \,\right]$ and $\boldsymbol{R}^{\top}(0) = \left[\, 1 \; 0 \cdots 0 \,\right]$.

Now we choose for the correction function the same expression than in the direct formulation (2.11). i.e. $\boldsymbol{H}_{\mathtt{r}} = \boldsymbol{H}_{\mathtt{d}}$

$$C_{\mathtt{r}} = \boldsymbol{H}_{\mathtt{r}}^{\top}(\boldsymbol{x}, \boldsymbol{x}_{\lambda}, \boldsymbol{x}_{\lambda} - \boldsymbol{x})\boldsymbol{b}_{\mathtt{r}} \tag{2.20}$$

introducing this expression into Eq. (2.19), it results in:

$$\left(\sum_{\lambda} \boldsymbol{R}(\boldsymbol{x}, \boldsymbol{x}_{\lambda}, \boldsymbol{x}_{\lambda} - \boldsymbol{x})\boldsymbol{H}_{\mathtt{r}}^{\top}(\boldsymbol{x}, \boldsymbol{x}_{\lambda}, \boldsymbol{x}_{\lambda} - \boldsymbol{x})\phi_{\lambda}\right)\boldsymbol{b}_{\mathtt{r}} = \boldsymbol{R}(0) \tag{2.21}$$

or being $\boldsymbol{R} \equiv \boldsymbol{H}_{\mathtt{r}} \equiv \boldsymbol{H}$.

$$\left(\sum_{\lambda} \boldsymbol{H}(\boldsymbol{x}, \boldsymbol{x}_{\lambda}, \boldsymbol{x}_{\lambda} - \boldsymbol{x})\boldsymbol{H}^{\top}(\boldsymbol{x}, \boldsymbol{x}_{\lambda}, \boldsymbol{x}_{\lambda} - \boldsymbol{x})\phi_{\lambda}\right)\boldsymbol{b}_{\mathtt{r}} = \boldsymbol{R}(0) \tag{2.22}$$

whose matrix form results in

$$\boldsymbol{M}_{\mathtt{r}}(\boldsymbol{x})\boldsymbol{b}_{\mathtt{r}}(\boldsymbol{x}) = \boldsymbol{R}(0) \tag{2.23}$$

where $\boldsymbol{M}_{\mathtt{r}}(\boldsymbol{x})$ is the enriched RKPM moment matrix defined by

$$\boldsymbol{M}_{\mathtt{r}}(\boldsymbol{x}) = \sum_{\lambda} \boldsymbol{H}(\boldsymbol{x}, \boldsymbol{x}_{\lambda}, \boldsymbol{x}_{\lambda} - \boldsymbol{x})\boldsymbol{H}^{\top}(\boldsymbol{x}, \boldsymbol{x}_{\lambda}, \boldsymbol{x}_{\lambda} - \boldsymbol{x})\phi_{\lambda} \tag{2.24}$$

Since Eqs. (2.9) and (2.19) are equivalent it directly follows that the vectors $\boldsymbol{b}_{\mathtt{d}}$ and $\boldsymbol{b}_{\mathtt{r}}$ are the same, and given $\boldsymbol{H}_{\mathtt{r}} = \boldsymbol{H}_{\mathtt{d}}$ we can conclude that Direct and the RKPM shape fonctions are the same.

2.4 MLS Shape Function

In this section we are going to prove that the MLS formulation [3] can be obtained by choosing the following form of the correction function

$$C = \boldsymbol{H}_{\mathtt{m}}^{\top}(\boldsymbol{x}_{\lambda})\boldsymbol{b}_{\mathtt{m}}(\boldsymbol{x}) \tag{2.25}$$

where $\boldsymbol{H}_{\mathtt{m}} = \boldsymbol{R}$. The reproducing conditions (2.9) can be rewritten

$$\sum_{\lambda} \boldsymbol{R}(\boldsymbol{x}_\lambda)\boldsymbol{H}_{\mathtt{m}}^\top(\boldsymbol{x}_\lambda)\boldsymbol{b}_{\mathtt{m}}(\boldsymbol{x})\phi_\lambda = \boldsymbol{R}(\boldsymbol{x}) \tag{2.26}$$

or

$$\boldsymbol{M}_{\mathtt{m}}(\boldsymbol{x})\boldsymbol{b}_{\mathtt{m}}(\boldsymbol{x}) = \boldsymbol{R}(\boldsymbol{x}) \tag{2.27}$$

where $\boldsymbol{M}_{\mathtt{m}}(\boldsymbol{x})$ is the MLS form of the moment matrix defined by

$$\boldsymbol{M}_{\mathtt{m}}(\boldsymbol{x}) = \sum_{\lambda} \boldsymbol{H}_{\mathtt{m}}(\boldsymbol{x}_\lambda)\boldsymbol{H}_{\mathtt{m}}^\top(\boldsymbol{x}_\lambda)\phi_\lambda \tag{2.28}$$

To derive the standard MLS shape function we consider Eq. (2.4) from which we obtain

$$u(\boldsymbol{x}) = \sum_{\lambda} \boldsymbol{H}_{\mathtt{m}}^\top(\boldsymbol{x}_\lambda)\boldsymbol{M}_{\mathtt{m}}^{-1}(\boldsymbol{x})\boldsymbol{R}(\boldsymbol{x})\phi_\lambda u(\boldsymbol{x}_\lambda) \tag{2.29}$$

Since $\boldsymbol{H}_{\mathtt{m}} = \boldsymbol{R}$, the moment matrix is symmetric, and Eq. (2.29) results in

$$u(\boldsymbol{x}) = \boldsymbol{H}_{\mathtt{m}}^\top(\boldsymbol{x})\boldsymbol{M}_{\mathtt{m}}^{-1}(\boldsymbol{x})\sum_{\lambda} \boldsymbol{H}_{\mathtt{m}}(\boldsymbol{x}_\lambda)\phi_\lambda u(\boldsymbol{x}_\lambda) \tag{2.30}$$

or

$$u(\boldsymbol{x}) = \boldsymbol{H}_{\mathtt{m}}^\top(\boldsymbol{x})\boldsymbol{a}_{\mathtt{m}} \tag{2.31}$$

where $\boldsymbol{a}_{\mathtt{m}}$ is computed from

$$\boldsymbol{M}_{\mathtt{m}}(\boldsymbol{x})\boldsymbol{a}_{\mathtt{m}} = \sum_{\lambda} \boldsymbol{H}_{\mathtt{m}}(\boldsymbol{x}_\lambda)\phi_\lambda u(\boldsymbol{x}_\lambda) \tag{2.32}$$

Now, we can prove that MLS shape functions coincide with the ones related to the direct procedure.

2.5 Direct and MLS Shape Function Equivalence

To prove the equivalence between both formulations, firstly, we consider the expression of the line i-th component of vector $\boldsymbol{M}_{\mathtt{m}}\boldsymbol{b}_{\mathtt{m}}$.

$$\begin{aligned}
[\boldsymbol{M}_{\mathtt{m}}\boldsymbol{b}_{\mathtt{m}}]_i &= b_{\mathtt{m}p_0} \sum_{\lambda} R_i(\boldsymbol{x}_\lambda)\phi_\lambda \\
&\quad + \sum_{0<|\alpha|\le m} b_{\mathtt{m}p_\alpha} \sum_{\lambda} R_i(\boldsymbol{x}_\lambda)(\boldsymbol{x}_\lambda)^\alpha \phi_\lambda \\
&\quad + \sum_{j=1}^{ne} b_{\mathtt{m}e_j} \sum_{\lambda} R_i(\boldsymbol{x}_\lambda)\chi(\boldsymbol{x}_\lambda)^j \phi_\lambda
\end{aligned} \tag{2.33}$$

R_i being the i-th component of vector $\boldsymbol{R}$.

In the same manner, the corresponding component related to vector $\boldsymbol{M}_{\mathtt{d}}\boldsymbol{b}_{\mathtt{d}}$ results in:

$$
\begin{aligned}
[\boldsymbol{M}_{\mathtt{d}}\boldsymbol{b}_{\mathtt{d}}]_i = {} & b_{\mathtt{d}p_0}\sum_{\lambda} R_i(\boldsymbol{x}_\lambda)\phi_\lambda \\
& + \sum_{0<|\alpha|\le m} b_{\mathtt{d}p_\alpha}\sum_{\lambda} R_i(\boldsymbol{x}_\lambda)(\boldsymbol{x}_\lambda-\boldsymbol{x})^\alpha\phi_\lambda \\
& + \sum_{j=1}^{ne} b_{\mathtt{d}e_j}\sum_{\lambda} R_i(\boldsymbol{x}_\lambda)\big(\chi_j(\boldsymbol{x}_\lambda)-\chi_j(\boldsymbol{x})\big)^j\phi_\lambda
\end{aligned}
\tag{2.34}
$$

Using the binomial theorem, we can write

$$
\begin{aligned}
& \sum_{j=1}^{ne} b_{\mathtt{d}e_j}\sum_{\lambda} R_i(\boldsymbol{x}_\lambda)\big(\chi(\boldsymbol{x}_\lambda)-\chi(\boldsymbol{x})\big)^j\phi_\lambda \\
& \quad = \sum_{j=0}^{ne} \tilde{b}_{\mathtt{d}e_j}(b_{\mathtt{d}e_1},\cdots,b_{\mathtt{d}e_{ne}},\chi(\boldsymbol{x}))\sum_{\lambda} R_i(\boldsymbol{x}_\lambda)\chi(\boldsymbol{x}_\lambda)^j\phi_\lambda
\end{aligned}
\tag{2.35}
$$

and

$$
\begin{aligned}
& \sum_{|\alpha|\le m} b_{\mathtt{d}p_\alpha}\sum_{\lambda} R_i(\boldsymbol{x}_\lambda)(\boldsymbol{x}_\lambda-\boldsymbol{x})^\alpha\phi_\lambda \\
& \quad = \sum_{|\alpha|\le m} \tilde{b}_{\mathtt{d}p_\alpha}(b_{\mathtt{d}p_{|\alpha|=0}},\cdots,b_{\mathtt{d}p_{|\alpha|=m}},\boldsymbol{x})\sum_{\lambda} R_i(\boldsymbol{x}_\lambda)\boldsymbol{x}_\lambda^\alpha\phi_\lambda
\end{aligned}
\tag{2.36}
$$

Using Eqs. (2.35) and (2.36), Eq. (2.34) results in

$$
\begin{aligned}
[\boldsymbol{M}_{\mathtt{d}}\boldsymbol{b}_{\mathtt{d}}]_i = {} & \tilde{b}_{\mathtt{d}p_0}\sum_{\lambda} R_i(\boldsymbol{x}_\lambda)\phi_\lambda \\
& + \sum_{0<|\alpha|\le m} \tilde{b}_{\mathtt{d}p_\alpha}\sum_{\lambda} R_i(\boldsymbol{x}_\lambda)(\boldsymbol{x}_\lambda)^\alpha\phi_\lambda \\
& + \tilde{b}_{\mathtt{d}e_0}\sum_{\lambda} R_i(\boldsymbol{x}_\lambda)\phi_\lambda \\
& + \sum_{j=1}^{ne} \tilde{b}_{\mathtt{d}e_j}\sum_{\lambda} R_i(\boldsymbol{x}_\lambda)\chi(\boldsymbol{x}_\lambda)^j\phi_\lambda
\end{aligned}
\tag{2.37}
$$

Subtracting Eqs. (2.37) and (2.33) leads to:

$$
\begin{aligned}
[\boldsymbol{M}_{\mathtt{m}}\boldsymbol{b}_{\mathtt{m}}]_i - [\boldsymbol{M}_{\mathtt{d}}\boldsymbol{b}_{\mathtt{d}}]_i = {} & \left(b_{\mathtt{m}p_0}-\tilde{b}_{\mathtt{d}p_0}-\tilde{b}_{\mathtt{d}e_0}\right)\sum_{\lambda} R_i(\boldsymbol{x}_\lambda)\phi_\lambda \\
& + \sum_{0<|\alpha|\le m}\left(b_{\mathtt{m}p_\alpha}-\tilde{b}_{\mathtt{d}p_\alpha}\right)\sum_{\lambda} R_i(\boldsymbol{x}_\lambda)(\boldsymbol{x}_\lambda)^\alpha\phi_\lambda \\
& + \sum_{j=1}^{ne}\left(b_{\mathtt{m}e_j}-\tilde{b}_{\mathtt{d}e_j}\right)\sum_{\lambda} R_i(\boldsymbol{x}_\lambda)\chi(\boldsymbol{x}_\lambda)^j\phi_\lambda
\end{aligned}
\tag{2.38}
$$

As Eqs. (2.27) and (2.13) imply $\boldsymbol{M}_{\mathrm{m}}\boldsymbol{b}_{\mathrm{m}} - \boldsymbol{M}_{\mathrm{d}}\boldsymbol{b}_{\mathrm{d}} = 0$, then Eq. (2.38) reduces to

$$\boldsymbol{M}_{\mathrm{m}} \begin{bmatrix} b_{\mathrm{m}p_0} - \tilde{b}_{\mathrm{d}p_0} - \tilde{b}_{\mathrm{de}_0} \\ \boldsymbol{b}^{\star}_{\mathrm{m}p} - \tilde{\boldsymbol{b}^{\star}}_{\mathrm{d}p} \\ \boldsymbol{b}^{\star}_{\mathrm{me}} - \tilde{\boldsymbol{b}^{\star}}_{\mathrm{de}} \end{bmatrix} = \begin{bmatrix} 0 \\ 0 \\ 0 \end{bmatrix} \tag{2.39}$$

As $\boldsymbol{M}_{\mathrm{m}}$ is non singular, we can conclude

$$\begin{bmatrix} b_{\mathrm{m}p_0} - \tilde{b}_{\mathrm{d}p_0} - \tilde{b}_{\mathrm{de}_0} \\ \boldsymbol{b}^{\star}_{\mathrm{m}p} - \tilde{\boldsymbol{b}^{\star}}_{\mathrm{d}p} \\ \boldsymbol{b}^{\star}_{\mathrm{me}} - \tilde{\boldsymbol{b}^{\star}}_{\mathrm{de}} \end{bmatrix} = \begin{bmatrix} 0 \\ 0 \\ 0 \end{bmatrix} \tag{2.40}$$

Using a similar reasoning, it is easy to prove that

$$\begin{aligned} \boldsymbol{H}_{\mathrm{m}}^{\top}\boldsymbol{b}_{\mathrm{m}} - \boldsymbol{H}_{\mathrm{d}}^{\top}\boldsymbol{b}_{\mathrm{d}} = {} & \left(b_{\mathrm{m}p_0} - \tilde{b}_{\mathrm{d}p_0} - \tilde{b}_{\mathrm{de}_0}\right) \\ & + \sum_{0<|\alpha|\leq m} \left(b_{\mathrm{m}p_\alpha} - \tilde{b}_{\mathrm{d}p_\alpha}\right) (\boldsymbol{x}_\lambda)^\alpha \\ & + \sum_{j=1}^{ne} \left(b_{\mathrm{me}_j} - \tilde{b}_{\mathrm{de}_j}\right) \chi(\boldsymbol{x}_\lambda)^j \end{aligned} \tag{2.41}$$

that implies, taking into account Eq. (2.40), that $\boldsymbol{H}_{\mathrm{m}}^{\top}\boldsymbol{b}_{\mathrm{m}} - \boldsymbol{H}_{\mathrm{d}}^{\top}\boldsymbol{b}_{\mathrm{d}} = 0$, which proves that both shape functions are the same.

Concluding Remarks

It is well known that Direct, MLS and RKPM shape functions are the same when the reproducing conditions are polynomial (see [8]). Here we have extended this result to general reproducing conditions.

3 Definition of the Continuous Shape Functions with Discontinuous Derivatives

Let Ω be the domain where the problem is defined and Γ_d a point, curve or surface (in 1D, 2D and 3D respectively) where the normal derivative of the problem solution becomes discontinuous. We assume that this discontinuity curve splits the domain in two subdomains Ω_0 and Ω_1 (see Fig. 1)

$$\begin{aligned} \Omega_0 \cup \Omega_1 \cup \Gamma_d = \Omega, \\ \Omega_0 \cap \Omega_1 = \emptyset, \end{aligned} \tag{3.42}$$

The enrichment function $\chi(\boldsymbol{x})$ which will be introduced in the reproduction vector $\boldsymbol{R}(\boldsymbol{x})$ must satisfy the discontinuity conditions. In that follows $\tilde{\boldsymbol{x}}$ represents a point located on the interface, u^0 (resp. u^1) is the function u defined

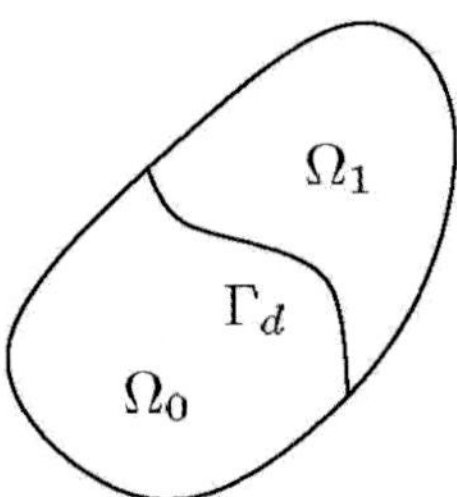

Figure 1. Problem domain containing an interface with a discontinuous normal derivative.

in any point in Ω_0 (resp. Ω_1). Thus, the transmission conditions related to a continuous approximation across the interface Γ_d with discontinuous normal derivative, result in:

$$u^0(\tilde{\boldsymbol{x}}) = u^1(\tilde{\boldsymbol{x}}) \tag{3.43}$$

$$u^0{}_{,i}(\tilde{\boldsymbol{x}}^-)n_i - u^1{}_{,i}(\tilde{\boldsymbol{x}}^+)n_i \neq 0 \tag{3.44}$$

It is also possible to add a discontinuity condition on the second derivative

$$u^0{}_{,ii}(\tilde{\boldsymbol{x}}^-) \neq u^1{}_{,ii}(\tilde{\boldsymbol{x}}^+) \tag{3.45}$$

To locate the interface Γ_d we make use of a "level-set" function $\Theta(\boldsymbol{x})$ defined as the signed distance from $\boldsymbol{x}$ to the interface Γ_d.

Thus, for satisfying the transmission conditions (3.43)-(3.44) the enrichment function $\chi(\boldsymbol{x})$ is assumed to be

$$\chi(\boldsymbol{x}) = H_0(\Theta(\boldsymbol{x}))\Theta(\boldsymbol{x}), \tag{3.46}$$

where $H_0(\boldsymbol{x})$ represents the usual Heaviside function

$$\begin{cases} H_0(\Theta(\boldsymbol{x})) = 1 \text{ if } \Theta(\boldsymbol{x}) \geq 0 \\ H_0(\Theta(\boldsymbol{x})) = 0 \text{ if } \Theta(\boldsymbol{x}) < 0 \end{cases}. \tag{3.47}$$

We are going to verify that this definition of $\chi(\boldsymbol{x})$ represents accurately those transmission conditions. According to the definition of $\chi(\boldsymbol{x})$, given by Eq. (3.46), our approximation reproduce in the domain Ω_0 the function u^0 defined by

$$u^0(\boldsymbol{x}) = \sum_\alpha a_\alpha \boldsymbol{x}^\alpha \tag{3.48}$$

and in the domain Ω_1

$$\begin{aligned} u^1(\boldsymbol{x}) =& \sum_\alpha a_\alpha \boldsymbol{x}^\alpha + e_1\Theta(\boldsymbol{x}) \\ =& u^0(\boldsymbol{x}) + e_1\Theta(\boldsymbol{x}) \end{aligned} \tag{3.49}$$

Since $\Theta(\tilde{\boldsymbol{x}}) = 0$, Eq. (3.49) implies that $u^1(\tilde{\boldsymbol{x}}) = u^0(\tilde{\boldsymbol{x}})$. Thus, the first transmission condition is satisfied.

Now, we evaluate the gradient of u^1 in the neighborhood of Γ_d from Eq. (3.49)

$$u^1_{,i}(\tilde{\boldsymbol{x}}^+)n_i = u^0_{,i}(\tilde{\boldsymbol{x}}^-)n_i + e_1\Theta_{,i}(\tilde{\boldsymbol{x}}^+)n_i \tag{3.50}$$

Since $\Theta_{,i}(\tilde{\boldsymbol{x}}^+)n_i \neq 0$ we verify that the second transmission condition (3.44) is also verified.

When $ne = 1$, that is, when the enrichment consists of a single function, the third condition (3.45) is verified if $\Theta_{,ii}(\tilde{\boldsymbol{x}}^+) \neq 0$. To avoid this difficulty one could consider $ne = 2$. In this case it is easy to prove that the last condition is verified for all Θ.

Thus, the vector $\boldsymbol{H}$ must be defined in the MLS framework as:

$$\boldsymbol{H}^\top_{\mathrm{m}}(\boldsymbol{x}) = \left[P_m(\boldsymbol{x}) \; \chi(\boldsymbol{x}) \; \chi^2(\boldsymbol{x}) \right] \tag{3.51}$$

and in the RKPA context as:

$$\boldsymbol{H}^\top_{\mathrm{r}}(\boldsymbol{x}, \boldsymbol{x}_\lambda, \boldsymbol{x}_\lambda - \boldsymbol{x}) = \left[P_m(\boldsymbol{x}_\lambda - \boldsymbol{x}) \; \chi(\boldsymbol{x}_\lambda) - \chi(\boldsymbol{x}) \; (\chi(\boldsymbol{x}_\lambda) - \chi(\boldsymbol{x}))^2 \right] \tag{3.52}$$

with $\chi(\boldsymbol{x}) = H_0(\Theta(\boldsymbol{x}))\Theta(\boldsymbol{x})$.

Figure (2(a)) (resp (2(b))) shows $\psi_\lambda(\boldsymbol{x})$ (resp. $\psi_\lambda(\boldsymbol{x})_{,x}$) evaluated for a node close to a circular interface.

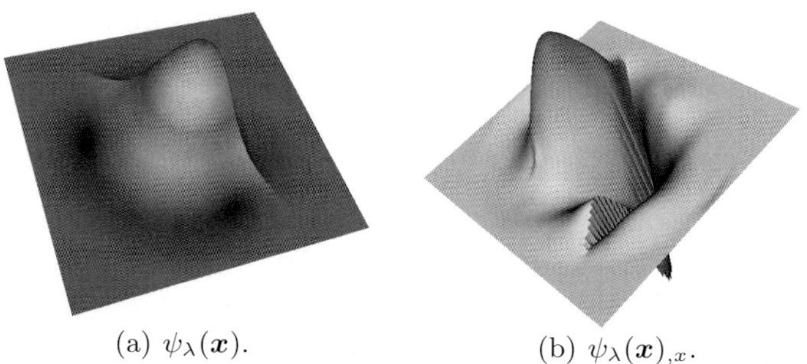

(a) $\psi_\lambda(\boldsymbol{x})$. (b) $\psi_\lambda(\boldsymbol{x})_{,x}$.

Figure 2. Enriched shape function and its derivative evaluated for a node close to a circular interface

4 Enrichment Function Expressed by a Polynomial

Consider the evaluation of the enriched shape function in the direct formulation $\psi_\lambda = (\boldsymbol{H}^\top_p \boldsymbol{b}_p + \boldsymbol{H}^\top_e \boldsymbol{b}_e)\phi_\lambda$. The matrix form of the reproducing conditions is

$$\begin{bmatrix} \boldsymbol{M}_{pp} & \boldsymbol{M}_{pe} \\ \boldsymbol{M}_{ep} & \boldsymbol{M}_{ee} \end{bmatrix} \begin{bmatrix} \boldsymbol{b}_p \\ \boldsymbol{b}_e \end{bmatrix} = \begin{bmatrix} \boldsymbol{R}_p \\ \boldsymbol{R}_e \end{bmatrix} \tag{4.53}$$

where the index p refers to the polynomial part and e to the non polynomial one. For the sake of simplicity, from now on, we omit the index referring the direct formulation.

Now consider a shape function computed only with the polynomial part $\hat{\psi}_\lambda = \boldsymbol{H}_p^\top \hat{\boldsymbol{b}}_p \phi_\lambda$. We have

$$\boldsymbol{M}_{pp} \hat{\boldsymbol{b}}_p = \boldsymbol{R}_p \tag{4.54}$$

From Eqs. (4.53) and (4.54) we can write

$$\boldsymbol{M}_{pp} \left(\hat{\boldsymbol{b}}_p - \boldsymbol{b}_p \right) = \boldsymbol{M}_{ep} \boldsymbol{b}_e \tag{4.55}$$

Now we consider a polynomial form of the enrichment functions:

$$\chi^j(\boldsymbol{x}) = \sum_{|\alpha| \le m} \boldsymbol{b}_{ej} (\boldsymbol{x}_\lambda - \boldsymbol{x})^\alpha (a_j)_\alpha \tag{4.56}$$

where $(a_j)_\alpha$ are the coefficients of the polynomials. Eq. (4.55) can be rewritten as

$$\sum_{|\alpha| \le m} \left(\hat{\boldsymbol{b}}_p - \boldsymbol{b}_p \right)_\alpha \sum_\lambda R_i(\boldsymbol{x}_\lambda)(\boldsymbol{x}_\lambda - \boldsymbol{x})^\alpha \phi_\lambda = \sum_{|\alpha| \le m} \sum_{j=1}^{ne} \boldsymbol{b}_{ej} (a_j)_\alpha \sum_\lambda R_i(\boldsymbol{x}_\lambda)(\boldsymbol{x}_\lambda - \boldsymbol{x})^\alpha \phi_\lambda \tag{4.57}$$

which implies

$$\left(\hat{\boldsymbol{b}}_p - \boldsymbol{b}_p \right)_\alpha = \sum_{j=1}^{ne} \boldsymbol{b}_{ej} (a_j)_\alpha \quad |\alpha| \le m \tag{4.58}$$

Now, we evaluate $\hat{\psi}_\lambda(\boldsymbol{x}) - \psi_\lambda(\boldsymbol{x})$, we obtain

$$\begin{aligned} \hat{\psi}_\lambda(\boldsymbol{x}) - \psi_\lambda(\boldsymbol{x}) &= \boldsymbol{H}_p^\top \hat{\boldsymbol{b}}_p \phi_\lambda - (\boldsymbol{H}_p^\top \boldsymbol{b}_p + \boldsymbol{H}_e^\top \boldsymbol{b}_e) \phi_\lambda \\ &= \phi_\lambda \sum_{|\alpha| \le m} \left(\left(\hat{\boldsymbol{b}}_p - \boldsymbol{b}_p \right)_\alpha - \sum_{j=1}^{ne} \boldsymbol{b}_{ej} (a_j)_\alpha \right) (\boldsymbol{x}_\lambda - \boldsymbol{x})^\alpha \end{aligned} \tag{4.59}$$

Using (4.58) in the previous equation whe conclude that $\psi_\lambda = \hat{\psi}_\lambda$.

From the equation (4.57) we notice that this result is valid only if the polynomial degree of χ is lower or equal than the degree of the polynomial part of the approximation, and if $(a_j)_\alpha$ do not depend on $\boldsymbol{x}_\lambda$.

In the enrichment just described, the presence of the Heaviside function implies that close to the interface $\psi_\lambda \neq \hat{\psi}_\lambda$ despite that the enrichment has a polynomial representation. ψ_λ will be equal to $\hat{\psi}_\lambda$ when all the nodes $\boldsymbol{x}_\lambda$

whose supports include the point $\boldsymbol{x}$ are located in the same subdomain (Ω_1 or Ω_0).

Thus, one could expect local enrichment of the approximation function in the neighborhood of the interface. However, the distance function previously proposed for enriching the approximation involving discontinuous normal derivatives across a fixed or moving interface, is rarely a polynomial. Thus, the resulting enrichment will result global which implies in the case of moving interfaces recompute at each time step all the shape functions. One possibility to circumvent this difficulty lies in the use of a Taylor expansion of the distance function, which results polynomial, according to:

$$\tilde{\Theta}^j(\boldsymbol{x}_\lambda) = \sum_{|\alpha| \le m} \frac{1}{\alpha!} \partial^\alpha \Theta^j(\boldsymbol{x})(\boldsymbol{x}_\lambda - \boldsymbol{x})^\alpha \tag{4.60}$$

5 Numerical Results

Let's consider the two-dimensional heat conduction problem in a bi-material cylinder depicted on figure (3(a)). We note Ω_0 and Ω_1 the domain of each material, and Γ the interface between both domains. Note that $\Omega_i \cap \Gamma = \emptyset$ and $\Omega = \Omega_0 \cup \Omega_0 \cup \Gamma$.

The governing equations are given by:

$$\lambda(\boldsymbol{x}) \left(T(\boldsymbol{x})_{,xx} + T(\boldsymbol{x})_{,yy} \right) = g \quad \text{for} \quad \boldsymbol{x} \in \Omega \tag{5.61}$$

with the conditions on the interface

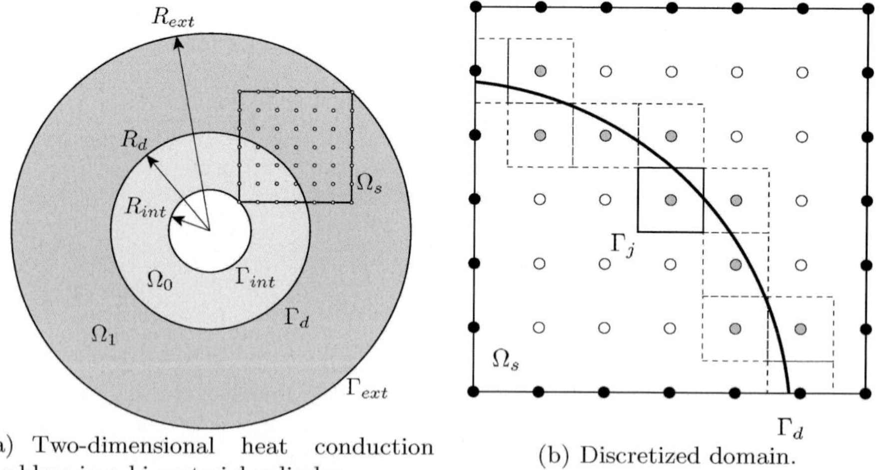

(a) Two-dimensional heat conduction problem in a bi-material cylinder.

(b) Discretized domain.

Figure 3. Domains definition

$$\lambda_0 T_{,i}(\boldsymbol{x}^-)\boldsymbol{n}_i(\boldsymbol{x}) = \lambda_1 T_{,i}(\boldsymbol{x}^+)\boldsymbol{n}_i(\boldsymbol{x}) \quad \text{for} \quad \boldsymbol{x} \in \Gamma \tag{5.62}$$

$$T(\boldsymbol{x}^-) = T(\boldsymbol{x}^+) \quad \text{for} \quad \boldsymbol{x} \in \Gamma \tag{5.63}$$

where $\boldsymbol{n}$ is the unit outwards vector defined on the interface. The boundary conditions are

$$T(\boldsymbol{x}) = T_{int} \quad \text{for} \quad \boldsymbol{x} \in \Gamma_{int} \tag{5.64}$$

$$T(\boldsymbol{x}) = T_{ext} \quad \text{for} \quad \boldsymbol{x} \in \Gamma_{ext} \tag{5.65}$$

The analytical solution of this problem can be found in [2].

5.1 Analysis of the Enriched Approximation

In this numerical test we analyze the representation of the analytical solution of the problem just described. The evaluation is done on the domain Ω_s depicted in figure (3(b)). Table (1) defines the different enriched functional approximations according to ne and $\Theta(\boldsymbol{x})$. In this table $\Theta_c(\boldsymbol{x})$ is the distance function associated with the circular interface which results $\Theta_c(\boldsymbol{x}) = \sqrt{x^2 + y^2} - R_d^2$, and $\tilde{\Theta}_c(\boldsymbol{x})$ is its Taylor expansion evaluated using Eq. (4.60). In all cases $m = 2$ and ϕ_λ is the classical cubic spline function (see [9]). The domain is discretized by a regular grid as shown in figure (3(b)). The support radius of ϕ_λ is $2.6 \times dx$, where dx is the internodal distance along the coordinate axes. The other problem parameters are $R_{int} = 0.1$, $R_{ext} = 1$, $R_d = 0.9$, $\lambda_2 = 100$, $g = 1$, $T_{int} = 0$, $T_{ext} = 100$.

For each simulation scenario (related to a different enriched approximation in table (1)) a convergence test is carried out with a number of nodes varying from 21×21 to 101×101. Convergence rates using the L^2 and the H^1 norms are given in table (2).

We can notice that the different enriched approximations exhibit the same behavior in term of rate of convergence and accuracy. They manifest a significant increase in the order of convergence (of approximatively one order) with respect to the standard (non enriched) RKPA approximation.

Table 1. Definition of the different enriched approximations: $P2$ refers to the second polynomial degree, $D1$ to the enrichment based on the distance function, $D2$ to the enrichment based on the distance function as well as on its square; and finally T refers to the Taylor's expansion of that distance function.

Approximation	ne	$\Theta(\boldsymbol{x})$
P2	0	
P2D1	1	$\Theta_c(\boldsymbol{x})$
P2D1T	1	$\tilde{\Theta}_c(\boldsymbol{x})$
P2D2	2	$\Theta_c(\boldsymbol{x})$
P2D2T	2	$\tilde{\Theta}_c(\boldsymbol{x})$

Table 2. Convergence analysis of the different enriched approximations: slope (coordinate at the origin) of the $log(Error)$ versus $log(1/\Delta x)$ curves

Functional approximation	λ_1	Analysis of the approximation L^2	Analysis of the approximation H^1	Analysis of the discretization L^2	Analysis of the discretization H^1
P2	0.01	-1.47(-0.64)	-0.49(-0.44)	-0.52(-0.86)	-0.41(-0.35)
	1	-1.47(-0.54)	-0.49(-0.29)	-0.41(-0.91)	-0.40(-0.22)
	10	-1.47(-0.79)	-0.49(-0.34)	-0.16(-1.41)	-0.30(-0.44)
P2D1	0.01	-2.51(-1.25)	-1.50(-1.15)	-1.96(-0.66)	-1.45(-0.74)
	1	-2.44(-0.66)	-1.43(-0.49)	-1.83(-0.30)	-1.45(-0.14)
	10	-2.44(-0.79)	-1.44(-0.38)	-1.37(-0.97)	-1.43(-0.17)
P2D1	0.01	-2.51(-1.25)	-1.50(-1.15)	-1.96(-0.66)	-1.45(-0.74)
	1	-2.44(-0.66)	-1.43(-0.49)	-1.83(-0.30)	-1.45(-0.14)
	10	-2.44(-0.79)	-1.44(-0.38)	-1.37(-0.97)	-1.43(-0.17)
P2D1T	0.01	-2.26(-0.58)	-1.52(-0.40)	-1.20(-0.85)	-1.17(-0.38)
	1	-2.43(-0.39)	-1.48(-0.09)	-1.25(-0.65)	-1.29(-0.05)
	10	-2.43(-0.57)	-1.47(-0.08)	-1.12(-1.06)	-1.22(-0.28)
P2D2	0.01	-2.47(-0.71)	-1.48(-0.58)	-2.03(-0.45)	-1.33(-0.64)
	1	-2.47(-0.19)	-1.48(-0.02)	-1.98(-0.03)	-1.38(-0.04)
	10	-2.47(-0.39)	-1.48(-0.02)	-1.44(-0.96)	-1.43(-0.05)
P2D2T	0.01	-2.44(-0.88)	-1.46(-0.72)	-2.06(-0.47)	-1.23(-0.77)
	1	-2.44(-0.36)	-1.46(-0.14)	-2.10(+0.09)	-1.28(-0.16)
	10	-2.44(-0.53)	-1.46(-0.11)	-1.37(-1.04)	-1.42(-0.10)

5.2 Analysis of the Discretization of a Poisson Problem

The problem previously described is now solved using the mixed point-subdomain collocation technique deeply described in [6]. The exact solution is prescribed on the boundary of Ω_s. We consider three sets of nodes. The first one, named Γ_{bc} contains the nodes located on the domain boundary where the temperature is prescribed. The second set, named Ω_d, consists of the nodes whose associated Voronoi cells intersect the interface (see figure (3(b))). The last one, named Ω_i, contains all the remaining nodes. The discretization is performed as follows:

- For each node $\boldsymbol{x}_j$ in Ω_i we consider the usual point collocation:

$$\lambda(\boldsymbol{x}_j)\left(\sum_{i=1}^{NP}(\psi_i^{[(2,0)]}(\boldsymbol{x}_j)+\psi_i^{[(0,2)]}(\boldsymbol{x}_j))T_i\right)=g(\boldsymbol{x}_j) \tag{5.66}$$

- For each node $\boldsymbol{x}_j$ in Ω_d proceed from a subdomain collocation

$$\int_{\Gamma_j}\lambda(\boldsymbol{x})(T_{,x}n_x+T_{,y}n_y)\,d\Gamma_j=g(\boldsymbol{x}_j)A_j \tag{5.67}$$

  where Γ_j is the boundary of the Voronoi cell associated with the node $\boldsymbol{x}_j$, A_j being its area. The integration on Γ_j is done using 5 gauss points on each edge.

- For each node $\boldsymbol{x}_j$ in Ω_d we enforce the known temperature on the domain boundary

$$\sum_{i=1}^{NP} \psi_i(\boldsymbol{x}_j) T_i = T(\boldsymbol{x}_j) \tag{5.68}$$

A convergence analysis is achieved for each simulation scenario. The results are grouped in table (2).

We can notice again that the different enriched approximations exhibit the same behavior in term of rate of convergence and accuracy. They manifest a significant increase in the order of convergence (of approximatively one order) with respect to the standard (non enriched) RKPA approximation.

6 Conclusion

In this paper we have proposed a new approximation technique within the context of meshless methods able to reproduce functions with discontinuous derivatives. This approach involves some concepts of the reproducing kernel particle method (RKPM), which have been extended in order to reproduce functions with discontinuous derivatives. The accuracy of the proposed technique has been compared with usual RKP approximations (which only reproduces polynomials) evidencing a significant increase in the order of convergence. Moreover, a Taylor's expansion of the distance function used to define the enrichment, allowed restricting the enrichment to the regions where the enrichment is required.

References

1. G. Touzot B. Nayrolles and P. Villon, *Generalizing the finite element method: Diffuse approximation and diffuse elements computational mechanics*, Journal of Computational Mechanics **10** (1992), no. 5, 307–318.
2. R.C. Batra, M. Porfiri, and D. Spinello, *Treatment of material discontinuity in two meshless local petrov-galerkin (MLPG) formulations of axisymmetric transient heat conduction*, Int. J. Numer. Meths. Eng. **61** (2004), 2461–2479.
3. T. Belytschko, Y. Kronggauz, D. Organ, and M. Fleming, *Meshless methods: an overview and recent developments*, Comput. Meths. Appl. Mech. Engrg. **139** (1996), 3–47.
4. M. Fleming, Y.A. Chu, B. Moran, and T. Belytschko, *Enriched element-free galerkin methods for crack tip fields*, Int. J. Numer. Meths. Eng. **40** (1997).
5. R.A. Gingold and J.J. Monaghan, *Smoothed particle hydrodynamics: Theory and application to non-spherical stars*, Monthly Notices Royal Astronomical Society **181** (1977), 375–389.
6. D. W. Kim and Y. Kim, *Point collocation methods using the fast moving least-square reproducing kernel approximation*, Int. J. Numer. Meths. Eng. **56** (2003), 1445–1464.

7. Y. Kronggauz and T. Belytschko, *Efg approximation with discontinuous derivatives*, Int. J. Numer. Meths. Eng. **41** (1998), 1215–1233.
8. S. Li and W.K. Liu, *Meshfree and particle methods and their applications*, Appl. Mech. Rev. **55** (2002), 1–34.
9. W. K. Liu, S. Li, and T. Belytshko, *Moving Least Square Reproducing Kernel Method. (I) Methodology and convergence*, Comput. Meths. Appl. Mech. Engrg. **143** (1997), 113–145.
10. W.K. Liu, S. Jun, S. Li, J. Adee, and T. Belytschko, *Reproducing kernel particle methods for structural dynamics*, Int. J. Numer. Meths. Eng. **38** (1995), 1655–1679.

Deterministic Particle Methods for High Dimensional Fokker-Planck Equations

M. Junk[1] and G. Venkiteswaran[2]

[1] Fachbereich Mathematik und Statistik, Fach D194, Universität Konstanz, Germany
michael.junk@uni-konstanz.de

[2] Department of Mathematics, Birla Institute of Technology and Science, Vidya Vihar Campus Pilani 333 031 Rajasthan, India
gvenki@bits-pilani.ac.in

Summary. We consider a mathematical model for polymeric liquids which requires the solution of high-dimensional Fokker-Planck equations related to stochastic differential equations. While Monte-Carlo (MC) methods are classically used to construct approximate solutions in this context, we consider an approach based on Quasi-Monte-Carlo (QMC) approximations. Although QMC has proved to be superior to MC in certain integration problems, the advantages are not as pronounced when dealing with stochastic differential equations. In this article, we illustrate the basic difficulty which is related to the construction of QMC product measures.

Key words: QMC, Fokker-Planck equation, stochastic differential equation, product measures

1 A Polymer Model

In order to understand the non-Newtonian behavior of polymeric liquids (see [1] for a list of phenomena), various polymer models have been investigated in the literature. Here, we consider the so called Rouse chain model [2], in which sub-strings of the polymer molecules are represented by beads (see figure 1) and interactions are indicated by connecting springs (even though a more complicated interaction potential is employed, as specified below). The geometrical configuration of such a chain is described by specifying all the connector vectors $\boldsymbol{Q}_i := \boldsymbol{r}_{i+1} - \boldsymbol{r}_i$, $i = 1, \dots, n-1$, where $\boldsymbol{r}_\nu$, $\nu = 1, \dots, n$ are the position vectors of the beads. Since each of the vectors $\boldsymbol{Q}_i$ has three components, the configuration space is $\mathbb{R}^s$ with $s = 3(n-1)$. For example, a chain with $n = 30$ beads requires $s = 87$ numbers to describe its geometrical configuration. We assume that many such bead-spring chains are immersed into a solvent liquid which undergoes a linear flow, for example a shear flow

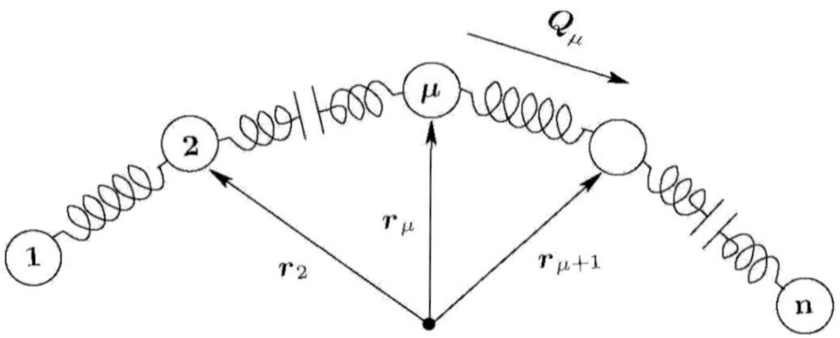

Figure 1. Rouse model of a bead-spring chain.

with velocity field

$$\boldsymbol{v}(\mathbf{x}) = \boldsymbol{\kappa}\,\mathbf{x}, \qquad \boldsymbol{\kappa}{:=}\beta \begin{pmatrix} 0\,1\,0 \\ 0\,0\,0 \\ 0\,0\,0 \end{pmatrix},$$

where β is the constant shear rate. The solvent is considered to exert a drag force, and a random Brownian force on the chain, and the chain is considered to interact with itself through a potential which consists of two contributions, a quadratic attractive part that prevents the beads of the chain from going very far apart, and a Gaussian repulsive part called the excluded volume potential [10], that resists any pair of beads from coming very close to each other. In the following, we list the momentum increment due to the various forces.

Potential Force

The momentum increase of bead ν due to potential forces is given by

$$d\boldsymbol{p}_\nu^{(\phi)}{:=} -\frac{\partial \phi}{\partial \boldsymbol{r}_\nu}\,dt, \tag{1.1}$$

where

$$\phi{:=}\frac{1}{2}\sum_i H\mathbf{Q}_i \cdot \mathbf{Q}_i + k_B T\,\frac{z}{d^3} \sum_{\substack{\mu,\nu=1 \\ \mu\neq\nu}}^{n} \exp\left(-\frac{H}{k_B T}\,\frac{r_{\mu\nu}^2}{2d^2}\right).$$

Here, H is the spring constant of the attractive part, T is the solvent temperature, k_B is Boltzmann's constant, $r_{\mu\nu}$ is the magnitude of the vector $\boldsymbol{r}_{\mu\nu}{:=}\boldsymbol{r}_\mu - \boldsymbol{r}_\nu$, connecting the pair of beads μ and ν, the parameter d controls the extent of the repulsive potential, and z describes its strength.

Hydrodynamic Drag Force

This is the force of resistance the bead experiences as it moves through the solvent. Under the assumption that the beads are spherical in shape, an expression for this force can be written using Stokes' law as

$$\boldsymbol{F}_\nu^{(h)} = -\xi \cdot (\dot{\boldsymbol{r}}_\nu - \boldsymbol{v}(\boldsymbol{r}_\nu)).$$

According to this law, the force on bead ν is directly proportional to the difference between the bead velocity $\dot{\boldsymbol{r}}_\nu$ and the velocity of the solution $\boldsymbol{v}(\boldsymbol{r}_\nu)$ at bead ν. The parameter ξ is the Stokes' friction coefficient. The corresponding momentum increment is

$$d\boldsymbol{p}_\nu^{(h)} := \boldsymbol{F}_\nu^{(h)}\, dt = -\xi \cdot (d\boldsymbol{r}_\nu - \boldsymbol{v}(\boldsymbol{r}_\nu)\, dt).$$

Brownian Force

Due to the thermal fluctuations of the solvent molecules, the bead experiences a random force and this force is modeled by a Wiener process. The momentum changes are thus given by

$$d\boldsymbol{p}_\nu^{(b)} := \sqrt{2k_B T \xi}\, d\boldsymbol{W}_\nu$$

where $\boldsymbol{W}_\nu$ are independent three dimensional Wiener processes. The factor $\sqrt{2k_B T \xi}$ signifies the fact that the energy of the solvent molecules is due to the temperature of the solvent, T, and this energy influences the collision with the beads.

Force Balance

In our force balance, we neglect interaction between different chains which amounts to the implicit assumption of a dilute polymeric solution. Also, the hydrodynamic interaction between the beads will be neglected, i.e. we assume that the flow field is given by $\boldsymbol{v}$, even though the presence of the beads leads to local perturbations. Finally, we assume that inertia forces play a negligible role in the process. Altogether, the force balance

$$d\boldsymbol{p}_\nu^{(\phi)} + d\boldsymbol{p}_\nu^{(h)} + d\boldsymbol{p}_\nu^{(b)} = 0$$

gives rise to a system of first order stochastic differential equations

$$d\boldsymbol{r}_\nu = \left[\boldsymbol{v}(\boldsymbol{r}_\nu) - \frac{1}{\xi}\frac{\partial \phi}{\partial \boldsymbol{r}_\nu}\right] dt + \sqrt{\frac{2k_B T}{\xi}}\, d\mathbf{W}_\nu, \qquad \nu = 1. \ldots, n.$$

Using the linear relation between the bead position vectors $\boldsymbol{r}_\nu$ and the connector vectors

$$\boldsymbol{Q}_k = \sum_\nu \bar{B}_{k\nu} \boldsymbol{r}_\nu, \qquad \bar{B}_{k\nu} = \delta_{k+1,\nu} - \delta_{k,\nu}$$

the system can be reformulated with $A = \bar{B}\bar{B}^T$ (see [12] for details)

$$d\boldsymbol{Q}_j = \left[\boldsymbol{v}(\boldsymbol{Q}_j) - \frac{1}{\xi}\sum_k A_{jk}\frac{\partial \phi}{\partial \boldsymbol{Q}_k}\right] dt + \sqrt{\frac{2k_B T}{\xi}}\left[\sum_\nu \bar{B}_{j\nu}\, d\mathbf{W}_\nu\right],\ j = 1, \ldots, n-1.$$

Combining all connector vectors to a single $\mathbb{R}^s$-vector $\boldsymbol{Q}$ and introducing obvious abbreviations, the system can be written in the compact form

$$d\boldsymbol{Q} = \boldsymbol{a}(\boldsymbol{Q})\, dt + D\, d\boldsymbol{W}. \tag{1.2}$$

Associated Fokker-Planck Equation

Assuming that the Itô process $\boldsymbol{Q}$ is a solution of (1.2) such that $\boldsymbol{Q}(t)$ possesses, for each $t \geq 0$, a smooth Lebesgue-density $\psi(t, \boldsymbol{q})$, we show formally, that ψ is a solution of a Fokker-Planck equation on $\mathbb{R}^s$. To this end, take any test function $f : [0, \infty) \times \mathbb{R}^s \to \mathbb{R}$. Itô's formula [5] implies (with summation convention for the indices i, j, k running from 1 to s)

$$\begin{aligned} f(t, \boldsymbol{Q}(t)) = f(0, \boldsymbol{Q}(0)) &+ \int_0^t D_{ij} \frac{\partial f}{\partial q_i}(s, \boldsymbol{Q}(s))\, dW_j(s) \\ &+ \int_0^t \left(\frac{\partial f}{\partial t} + a_i \frac{\partial f}{\partial q_i} + \frac{1}{2} D_{ik} D_{jk} \frac{\partial^2 f}{\partial q_i \partial q_j} \right)(s, \boldsymbol{Q}(s))\, ds. \end{aligned}$$

Computing the expected value of $f(t, \boldsymbol{Q}(t))$ using the density of $\boldsymbol{Q}(t)$ and noting that the expected value of the stochastic integral vanishes, we obtain

$$\begin{aligned} \int_{\mathbb{R}^s} f(t, \boldsymbol{q}) \psi(t, \boldsymbol{q})\, d\boldsymbol{q} = \int_{\mathbb{R}^s} f(0, \boldsymbol{q}) \psi(0, \boldsymbol{q})\, d\boldsymbol{q} \\ + \int_{\mathbb{R}^s} \int_0^t \left(\frac{\partial f}{\partial t} + a_i \frac{\partial f}{\partial q_i} + \frac{1}{2} D_{ik} D_{jk} \frac{\partial^2 f}{\partial q_i \partial q_j} \right)(s, \boldsymbol{q}) \psi(s, \boldsymbol{q})\, ds\, d\boldsymbol{q}. \end{aligned}$$

Integration by parts allows us to move the derivatives over to the density

$$\int_{\mathbb{R}^s} \int_0^t \left(\frac{\partial \psi}{\partial t} + \frac{\partial}{\partial q_i}(a_i \psi) - \frac{1}{2} D_{ik} D_{jk} \frac{\partial^2 \psi}{\partial q_i \partial q_j} \right)(s, \boldsymbol{q}) f(s, \boldsymbol{q})\, ds\, d\boldsymbol{q} = 0$$

and since f was an arbitrary test function, we see that ψ is a solution of the Fokker-Planck equation

$$\frac{\partial \psi}{\partial t} + \frac{\partial}{\partial q_i}(a_i \psi) = \frac{1}{2} D_{ik} D_{jk} \frac{\partial^2 \psi}{\partial q_i \partial q_j}$$

depending on the high-dimensional variable $\boldsymbol{q} = (q_1, \ldots, q_s) \in \mathbb{R}^s$. In our particular case, the equation has the form

$$\frac{\partial \psi}{\partial t} = -\sum_{j=1}^{n-1} \frac{\partial}{\partial \boldsymbol{Q}_j} \cdot \left(\boldsymbol{\kappa}\, \boldsymbol{Q}_j - \frac{1}{4} \sum_{k=1}^{n-1} A_{jk} \frac{\partial \phi}{\partial \boldsymbol{Q}_k} \right) \psi + \frac{1}{4} \sum_{j,\,k=1}^{n-1} A_{jk} \frac{\partial}{\partial \boldsymbol{Q}_j} \cdot \frac{\partial \psi}{\partial \boldsymbol{Q}_k}$$

where, $\partial / \partial \boldsymbol{Q}_j \cdot$ denotes divergence with respect to $\boldsymbol{Q}_j$ and $\partial \psi / \partial \boldsymbol{Q}_k$ the $\boldsymbol{Q}_k$ gradient.

The Target Quantity

The quantity we are ultimately interested in is the stress tensor $\boldsymbol{\tau}^s + \boldsymbol{\tau}^p$ which characterizes the flow behavior of polymeric liquids. It consists of two contributions namely, one from the solvent $\boldsymbol{\tau}^s$, and the other from the polymer

$\boldsymbol{\tau}^p$. The rheological properties of the polymer solution can be obtained by calculating the polymer contribution to the stress tensor, which is given by Kramers expression [2],

$$\boldsymbol{\tau}^p := \int_{\mathbb{R}^s} \Gamma(\boldsymbol{q})\psi_\infty(\boldsymbol{q})\, d\boldsymbol{q}, \qquad \Gamma(\boldsymbol{q}) := -\sum_{j=1}^{n-1} \boldsymbol{q}_j \otimes \frac{\partial \phi}{\partial \boldsymbol{q}_j}(\boldsymbol{q}). \tag{1.3}$$

where ψ_∞ is the stationary solution of the Fokker-Planck equation.

Mathematical Problem

In view of the above, the mathematical task can be summarized as follows. We have to solve a system of s stochastic differential equations

$$d\boldsymbol{Q} = \boldsymbol{a}(\boldsymbol{Q})\, dt + D\, d\boldsymbol{W}, \qquad \boldsymbol{Q}(0) = \boldsymbol{Q}_0 \tag{1.4}$$

and compute the expected value $E(\Gamma(\boldsymbol{Q}(t)))$ for $t \to \infty$. Equivalently, we can solve the Fokker-Planck initial value problem on $\mathbb{R}^s$

$$\frac{\partial \psi}{\partial t} + \frac{\partial}{\partial q_i}(a_i \psi) = \frac{1}{2} D_{ik} D_{jk} \frac{\partial^2 \psi}{\partial q_i \partial q_j}, \qquad \psi|_{t=0} = \psi_0 \tag{1.5}$$

for the density corresponding to $\boldsymbol{Q}$ and compute, for large t,

$$\boldsymbol{\tau}(t) = \int_{\mathbb{R}^s} \Gamma(\boldsymbol{q})\psi(t, \boldsymbol{q})\, d\boldsymbol{q}. \tag{1.6}$$

Model Problem

To develop ideas, we study an important but simplified version of problem (1.4) resp. (1.5) where the deterministic part $\boldsymbol{a}$ is set to zero. Specifically, we consider a system of s stochastic differential equations

$$d\boldsymbol{z}_t = d\boldsymbol{W}_t, \qquad \boldsymbol{z}_0 = 0 \tag{1.7}$$

with associated Fokker-Planck equation

$$\frac{\partial u}{\partial t} = \frac{1}{2}\Delta u, \qquad u|_{t=0} = \delta_0 \tag{1.8}$$

which is the diffusion equation with Dirac-delta initial value. The target value is assumed to be of the form

$$\mu_t = E(g(\boldsymbol{z}_t)) = \int_{\mathbb{R}^s} g(\boldsymbol{y}) u(t, \boldsymbol{y})\, d\boldsymbol{y} \tag{1.9}$$

where g is some given function which generalizes Γ in (1.6).

2 Meshfree Solution Methods

Due to the high dimension of the configuration space $\mathbb{R}^s$, traditional methods like quadrature rules, finite differences or finite elements, are not useful for the discretization of integral functionals of type (1.9) or parabolic equations like (1.5) or (1.8). Indeed, the number of node points needed to achieve a prescribed accuracy with these methods grows exponentially with the dimension which is referred to as the curse of dimension [9]. The famous approach which breaks this curse is the Monte Carlo (MC) method. In our context, it is used to approximately evaluate the integral (1.9) and to solve (1.7). Using MC integration, we obtain

$$\mu_T = E(g(\boldsymbol{z}_T)) \approx \frac{1}{N}\sum_{i=1}^{N} g(\tilde{\boldsymbol{z}}_T(i)) \tag{2.10}$$

provided $\tilde{\boldsymbol{z}}_T(1), \tilde{\boldsymbol{z}}_T(2), \dots$ are independent pseudo random vectors which are distributed approximately like the solution $\boldsymbol{z}_T$ of (1.7). Such vectors can be obtained, for example, with the Euler-Maruyama MC method [5]. This straight forward solution algorithm is based on replacing the differentials d in (1.7) with corresponding time differences. Introducing a regular time discretization $t_m = mh$ with $m \in \mathbb{N}_0$, the Wiener increments $\boldsymbol{W}_{t_{m+1}} - \boldsymbol{W}_{t_m}$ are normal distributed with variance equal to the time increment $t_{m+1} - t_m = h$. If $\tilde{\boldsymbol{y}}_{t_m}(1), \tilde{\boldsymbol{y}}_{t_m}(2), \dots$ are independent pseudo random vectors with the same distribution as the Wiener increments, the Euler-Maruyama method for (1.7) reads

$$\tilde{\boldsymbol{z}}_{t_{m+1}(i)} = \tilde{\boldsymbol{z}}_{t_m}(i) + \tilde{\boldsymbol{y}}_{t_m}(i), \quad \tilde{\boldsymbol{z}}_0(i) = 0, \qquad i = 1, \dots, N. \tag{2.11}$$

An alternative to the classical MC methods described above is given by Quasi-Monte-Carlo (QMC) methods which are in the focus of this article. Particularly for the integration problem, QMC methods have shown to be superior to the MC approach under certain assumptions on the integrand. While the QMC integral approximation also has the form (2.10), the pseudo random vectors $\tilde{\boldsymbol{z}}_T(i)$ are replaced by points $\boldsymbol{Z}_T(1), \dots, \boldsymbol{Z}_T(N) \in \mathbb{R}^s$ with different properties. To obtain a good approximation by

$$\frac{1}{N}\sum_{i=1}^{N} g(\boldsymbol{Z}_T(i)) \approx E(g(\boldsymbol{z}_T)) = \int_{\mathbb{R}^s} g(\boldsymbol{y}) u(T, \boldsymbol{y})\, d\boldsymbol{y},$$

the points $\boldsymbol{Z}_T(i)$ should be constructed in such a way that the associated point measure

$$\Pi_{\boldsymbol{Z}_T} := \frac{1}{N}\sum_{i=1}^{N} \delta_{\boldsymbol{Z}_T(i)}$$

is a good approximation of the measure $u(t, \boldsymbol{y})\, d\boldsymbol{y}$. The approximation error is quantified deterministically in terms of the *discrepancy* between the two measures (for details, see [8] and the following section) which is an advantage compared to MC methods where the quality of the pseudo random vectors is checked with less reliable statistical tests. On the other hand, the basic MC concept of *independence* is typically *not* available in the QMC context which turns out to be a drawback in the construction of the measure approximation $\Pi_{\boldsymbol{Z}_T}$. Following [6, 14], for example, the required points $\boldsymbol{Z}_T(i)$ are constructed from a measure approximation of the solution to (1.8) which uses the integral representation in terms of the fundamental solution. The final algorithm has a form similar to the Euler-Maruyama method (2.11)

$$\boldsymbol{Z}_{t_{m+1}}(i) \;=\; \boldsymbol{Z}_{t_m}(i) + \boldsymbol{Y}_{t_m}(P(i)), \quad \boldsymbol{Z}_0(i) = 0, \qquad i = 1, \ldots, N. \tag{2.12}$$

In this recursion formula, the vectors $\boldsymbol{Y}_{t_m}(1), \ldots, \boldsymbol{Y}_{t_m}(N)$ should constitute a point measure approximation $\Pi_{\boldsymbol{Y}_{t_m}}$ of the measure $G_h(\boldsymbol{y})\,d\boldsymbol{y}$ related to the fundamental solution

$$G_h(\boldsymbol{y}) := \frac{1}{(4\pi h)^{s/2}} \exp\left(-\frac{|\boldsymbol{y}|^2}{4h}\right), \quad \boldsymbol{y} \in \mathbb{R}^s \tag{2.13}$$

of the diffusion equation at time $h > 0$. Note that this requirement parallels the assumption on the pseudo random vectors $\tilde{\boldsymbol{y}}_{t_m}(i)$ in (2.11) which must be normal distributed with the same density (2.13). Also the *independence* of the increments $\tilde{\boldsymbol{y}}_{t_m}(i)$ from the states $\tilde{\boldsymbol{z}}_{t_m}(i)$ has its parallel in the QMC algorithm. To explain this point, we recall that independence of $\tilde{\boldsymbol{y}}_{t_m}(i)$ and $\tilde{\boldsymbol{z}}_{t_m}(i)$ is equivalent to the condition that the distribution of the pairs $(\tilde{\boldsymbol{z}}_{t_m}(i), \tilde{\boldsymbol{y}}_{t_m}(i)) \in \mathbb{R}^{2s}$ is the *product* of the separate distributions. In view of this, it is not surprising that the permutation $P : \{1, \ldots, N\} \to \{1, \ldots, N\}$ in (2.12) has to ensure that the pairs $(\boldsymbol{Z}_{t_m}(i), \boldsymbol{Y}_{t_m}(P(i))) \in \mathbb{R}^{2s}$ give rise to a point measure approximation of the *product measure* $\Pi_{\boldsymbol{Z}_{t_m}} \otimes \Pi_{\boldsymbol{Y}_{t_m}}$.

The two requirements on the increments $\boldsymbol{Y}_{t_m}(P(1)), \ldots, \boldsymbol{Y}_{t_m}(P(N))$ to be correctly distributed according to (2.13) and to be suitably independent from the vectors $\boldsymbol{Z}_{t_m}(1), \ldots, \boldsymbol{Z}_{t_m}(N)$ are essential in the derivation presented in [6, 14]. Obviously, the main difference to the MC approach is the required construction of the permutation P to ensure independence, or equivalently, to ensure the product measure approximation. Unfortunately, this crucial construction is both time consuming (P involves high-dimensional sorting and quasi-random mixing) and unfavorable for the accuracy of the QMC approach. In fact, the discrepancy estimate given in [6] indicates a convergence order well below $\sqrt{N}$ in high dimensions, although in practice, the QMC approach is still somewhat more accurate than the MC version.

A similar observation has been made in [13], and in [14], where the importance of sorting the QMC points before adding the increments $\boldsymbol{Y}_{t_m}(i)$ has been demonstrated experimentally. Similarly, in [7], the role of sorting the QMC points is highlighted.

These observations have motivated us to take a closer look at the important question, how to construct product measures from given QMC point sets.

3 QMC Product Measures

In this section, we study the fundamental problem how to construct a point measure approximation of the Lebesgue measure on the s dimensional unit cube from two lower dimensional point measure approximations. Our construction is guided by a discrepancy estimate for the resulting product measure in terms of the two participating measures. The discussion in the previous section suggests that changing the numbering of the QMC points (using permutations) plays an important role in the construction. This will also become obvious from the results below.

To introduce the relevant concepts, we first define generalized intervals $[\boldsymbol{a}, \boldsymbol{b})$ with $\boldsymbol{a}, \boldsymbol{b} \in \mathbb{R}^s$ according to

$$[\boldsymbol{a}, \boldsymbol{b}) := \prod_{i=1}^{s} [a_i, b_i).$$

If $\mathbf{0}$ denotes the zero vector and $\mathbf{1}$ the vector having a one in each component, the s-dimensional unit cube I^s can be written in the form $I^s = [\mathbf{0}, \mathbf{1})$. The set of all sub intervals $[\mathbf{0}, \boldsymbol{u})$, $\boldsymbol{u} \in I^s$ of the unit interval with $\mathbf{0}$ as a corner is denoted $\mathcal{R}_s^*$.

Since we restrict our attention to point measures with equal weights, it suffices to specify the point locations to uniquely define the measure. Hence, a point measure on I^s is characterized by a mapping $X : \mathcal{N} \to I^s$ on a finite index set $\mathcal{N} = \{1, \ldots, N\}$. We will employ the usual notation $X \in J$ resp. $\{X \in J\}$ to denote the set of indices $\{i \in \mathcal{N} : X(i) \in J\}$. If Cnt is the counting measure, $\delta_{X(i)}$ the Dirac measure located at $X(i)$, and $\mathcal{B}$ the family of Borel sets in I^s, the point measure associated to X is defined by

$$\Pi_X(J) := \frac{1}{N} \sum_{i=1}^{N} \delta_{X(i)}(J) = \frac{\mathrm{Cnt}(X \in J)}{\mathrm{Cnt}(X \in I^s)}, \qquad J \in \mathcal{B}.$$

Notice that Π_X is invariant under permutations, i.e. $\Pi_X = \Pi_{X \circ P}$ for every permutation $P : \mathcal{N} \to \mathcal{N}$.

The point measure of a set J approximates its s-dimensional Lebesgue measure $\lambda_s(J)$ if the difference

$$\Delta(X, J) := |\Pi_X(J) - \lambda_s(J)|$$

is small. If $\Delta(X, J)$ is small for *all* intervals $J \in \mathcal{R}_s^*$, then the point measure is a good approximation of the Lebesgue measure. The approximation quality can be measured with the star-discrepancy of X

$$D^*(X) := \sup_{J \in \mathcal{R}_s^*} \Delta(X, J).$$

Since Π_X is invariant under renumbering, we conclude that also the discrepancy satisfies $D^*(X) = D^*(X \circ P)$ for every permutation $P : \mathcal{N} \to \mathcal{N}$.

The problem we address is the following: Assume X_1, X_2 are two functions from $\mathcal{N}$ into the unit cubes I^{s_1} respectively I^{s_2} and assume we have some information about the discrepancies $D^*(X_1)$ and $D^*(X_2)$. What can we say about the discrepancy $D^*(X)$ of the measure defined by $X = (X_1, X_2)$ on the product cube I^s, $s = s_1 + s_2$?

The special case $X_2 = X_1$ immediately shows that the quality of the product measure may be arbitrarily bad unless we specify further conditions on X_1 and X_2 (in the case $X_2 = X_1$ all points will be on the s_1-dimensional diagonal in the $2s_1$-dimensional product cube, which is far from the Lebesgue measure, even if X_1 has a very small star-discrepancy).

Our assumptions are motivated by the observation that low discrepancy point sets $\{X_2(1), \ldots, X_2(N)\}$ can be constructed with a discrepancy estimate of the form

$$D^*(X_2) \leq \frac{l_2(N)}{N}, \qquad N = \mathrm{Cnt}(X_2 \in I^{s_2}) \tag{3.14}$$

where $l_2(N)$ is of the form $c \log^s(N)$ with some constant $c > 0$. However, our estimates will only use that l_2 is a non-negative and non-decreasing function.

Our assumptions on X_1 are a generalization of the situation which arises if we take $X_1(1), \ldots, X_1(N)$ as the first N points of a Faure sequence [3]. This low discrepancy *sequence* $\boldsymbol{y}_1, \boldsymbol{y}_2, \ldots$ has the property that any consecutive subset $\boldsymbol{y}_{k+1}, \boldsymbol{y}_{k+2}, \ldots, \boldsymbol{y}_{k+m}$ gives rise to a point measure with a discrepancy estimate depending only on the length m of the subset and *not* on the starting index $k+1$. The estimate is again of the form $l_1(m)/m$ with a logarithmic factor l_1. To state our assumptions more precisely, we need

Definition 1. *A finite set $A \subset \mathbb{N}$ is called consecutive if it contains all $n \in \mathbb{N}$ between* $\min A$ *and* $\max A$.

Then we assume that

$$D^*(X_1|_C) \leq \frac{l_1(m)}{m}, \qquad m = \mathrm{Cnt}(C),\ C \subset \mathcal{N} \text{ consecutive} \tag{3.15}$$

where l_1 is a non-negative and non-decreasing function.

In order to formulate our main discrepancy estimate, we now introduce a concept which eventually demonstrates why and how the numbering of the points $X_2(1), \ldots, X_2(N)$ affects the quality of the product measure. To explain the idea, let us take some interval $J_2 \in \mathcal{R}^*_{s_2}$ and assume that the associated index set $\{X_2 \in J_2\}$ has the form

$$\{X_2 \in J_2\} = \{12, 13, 25, 28, 29, 30\}.$$

This set has several *consecutive representations*, for example

$$\{12, 13, 25, 28, 29, 30\} = \{12\} \cup \{13\} \cup \{25\} \cup \{28\} \cup \{29, 30\},$$

but only one *smallest* representation

$$\{12, 13, 25, 28, 29, 30\} = \{12, 13\} \cup \{25\} \cup \{28, 29, 30\},$$

with $\mathrm{cc}(X_2 \in J_2) = 3$ connected components. In this way, we can associate to each interval J_2 the number $\mathrm{cc}(X_2 \in J_2)$ of connected components and the largest number that appears in this process is denoted $\mathrm{cc}^*(X_2)$. It should be clear that this value depends on the numbering, i.e. we typically have $\mathrm{cc}^*(X_2) \neq \mathrm{cc}^*(X_2 \circ P)$ if $P : \mathcal{N} \to \mathcal{N}$ is a permutation.

In general, it is clear that if A and A' are consecutive and if $A \cap A' \neq \emptyset$ then $A \cup A'$ is also consecutive. Using this observation, we can conclude that, for every finite set $A \subset \mathbb{N}$, there exists a *unique* smallest consecutive representation of pairwise disjoint consecutive sets (if the sets were not pairwise disjoint, one could merge at least two of them violating the minimality; if there was a different representation with the same number of sets, at least one of them would intersect two other sets of the original representation which again violates minimality). This observation allows us to state

Definition 2. *The sets of the smallest consecutive representation of a finite set $A \subset \mathbb{N}$ are called consecutive components. The number of consecutive components is denoted* $\mathrm{cc}(A)$. *For $X_2 : \{1, \ldots, N\} \to I^{s_2}$ with $N, s_2 \in \mathbb{N}$ we denote*

$$\mathrm{cc}^*(X_2) := \sup_{J_2 \in \mathcal{R}^*_{s_2}} \mathrm{cc}(X_2 \in J_2).$$

The proof of the following main result can be found in [4].

Theorem 1. *Let $X_i : \mathcal{N} \to I^{s_i}$, $i = 1, 2$ be mappings from $\mathcal{N} = \{1, \ldots, N\}$ into the unit cubes I^{s_1} and I^{s_2} with $N, s_1, s_2 \in \mathbb{N}$ which satisfy* (3.14) *and* (3.15). *Then the star-discrepancy of $X = (X_1, X_2)$ can be estimated by*

$$D^*(X) \leq \frac{l_2(N)}{N} + \mathrm{cc}^*(X_2)\frac{l_1(N)}{N}.$$

Obviously, the discrepancy estimate of Theorem 1 is optimal, if the worst number of consecutive components $\mathrm{cc}^*(X_2)$ is as small as possible. Since $\mathrm{cc}^*(X_2)$ depends on the numbering of the points, it is also clear that a suitable permutation of the point numbers may improve the quality of the product measure.

3.1 The One-Dimensional Case

To illustrate the role of sorting in the construction of product measures the case $s_2 = 1$ is most enlightening.

For example, if $X_2(1), \ldots, X_2(N)$ are the first N members of the Sobol sequence [11]

$$0, \frac{1}{2}, \frac{1}{4}, \frac{3}{4}, \frac{3}{8}, \frac{7}{8}, \frac{1}{8}, \frac{5}{8}, \frac{5}{16}, \frac{13}{16}, \ldots$$

we find for $J_2 = [0, 1/2)$ and $N = 10$

$$\{X_2 \in J_2\} = \{1, 3, 5, 7, 9\}$$

so that the number of consecutive components is $\mathrm{cc}(X_2 \in J_2) = 5 = N/2$. A similar behavior is observed for larger N in the case $J_2 = [0, 1/2)$, i.e. the number $\mathrm{cc}(X_2 \in J_2)$ of consecutive subsets is essentially $N/2$ which ruins the estimate of Theorem 1.

However, if we *sort* the leading N values of the Sobol sequence before assigning them as values of X_2, the situation is much better. In our example above, the sorted values are

$$0, \frac{1}{8}, \frac{1}{4}, \frac{5}{16}, \frac{3}{8}, \frac{1}{2}, \frac{5}{8}, \frac{3}{4}, \frac{13}{16}, \frac{7}{8}$$

so that

$$\{X_2 \in J_2\} = \{1, 2, 3, 4, 5\}$$

which is consecutive, i.e. $cc(X_2 \in J_2) = 1$. More generally, if the points $X_2(1), \ldots, X_2(N)$ are sorted, or in other words if X_2 is a monotone function, we can find for each $J_2 = [0, u_2)$ a unique index m such that

$$\{X_2 \in J_2\} = \{1, \ldots, m\}.$$

and hence $\mathrm{cc}^*(X_2 \in J_2) = 1$. Using Theorem 1, we conclude

Corollary 1. *Let X_2 be an increasing mapping from $\mathcal{N} = \{1, \ldots, N\}$, $N \in \mathbb{N}$ to the unit interval I with a discrepancy estimate of the form* (3.14). *Assume further that $X_1 : \mathcal{N} \to I^{s_1}$ satisfies* (3.15). *Then the discrepancy of $X = (X_1, X_2)$ can be estimated by*

$$D^*(X) \leq \frac{l_1(N) + l_2(N)}{N}.$$

We remark that this result parallels the case mentioned in [14] where the case $s_1 = s_2 = 1$ has been considered with a slightly weaker assumption on the points X_1.

3.2 The General Case

We describe the two-dimensional case $s_2 = 2$ to explain the effect of sorting in higher dimensional situations.

To give a first example, we consider $N = 16$ points of the two-dimensional Halton sequence with bases two and three. If we connect consecutive points with lines, the curve on the left of figure 2 appears. Given an interval of the form $J_2 = [\mathbf{0}, \boldsymbol{u}_2)$, the consecutive components of $\{X_2 \in J_2\}$ can then be visualized in the following way: as before, we connect consecutive points but now only as long as they are in J_2. Whenever a node $X_2(i) \in J_2$ has a successor outside J_2, the curve terminates at $X_2(i)$ and a new curve is started at the smallest index $j > i$ for which again $X_2(j) \in J_2$. Each snippet obtained in this way is a connected subset of the original curve and corresponds directly to

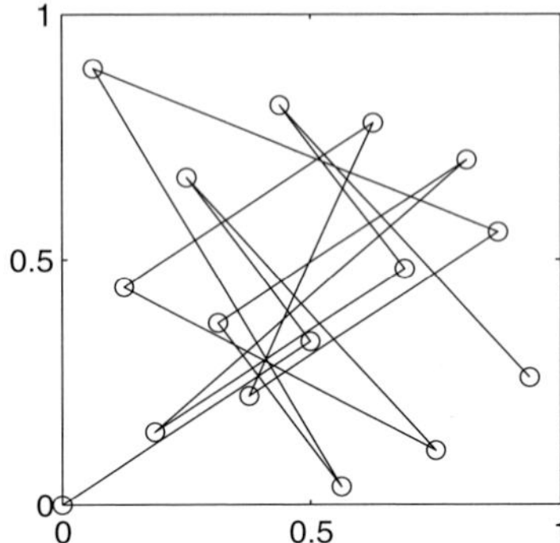

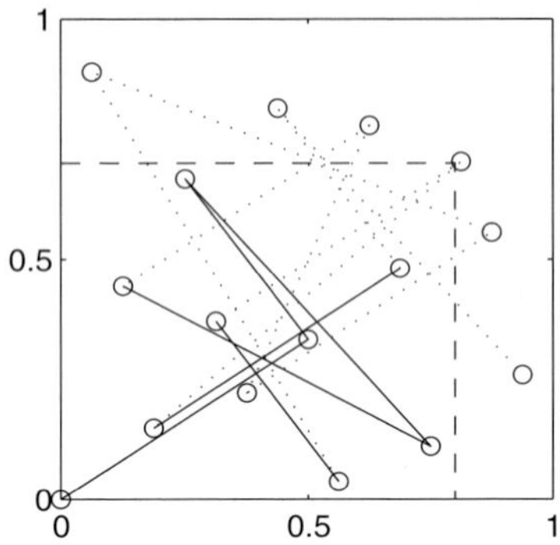

Figure 2. Left: the first 16 points of the Halton sequence with bases two and three connected in the order of appearance. Right: the interval J_2 is indicated by the dashed line and the four consecutive subsets of $\{X_2 \in J_2\}$ are represented by solid lines. The snippets of the original curve correspond to the consecutive index sets $\{1,\ldots,5\},\{7\},\{10,11\},\{13,14\}$.

a consecutive component of $\{X_2 \in J_2\}$. For our example above, the snippets corresponding to $\boldsymbol{u}_2 = \binom{0.8}{0.7}$ are visualized on the right of figure 2.

From these considerations it should intuitively be clear that a very irregular curve is generally cut into more snippets by intersection with intervals than a more regular curve. The idea is therefore to keep the node locations which assure the required low discrepancy of X_2 but to connect them in a different order, resulting in a more regular curve which leads to less snippets when cut with an interval. In other words, if $\tilde{X}_2 : \mathcal{N} \to I^2$ refers to the original node distribution, we try to construct a permutation $P : \mathcal{N} \to \mathcal{N}$ such that $X_2 = \tilde{X}_2 \circ P$ gives rise to a better behavior.

Again, this can be achieved by a suitable sorting. To this end, we choose some $m \in \mathbb{N}$, split the unit square into m equal sized columns (m-bins)

$$B_k := \Big[\frac{k}{m}, \frac{k+1}{m}\Big) \times \big[0,1\big), \qquad k = 0,\ldots,m-1$$

and sort the points in each bin separately with respect to the last component. Similarly, in the general case, m-bins are sub-intervals which have side length $1/m$ in the first $s_2 - 1$ directions and length 1 in direction s_2. Sorting is carried out in each bin with respect to the last component. Connecting the sorted subsets according to the bin numbers, we obtain a new global numbering of the points. In our previous example, the corresponding curve is visualized in figure 3 for the case $m = 4$. Of course, there are many other ways to connect the given nodes with reasonable curves. Some more examples are presented and analyzed in [4].

Using Theorem 1, we can prove that sorting in m-bins allows to produce reasonable QMC product measures, by estimating $\mathrm{cc}^*(X_2)$. In other words, we have to find an upper bound for the number of consecutive components of

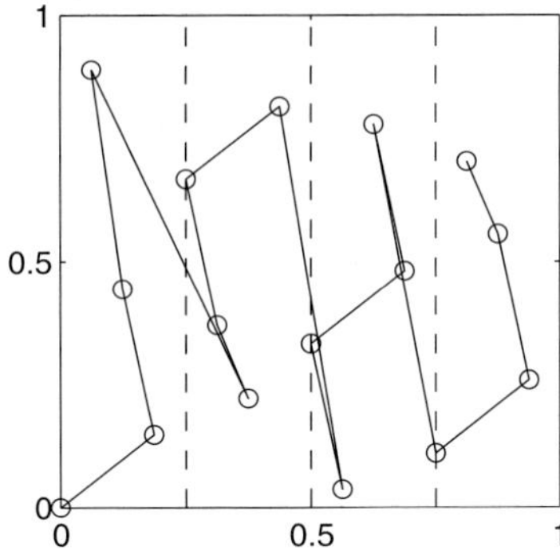

Figure 3. The first 16 points of the Halton sequence with bases two and three connected according to the numbering which follows from sorting the second coordinate in each bin separately (the bins are indicated by vertical dashed lines).

sets $\{X_2 \in J\}$ with $J = [\mathbf{0}, \boldsymbol{u})$ and $\boldsymbol{u} \in I^{s_2}$. We develop the idea in the case $s_2 = 2$ and refer to [4] for a rigorous investigation. The idea of the estimate is the following: if k is the number of the bin which contains $\boldsymbol{u}$, all the bins to the left of bin B_k are cut by J. In each of these bins B_i, the intersection $J \cap B_i$ contains not more than one consecutive component. Hence, the number of consecutive components related to $J \cap B$ with $B = \cup_{i=1}^{k-1} B_i$ is bounded by $k \leq m$ (respectively by $m^{s_2 - 1}$ in the general case so that m should not be too large). In the remaining set $J \backslash B$ we could face the worst case that the curve connecting the points is zig-zagging around the vertical line through u_1. This would give rise to many single-point components (essentially half the number of points contained in $J \backslash B$). Hence, m should be as large as possible because small bins contain fewer points. A *balanced* estimate is obtained if m is chosen of the order of $N^{\frac{1}{s_2}}$. For the two-dimensional case, this is easy to see because each of the $m = \sqrt{N}$ bins contains essentially $N/m = \sqrt{N}$ points if the points are uniformly distributed, i.e. if the discrepancy of the point set is reasonably small. Hence the number of consecutive components can be estimated by $\mathrm{cc}^*(X_2) \leq m + N/m = O(\sqrt{N})$. A careful version of this argument is given in [4] which proves

Theorem 2. *Let $X_i : \mathcal{N} \to I^{s_i}$, $i = 1, 2$ be mappings from $\mathcal{N} = \{1, \ldots, N\}$ into the unit cubes I^{s_1} and I^{s_2} with $N, s_1, s_2 \in \mathbb{N}$. Assume further that X_1, X_2 satisfy (3.14) and (3.15) and that that the points $X_2(i)$ are sorted in m-bins with $m = \lfloor N^{\frac{1}{s_2}} \rfloor + 1$. Then the star-discrepancy of $X = (X_1, X_2)$ satisfies*

$$D^*(X) \leq \frac{l_1(N)(1 + s_2)}{N^{\frac{1}{s_2}}} + \frac{l_2(N)(1 + 2l_1(N))}{N}.$$

In summary, we can say that sorting plays an essential role in the construction of QMC product measures. The obtained discrepancy estimate, however, is quite poor in high dimensions. Among other things, this is due to the fact that we estimate the worst case error which may not play a big role in specific applications. Therefore, numerical investigations are necessary to judge about the practical usefulness of the QMC approach. In the following section, such questions are addressed.

4 QMC Simulations

4.1 The Diffusion Equation

In order to construct approximate solutions of the model problem (1.8), we use a QMC algorithm of the form

$$\boldsymbol{Z}_{t_{m+1}}(i) = \boldsymbol{Z}_{t_m}(i) + \boldsymbol{Y}_{t_m}(P(i)), \qquad i = 1, \ldots, N.$$

The increment vectors $\boldsymbol{Y}_{t_m}$ are constructed from a suitable low discrepancy sequence $\boldsymbol{u}_1, \boldsymbol{u}_2, \boldsymbol{u}_3, \ldots$ according to the rule

$$\boldsymbol{Y}_{t_m}(i) = \sqrt{h}\boldsymbol{H}^{-1}(\boldsymbol{u}_{mN+i})$$

where $\boldsymbol{H}(\boldsymbol{z}) = (H(z_1), \ldots, H(z_s))$ with H given by

$$H(x) := \frac{1}{2}\left(1 + \operatorname{erf}\left(\frac{x}{\sqrt{2}}\right)\right) \quad \text{where} \quad \operatorname{erf}(z) = \frac{2}{\sqrt{\pi}} \int_0^z e^{-t^2} dt.$$

This construction ensures that the increments are properly distributed according to (2.13) and that the condition (3.15) can be met with $\boldsymbol{Y}_{t_m}$ in place of X_1 (for example, when $(\boldsymbol{u}_k)$ is a Faure sequence). As a consequence, the points $\boldsymbol{Z}_{t_m}$ play the role of X_2, so that a renumbering of these points is required to ensure a reasonable approximation of the product measure. This renumbering in the form of a permutation $P^{-1} : \{1, \ldots, N\} \to \{1, \ldots, N\}$ can be constructed by various sorting and mixing procedures based on the components of $\boldsymbol{Z}_{t_m}$. The resulting pairs $(\boldsymbol{Z}_{t_m}(P^{-1}(i)), \boldsymbol{Y}_{t_m}(i))$, or equivalently $(\boldsymbol{Z}_{t_m}(i), \boldsymbol{Y}_{t_m}(P(i)))$ are then used to define $\boldsymbol{Z}_{t_{m+1}}$ recursively. In [14], a family of algorithms (denoted QMC(m, r)) has been introduced which differ only in the construction of the permutation P. For example, QMC$(0, r)$ is based on a permutation which is constructed by sorting the components of $\boldsymbol{Z}_{t_m}(i)$ with respect to a bin-structure in the subspace of the first r coordinates. In the algorithm QMC$(m, 0)$, the sequence $\boldsymbol{u}_1, \boldsymbol{u}_2, \ldots$ of uniformly distributed quasi random numbers is taken $(s+m)$-dimensional instead of s-dimensional, where the first s components are used to define $\boldsymbol{Y}_{t_m}$ and the last m components are taken to mix $\boldsymbol{Z}_{t_m}(i), \boldsymbol{Y}_{t_m}(i)$ in a quasi-random way (for details see [14]). Finally, QMC(m, r) represents the algorithm where both mixing and sorting is performed. The specific case with maximal values for m, r, i.e. QMC(s, s), is the algorithm proposed in [6]. The higher the values for m, r, the more expensive the construction of the permutation. Moreover, high values of m, r

restrict the applicability in high dimensions because of hardware limitations. Therefore, it is important to check whether the algorithms with small m, r work satisfactorily.

For a numerical investigation, we consider the diffusion equation (1.8) with initial value $u_0(\boldsymbol{z}) = \pi^{-s/2} \exp(-|\boldsymbol{z}|^2)$. The exact solution for this problem is

$$u(\boldsymbol{z}, t) = \frac{1}{(\pi(1+2t))^{s/2}} \exp\left(-\frac{|\boldsymbol{z}|^2}{1+2t}\right), \qquad \boldsymbol{z} \in \mathbb{R}^s. \tag{4.16}$$

Since the discrepancy between the point measure $\Pi_{\boldsymbol{Z}_{t_m}}$ associated to the approximate solution and the exact measure $u(t_m, \boldsymbol{z})d\boldsymbol{z}$ is difficult to compute exactly, we use the following approximation. Instead of comparing the two measures on *all* s-dimensional intervals of the form $(-\infty, \boldsymbol{\omega})$, we only choose a large number (10000) of such intervals where $\boldsymbol{\omega}$ is normally distributed with mean 3 and variance 1. Based on this error measure, we compute numerical convergence rates. For various methods mentioned above, we calculate the numerical convergence rate for different dimensions taking 100 time steps with $h = 0.0001$. The results are given in table 1. The Method MC refers to the Euler Maruyama method introduced in section 2 From table 1, we conclude

Table 1. Numerical order of convergence for the various methods. NA refers to non applicability due to memory restrictions.

Method	$s = 3$	$s = 6$	$s = 9$	$s = 12$	$s = 50$
MC	-0.2698	-0.2538	-0.2862	-0.2765	-0.2946
QMC$(1, 1)$	-0.4517	-0.4162	-0.5299	-0.4757	-0.6106
QMC$(0, 1)$	-0.5662	-0.5407	-0.5783	-0.5542	-0.6070
QMC$(1, 0)$	-0.00002	-0.0002	-0.0001	-0.0001	-0.00008
QMC(s, s)	-0.4555	NA	NA	NA	NA
QMC$(0, s)$	-0.2149	NA	NA	NA	NA
QMC$(s, 0)$	-0.1715	NA	NA	NA	NA
QMC$(3, 2)$	-0.1699	-0.4100	-0.2268	-0.4627	NA

that algorithm QMC$(0, 1)$ outperforms the others. One can also observe that algorithm QMC$(1, 1)$ performs well with the only disadvantage of extra mixing time. Algorithm QMC$(1, 0)$ does not converge at all implying that sorting is essential for convergence and is in accordance with [7]. We stress that the numerical order of convergence is only one indication for the performance of the algorithm but one should also consider the absolute errors. Typical error plots corresponding to the results presented in table 1 are shown in

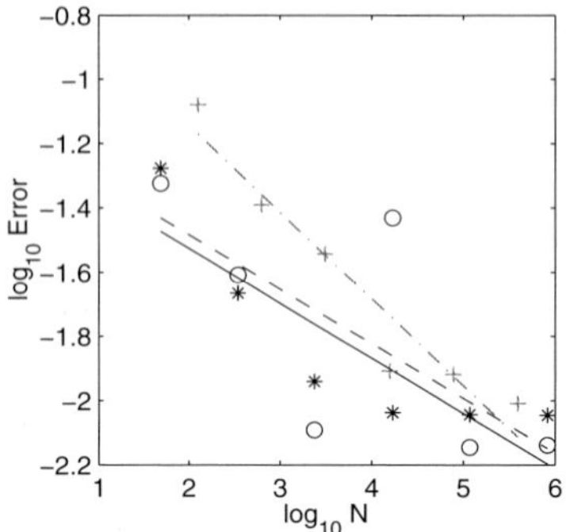

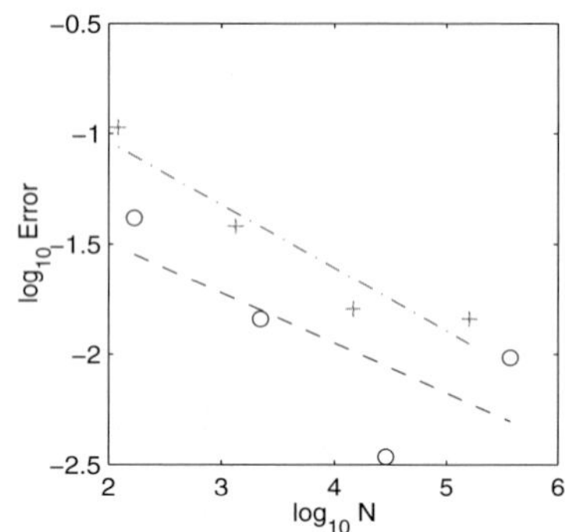

Figure 4. Even though the methods QMC$(s, 0)$ (stars), QMC$(3, 2)$ (circles) show a poor convergence order compared to the MC method (plus), the absolute error is still lower for particle numbers N below 10^6. The results are shown for dimensions $s = 3$ (left) and $s = 9$ (right). The convergence order is obtained as the slope of the least squares fit: solid line for QMC$(s, 0)$, dashed line for QMC$(3, 2)$, and dashed dot line for MC.

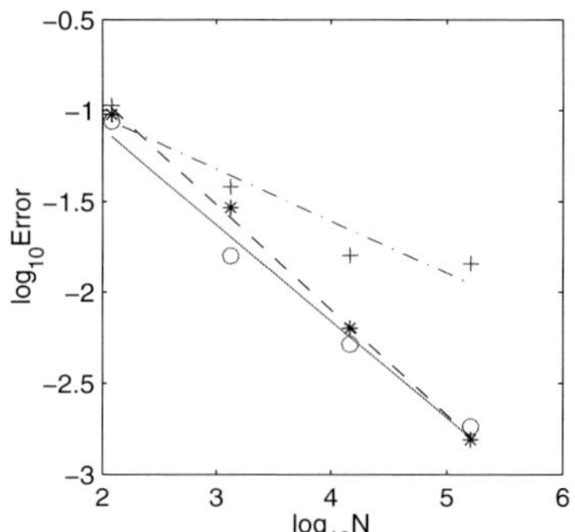

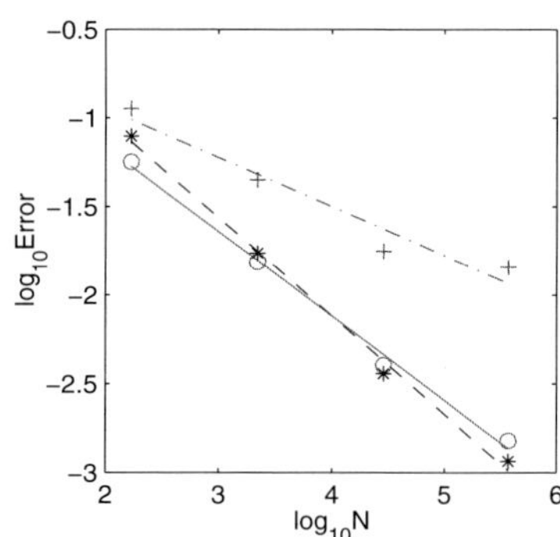

Figure 5. The methods QMC$(1, 1)$ (circles) and QMC$(0, 1)$ (stars) clearly outperform the Monte Carlo method both in convergence order and in absolute error. The left plot refers to dimension $s = 9$ and the right one to $s = 12$.

figures 4 and 5. In conclusion, we can say that the methods QMC$(0, 1)$ and QMC$(1, 1)$ seem to be interesting alternatives for the Monte Carlo approach. Of course, the construction of the required permutation requires additional computational time but the increased accuracy allows to use lower particle numbers which balances this disadvantage.

4.2 The Polymer Model

Based on the previous considerations, we are able to construct approximate solutions of the polymer problem (1.5). The QMC algorithm has the form

$$\boldsymbol{Q}_{t_{m+1}}(i) = \boldsymbol{Q}_{t_m}(i) + \boldsymbol{a}(\boldsymbol{Q}_{t_m})h + \boldsymbol{D}\boldsymbol{Y}_{t_m}(P(i)), \qquad i = 1, \ldots, N.$$

where the increments $\boldsymbol{Y}_{t_m}$ are defined as specified in the previous section. The permutation P is constructed according to the algorithm QMC(1, 1). As initial configuration distribution ψ_0, we compare three different choices

- (SN) $\psi_0(\boldsymbol{q}) = (2\pi)^{-s/2} \exp(-|\boldsymbol{q}|^2/2)$
- (DD) $\psi_0(\boldsymbol{q}) = \delta_{\boldsymbol{0}}(\boldsymbol{q})$
- (EQ) $\psi_0(\boldsymbol{q}) = N_{eq} \exp(-\phi(\boldsymbol{q}))$, with ϕ as in (1.1) (refer [2]).

The abbreviations SN, DD and EQ stand for standard normal, Dirac delta and equilibrium distribution respectively. As a result of the simulation, we consider a component of the integral functional $\boldsymbol{\tau}(t)$ defined in (1.6), where we replace the exact measure $\psi(t_m, \boldsymbol{q})\, d\boldsymbol{q}$ by the approximate point measure $\Pi_{\boldsymbol{Q}_{t_m}}$. A typical time evolution is shown in figure 6. If we focus on the

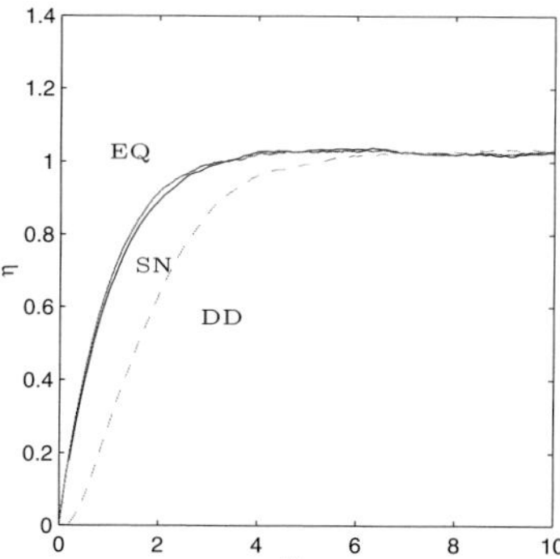

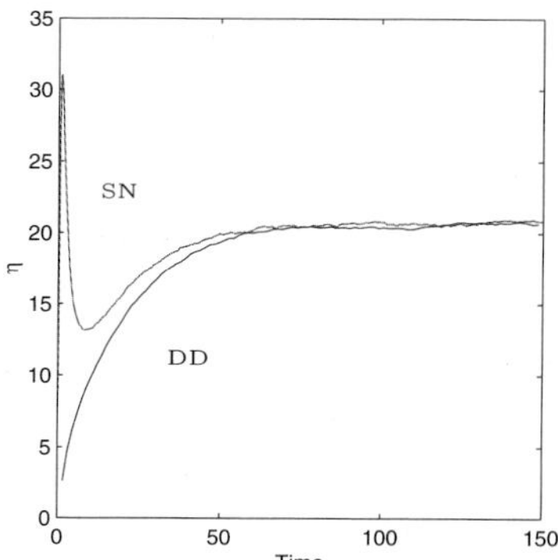

Figure 6. One component η of the functional $\boldsymbol{\tau}(t)$ for different choices of the initial condition. Left: the case of two beads $s = 3$ with $z = 0.1$, $\beta = 1.0$ and $d = 0.5$. Right: the case of eight beads $s = 21$ with the same force parameters. Obviously, the physically relevant stationary value is not influenced by the choice of the initial value.

stationary part of the curve which yields the required stationary value $\boldsymbol{\tau}^p$ defined in (1.3), we see that the result is noisy with a variation of the order of a few percent of the stationary value. The noise clearly reduces if we choose more particles (see figure 7). Compared to the MC result, the QMC algorithm shows considerably less oscillations. This can be seen in figure 8 where we compare the results of the algorithms for large t. It is clear that the QMC trajectory has less oscillations compared to the MC trajectory, meaning that one has to average MC over several runs to obtain a similar result. The average values obtained from the three separate runs of MC is shown in figure 8 (right). At this stage, we have the following situation: MC and QMC both work for high dimensions, the former can be implemented in a straightforward manner whereas the latter requires sorting the particle positions at each time

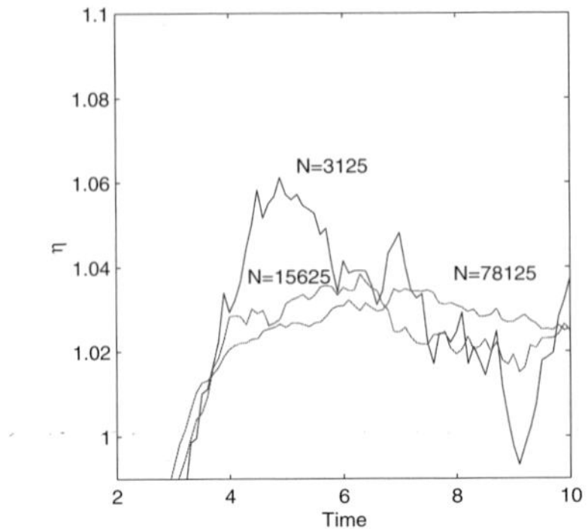

Figure 7. Variation of the component η of $\boldsymbol{\tau}(t)$ for large t decreases with increasing particle number. We indicate the case $s = 3$ with $z = 0.1$, $\beta = 1.0$ and $d = 0.5$.

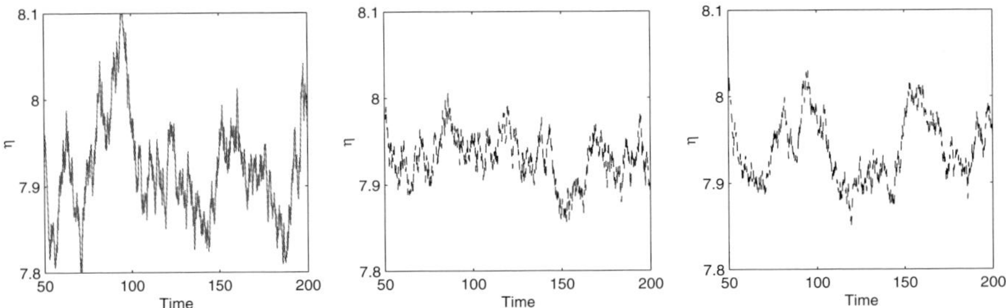

Figure 8. Component of $\boldsymbol{\tau}(t)$ with a single run of MC (left), QMC(1, 1) (middle) and the average of three separate MC simulations (right) for the case of five beads $s = 12$ using 28561 particles with $z = 0.1$, $\beta = 1.0$ and $d = 1.0$.

step. The advantage with QMC is that the results have less noise compared to MC, but the extra processes required to properly represent the diffusion take up additional time. For the polymer problem, however, this does not contribute significantly as the evaluation of the function $\boldsymbol{a}(\boldsymbol{q})$ dominates the total computational time and this is required in both MC and QMC simulations. For further comparisons, we refer to [13].

References

1. R. Byron Bird, Charles. F. Curtiss, Robert. C. Armstrong, and Ole Hassager, *Dynamics of polymeric liquids*, vol. 1, Wiley Interscience, 1987.
2. ———, *Dynamics of polymeric liquids*, vol. 2, Wiley Interscience, 1987.
3. H. Faure, *Discrépance de suites associées à un systéme de numération (en dimension s)*, Acta Arith. **42** (1982), 337–351.
4. M. Junk and G. Venkiteswaran, *Constructing product measures with Quasi-Monte Carlo points*, preprint.

5. E. Kloeden and E. Platen, *Numerical solution of stochastic differential equations*, Springer, 1999.
6. C. Lécot and F. E. Khettabi, *Quasi-Monte Carlo simulation of diffusion*, Journal of Complexity **15** (1999), 342–359.
7. W. J. Morokoff and R. E. Caflisch, *A quasi-Monte Carlo approach to particle simulation of the heat equation*, Siam J. Numer. Anal. **30** (1993), no. 6, 1558–1573.
8. H. Niederreiter, *Random number generation and quasi-Monte Carlo methods*, vol. 6, Society for Industrial and Applied Mathematics, 1992.
9. E. Novak and K. Ritter, *The curse of dimension and a universal method for numerical integration*, Multivariate approximation and splines (G. Nüneberger, J.W. Schmidt, and G. Walz, eds.), ISNM, 1997, pp. 177–188.
10. H. C. Öttinger, *Stochastic processes in polymeric fluids*, Springer, 1996.
11. I. M. Sobol, *On the distribution of points in a cube and the approximate evaluation of integrals*, USSR Comput. Math. Math. Phys (1967).
12. G. Venkiteswaran, *A particle method for Fokker-Planck equations in high dimensions*, PhD dissertation, University of Kaiserslautern, Department of Mathematics, January 2003.
13. G. Venkiteswaran and M. Junk, *A QMC approach for high dimensional Fokker-Planck equations modelling polymeric liquids*, Math. Comp. Simul. **68** (2005), 23–41.
14. ———, *Quasi-Monte Carlo algorithms for diffusion equations in high dimensions*, Math. Comp. Simul. **68** (2005), 43–56.

Bridging Scale Method and Its Applications

W.K. Liu[1], H.S. Park[2], E.G. Karpov[1], and D. Farrell[1]

[1] Northwestern University, Mechanical Engineering Department, 2145 Sheridan Rd., Evanston, IL 60208, USA
w-liu@northwestern.edu

[2] Vanderbilt University, School of Engineering, Civil and Environmental Engineering, Nashville, TN 37235-1831, USA
harold.park@vanderbilt.edu

Summary. This paper is a review of the bridging scale method, which was recently proposed to couple atomistic and continuum simulation methods. The theory will be shown in a fully generalized three-dimensional setting, including the numerical calculation of the time history kernel in multiple dimensions, such that a two-way coarse/fine coupled non-reflecting molecular dynamics boundary condition can be found. We present numerical examples in three dimensions validating the bridging scale methodology. The bridging scale method is tested on highly nonlinear dynamic fracture examples, and the ability of the numerically calculated time history kernel in eliminating high frequency wave reflection at the MD/FE interface is shown. All results are compared to benchmark full MD simulations for verification and validation.

Key words: Bridging scale method, multiscale simulation, fracture.

1 Bridging Scale Fundamentals

The results in this paper are a summary of the one-dimensional work of [1], and the two and three-dimensional work [2, 3]. The bridging scale was recently proposed in [4] to couple atomistic and continuum simulations. The fundamental idea is to decompose the total displacement field $\mathbf{u}(\mathbf{x})$ into coarse and fine scales

$$\mathbf{u}(\mathbf{x}) = \bar{\mathbf{u}}(\mathbf{x}) + \mathbf{u}'(\mathbf{x}) \tag{1.1}$$

The coarse scale $\bar{\mathbf{u}}$ is that part of the solution which can be represented by a set of basis functions, i.e. finite element (FE) shape functions. The fine scale $\mathbf{u}'$ is defined as the part of the solution whose projection onto the coarse scale basis functions is zero; this implies orthogonality of the coarse and fine scale solutions.

In order to describe the bridging scale, first imagine a body in any dimension which is described by N_a atoms. The notation used here will mirror that used in [4]. The total displacement of an atom α is written as $\mathbf{u}_\alpha$. The coarse scale displacement is a function of the initial positions $\mathbf{X}_\alpha$ of the atoms. It should be noted that the coarse scale would at first glance be thought of as a continuous field, since it can be interpolated between atoms. However, because the fine scale is defined only at atomic positions, the total displacement and thus the coarse scale are discrete functions that are defined only at atomic positions. For consistency, Greek indices $(\alpha, \beta, ...)$ will define atoms for the remainder of this work, and uppercase Roman indices $(I, J, ...)$ will define coarse scale nodes.

Introduce $N_I^\alpha = N_I(\mathbf{X}_\alpha)$ to be the shape function of node I evaluated at the initial atomic position $\mathbf{X}_\alpha$, and $\mathbf{d}_I$ the FE nodal displacements associated with node I. Then the coarse scale is defined to be

$$\bar{\mathbf{u}}(\mathbf{X}_\alpha) = \sum_I N_I^\alpha \mathbf{d}_I \tag{1.2}$$

The fine scale in the bridging scale decomposition is simply that part of the total displacement that the coarse scale cannot represent. Thus, the fine scale is defined to be the projection of the total displacement $\mathbf{u}$ onto the FE basis functions subtracted from the total solution $\mathbf{u}$. We will select this projection operator to minimize the mass-weighted square of the fine scale, which we call J and can be written as

$$J = \sum_\alpha m_\alpha \left(\mathbf{u}_\alpha - \sum_I N_I^\alpha \mathbf{w}_I \right)^2 \tag{1.3}$$

m_α is the atomic mass of an atom α and $\mathbf{w}_I$ are temporary nodal (coarse scale) degrees of freedom, which are sought as close to the total solution, as possible at given choice of the shape function set. It should be emphasized that (1.3) is only one of many possible ways to define an error metric. In order to solve for $\mathbf{w}$, the error is minimized with respect to $\mathbf{w}$, yielding the following result:

$$\mathbf{w} = \mathbf{M}^{-1}\mathbf{N}^T\mathbf{M}_A\mathbf{u} \tag{1.4}$$

where the coarse scale mass matrix $\mathbf{M}$ is defined as

$$\mathbf{M} = \mathbf{N}^T\mathbf{M}_A\mathbf{N} \tag{1.5}$$

In (1.4) and (1.5), $\mathbf{M}_A$ is a diagonal matrix with the atomic masses on the diagonal, and $\mathbf{N}$ is a matrix containing the values of the FE shape functions evaluated at all the atomic positions. In general, the size of $\mathbf{N}$ is $N_{a1} \times N_{n1}$, where N_{n1} is the number of finite element nodes whose support contains an atomic position, and N_{a1} is the total number of atoms. The fine scale $\mathbf{u}'$ can thus be written as

$$\mathbf{u}' = \mathbf{u} - \mathbf{N}\mathbf{w} \tag{1.6}$$

or

$$\mathbf{u}' = \mathbf{u} - \mathbf{Pu} \tag{1.7}$$

where the projection matrix $\mathbf{P}$ is defined to be

$$\mathbf{P} = \mathbf{N}\mathbf{M}^{-1}\mathbf{N}^T\mathbf{M}_A \tag{1.8}$$

The total displacement $\mathbf{u}_\alpha$ can thus be written as the sum of the coarse and fine scales as

$$\mathbf{u} = \mathbf{Nd} + \mathbf{u} - \mathbf{Pu} \tag{1.9}$$

The final term in the above equation is called the bridging scale. It is the part of the solution that must be removed from the total displacement so that a complete separation of scales is achieved, i.e. the coarse and fine scales are orthogonal to each other. This bridging scale approach was first used by Liu et al. [5] to enrich the finite element method with meshfree shape functions. Wagner and Liu [6] used this approach to consistently apply essential boundary conditions in meshfree simulations. Zhang et al. [7] applied the bridging scale in fluid dynamics simulations. Qian et al. [8] recently used the bridging scale in quasi-static simulations of carbon nanotube buckling. The bridging scale was also used in conjunction with a multiscale constitutive law to simulate strain localization by Kadowaki and Liu [9].

Now that the details of the bridging scale have been laid out, some comments are in order. In equation (1.3), the fact that an error measure was defined implies that $\mathbf{u}_\alpha$ is the "exact" solution to the problem. In our case, the atomistic simulation method we choose to be our "exact" solution is molecular dynamics (MD). After determining that the MD displacements shall be referred to by the variable $\mathbf{q}$, equation (1.3) can be re-written as

$$J = \sum_\alpha m_\alpha \left(\mathbf{q}_\alpha - \sum_I N_I^\alpha \mathbf{w}_I \right)^2 \tag{1.10}$$

where the MD displacements $\mathbf{q}$ now take the place of the total displacements $\mathbf{u}$. The equation for the fine scale $\mathbf{u}'$ can now be re-written as

$$\mathbf{u}' = \mathbf{q} - \mathbf{Pq} \tag{1.11}$$

The fine scale is now clearly defined to be the difference between the MD solution and its projection onto a pre-determined coarse scale basis space. This implies that the fine scale can thus be interpreted as a built in error estimator to the quality of the coarse scale approximation. Finally, the equation for the total displacement $\mathbf{u}$ can be re-written as

$$\mathbf{u} = \mathbf{Nd} + \mathbf{q} - \mathbf{Pq} \tag{1.12}$$

Because of the equality of $\mathbf{q}$ and $\mathbf{u}$, it would appear that solving the FE equation of motion is unnecessary, since the coarse scale can be calculated directly as the projection of $\mathbf{q}$, i.e. $\mathbf{Nd} = \mathbf{Pq}$. However, because the goal is to eliminate the fine scale from large portions of the domain, the MD displacements $\mathbf{q}$ are not defined over the entire domain, and thus it is not possible to calculate the coarse scale solution everywhere via direct projection of the MD displacements. Thus, the solution of the FE equation of motion everywhere ensures a continuous coarse scale displacement field.

1.1 Multiscale Equations of Motion

The next step in the multiscale process is to derive the coupled MD and FE equations of motion. This is done by first constructing a Lagrangian $\mathcal{L}$, which is defined to be the kinetic energy minus the potential energy

$$\mathcal{L}(\mathbf{u}, \dot{\mathbf{u}}) = \mathcal{K}(\dot{\mathbf{u}}) - V(\mathbf{u}) \tag{1.13}$$

Ignoring external forces, (1.13) can be written as

$$\mathcal{L}(\mathbf{u}, \dot{\mathbf{u}}) = \frac{1}{2}\dot{\mathbf{u}}^T \mathbf{M}_A \dot{\mathbf{u}} - U(\mathbf{u}) \tag{1.14}$$

where $U(\mathbf{u})$ is the interatomic potential energy. Differentiating the total displacement $\mathbf{u}$ in (1.12) with respect to time gives

$$\dot{\mathbf{u}} = \mathbf{N}\dot{\mathbf{d}} + \mathbf{Q}\dot{\mathbf{q}} \tag{1.15}$$

where the complimentary projection operator $\mathbf{Q} \equiv \mathbf{I} - \mathbf{P}$. Substituting (1.15) into the Lagrangian (1.14) gives

$$\mathcal{L}(\mathbf{d}, \dot{\mathbf{d}}, \mathbf{q}, \dot{\mathbf{q}}) = \frac{1}{2}\dot{\mathbf{d}}^T \mathbf{M}\dot{\mathbf{d}} + \frac{1}{2}\dot{\mathbf{q}}^T \mathcal{M}\dot{\mathbf{q}} - U(\mathbf{d}, \mathbf{q}) \tag{1.16}$$

where the fine scale mass matrix $\mathcal{M}$ is defined to be $\mathcal{M} = \mathbf{Q}^T \mathbf{M}_A$. One elegant feature of (1.16) is that the total kinetic energy has been decomposed into the sum of the coarse scale kinetic energy plus the fine scale kinetic energy.

The multiscale equations of motion are obtained from the Lagrangian by following the Lagrange equations

$$\frac{d}{dt}\left(\frac{\partial \mathcal{L}}{\partial \dot{\mathbf{d}}}\right) - \frac{\partial \mathcal{L}}{\partial \mathbf{d}} = 0 \tag{1.17}$$

$$\frac{d}{dt}\left(\frac{\partial \mathcal{L}}{\partial \dot{\mathbf{q}}}\right) - \frac{\partial \mathcal{L}}{\partial \mathbf{q}} = 0 \tag{1.18}$$

Substituting the Lagrangian (1.16) into (1.17) and (1.18) gives

$$\mathbf{M}\ddot{\mathbf{d}} = -\frac{\partial U(\mathbf{d}, \mathbf{q})}{\partial \mathbf{d}} \tag{1.19}$$

$$\mathcal{M}\ddot{\mathbf{q}} = -\frac{\partial U(\mathbf{d}, \mathbf{q})}{\partial \mathbf{q}} \tag{1.20}$$

The two equations (1.19) and (1.20) are coupled through the derivative of the potential energy U, which can be expressed as functions of the interatomic force $\mathbf{f}$ as

$$\mathbf{f} = -\frac{\partial U(\mathbf{u})}{\partial \mathbf{u}} \tag{1.21}$$

Expanding the right hand sides of (1.19) and (1.20) with a chain rule and using (1.21) together with (1.12) gives

$$\mathbf{M}\ddot{\mathbf{d}} = \frac{\partial U}{\partial \mathbf{u}}\frac{\partial \mathbf{u}}{\partial \mathbf{d}} = \mathbf{N}^T\mathbf{f} \tag{1.22}$$

$$\mathcal{M}\ddot{\mathbf{q}} = -\frac{\partial U}{\partial \mathbf{u}}\frac{\partial \mathbf{u}}{\partial \mathbf{q}} = \mathbf{Q}^T\mathbf{f} \tag{1.23}$$

Using the fact that $\mathcal{M} = \mathbf{Q}^T\mathbf{M}_A$, (1.23) can be rewritten as

$$\mathbf{Q}^T\mathbf{M}_A\ddot{\mathbf{q}} = \mathbf{Q}^T\mathbf{f} \tag{1.24}$$

Because $\mathbf{Q}$ can be proven to be a singular matrix [4], there are many unique solutions to (1.24). However, one solution which does satisfy (1.24) and is beneficial to us is (including the coarse scale equation of motion):

$$\mathbf{M}_A\ddot{\mathbf{q}} = \mathbf{f}(\mathbf{q}) \tag{1.25}$$

$$\mathbf{M}\ddot{\mathbf{d}} = \mathbf{N}^T\mathbf{f}(\mathbf{u}) \tag{1.26}$$

Now that the coupled multiple scale equations of motion have been derived, we make some relevant comments

(i) The fine scale equation of motion (1.25) is simply the MD equation of motion. Therefore, a standard MD solver can be used to obtain the MD displacements $\mathbf{q}$, while the MD forces $\mathbf{f}$ can be found by using any relevant potential energy function.
(ii) The coarse scale equation of motion (1.26) is simply the finite element momentum equation. Therefore, we can use standard finite element methods to find the solution to (1.26), while noting that the finite element mass matrix $\mathbf{M}$ is defined to be a consistent (generally, non-diagonal) mass matrix.
(iii) The coupling between the two equations is through the coarse scale internal force $\mathbf{N}^T\mathbf{f}(\mathbf{u})$, which is a direct function of the MD internal force $\mathbf{f}$. In the region in which MD exists, the coarse scale force is calculated by interpolating the MD force using the finite element shape functions $\mathbf{N}$. In the region in which MD has been eliminated, the coarse scale force can be calculated in multiple ways. The elimination of unwanted MD degrees of freedom is discussed in the next section.
(iv) We note that the total solution $\mathbf{u}$ satisfies the same equation of motion as $\mathbf{q}$, i.e.

$$\mathbf{M}_A\ddot{\mathbf{u}} = \mathbf{f} \tag{1.27}$$

This result is due to the fact that $\mathbf{q}$ and $\mathbf{u}$ satisfy the same initial conditions, and will be utilized in deriving the boundary conditions on the MD simulation in a later section.

(v) Due to the kronecker-delta property of the finite element shape functions, for the case in which the FE nodal positions correspond exactly to the MD atomic positions, the FE equation of motion (1.26) converges to the MD equation of motion (1.25).

(vi) The FE equation of motion is redundant for the case in which the MD and FE regions both exist everywhere in the domain, because the FE equation of motion is simply an approximation to the MD equation of motion, with the quality of the approximation controlled by the finite element shape functions $\mathbf{N}$. This redundancy will be removed by eliminating the fine scale from large portions of the domain.

(vii) We note that the right hand side of (1.25) constitutes an approximation; the internal force $\mathbf{f}$ should be a function of the total displacement $\mathbf{u}$. We utilize the MD displacements $\mathbf{q}$ for two reasons. The first reason, as stated above, is the equality of $\mathbf{q}$ and $\mathbf{u}$. The second reason relates to computational efficiency, as determining $\mathbf{u}$ at each MD time step requires the calculation of the coarse scale solution $\mathbf{Nd}$, which would defeat the purpose of keeping a coarse FE mesh over the entire domain.

The last comment above motivates the developments in the next section. Because of the redundance of the FE equation of motion, we now proceed to eliminate the MD region from a large portion of the domain, such that the redundance of the FE equation of motion is removed.

2 Reduction of the MD Domain

In this section, we apply the arguments of lattice mechanics to the bridging scale. In this approach, we assume that the atomistic region can be subdivided into two regions. In the first region, i.e. around a crack tip, defects or other locally interesting physical phenomena, an anharmonic, or nonlinear potential is necessary to accurately represent the atomic interactions. However, at some distance away from the process of interest, an anharmonic representation of the atomistic physics no longer becomes necessary. At this point, the atomic interactions are sufficiently captured using a harmonic approximation. Therefore, the remainder of this section will summarize the methodology used to eliminate those atoms which we assume to interact harmonically; the eliminated fine scale degrees of freedom will be accounted for via the time history kernel $\boldsymbol{\theta}(t)$.

The first step in this process is to linearize the MD equation of motion (1.25), while using the equality of $\mathbf{q}$ and $\mathbf{u}$ to obtain

$$\mathbf{M}_A\ddot{\mathbf{q}} = \mathbf{M}_A\ddot{\bar{\mathbf{u}}} + \mathbf{M}_A\ddot{\mathbf{u}}' = \mathbf{f}(\bar{\mathbf{u}}) + \mathbf{K}\mathbf{u}' \tag{2.28}$$

where

$$\mathbf{K} = \frac{\partial \mathbf{f}}{\partial \mathbf{u}}|_{\mathbf{u}'=0} \tag{2.29}$$

The goal in performing the linearization is to decompose the MD equation of motion into coarse and fine scale components. In doing so, we will work exclusively with the fine scale equation such that we can achieve our stated goal of limiting the fine scale to a small region of the domain, while keeping the coarse scale everywhere in the domain. The major assumption in our derivation is that we can write the fine scale equation of motion neglecting contributions from the coarse scale This assumption is justified by the orthogonality of the coarse and fine scales, and because the time scale for the coarse scale is much larger than that of the atomic vibrations present in the fine scale. Thus, we re-write the fine scale portion of the linearized MD equation of motion (2.28) while including the effects of external forces as

$$\mathbf{M}_A\ddot{\mathbf{u}}' - \mathbf{K}\mathbf{u}' = \mathbf{f}^{ext} \tag{2.30}$$

We note that the complete nonlinear force $\mathbf{f}(\mathbf{u})$ has been decomposed as

$$\mathbf{f}(\mathbf{u}) = \mathbf{f}(\bar{\mathbf{u}}) + \mathbf{K}\mathbf{u}' \tag{2.31}$$

Before we can proceed to eliminate the fine scale, we first need to recall some essentials of Laplace and discrete Fourier transform techniques. Assuming that the function x can be defined at all atomic positions n, we denote the value of x and position n as x_n. For the convenience or the discussion below, the DFT of x is written in the following form,

$$\hat{x}(p) = \mathcal{F}_{n\to p}\{x_n\} \equiv \sum_{n=-N/2}^{N/2-1} x_n e^{-i2\pi pn/N} \tag{2.32}$$

Here, N denotes the number of lattice sites, and p can take any integer value between $-N/2$ and $N/2-1$. The inverse Fourier transform (IFT) is then written as

$$x_n = \mathcal{F}^{-1}_{p\to n}\{\hat{x}(p)\} \equiv \frac{1}{N}\sum_{p=-N/2}^{N/2-1} \hat{x}(p)e^{i2\pi pn/N} \tag{2.33}$$

The Laplace transform and the second time derivative rule will by utilized in their standard forms. The inversion of Laplace transform will be done numerically by utilizing the numeral algorithm of [10].

2.1 Elimination of Fine Scale Degrees of Freedom: 3D Generalization

We now generalize the ideas presented in the previous section to three dimensions. The periodic lattice now consists of spatially repeated unit cells

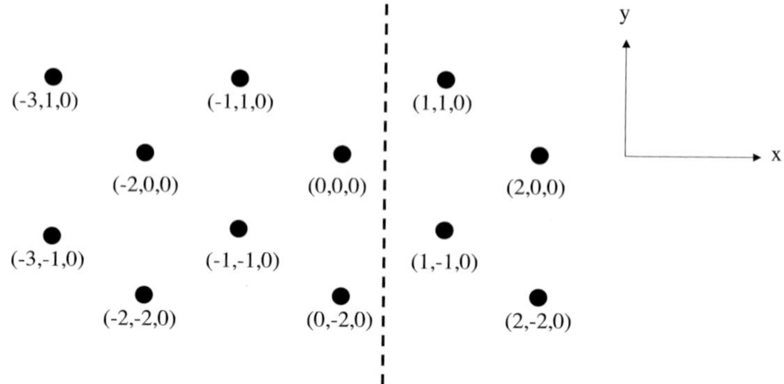

Figure 1. Periodic two-dimensional hexagonal lattice structure numbered using unit cell nomenclature. The solid line represents the boundary between the MD region to be simulated (bottom), and the MD region to be eliminated (top).

which are repeated in three directions. Each repeated cell has n_a atoms, each of which can move in n_{sd} spatial directions. The total number of degrees of freedom in each unit cell is then $n_{dof} = n_a \times n_{sd}$. Each unit cell can be labeled with three indices, l, m and n, indicating the position along axes in the direction of the three primitive vectors of the crystal structure. A two-dimensional example for a hexagonal lattice labelled using unit cell nomenclature is illustrated in Figure (1). Let μ and ν represent the range of the forces in the m and n coordinate directions; then we arrive at

$$\ddot{\mathbf{u}}'_{l,m,n}(t) = \sum_{l'=l-1}^{l+1} \sum_{m'=m-\mu}^{m+\mu} \sum_{n'=n-\nu}^{n+\nu} \mathbf{M}_A^{-1} \mathbf{K}_{l-l',m-m',n-n'} \mathbf{u}'_{l',m',n'}(t) + \mathbf{M}_A^{-1} \mathbf{f}^{ext}_{l,m,n}(t) \quad (2.34)$$

where $\mathbf{f}^{ext}_{l,m,n}(t)$ is the external force acting upon unit cell (l, m, n), the matrices $\mathbf{K}$ relate the displacements in cell $(-l', -m', -n')$ to the forces in cell (l, m, n), and $\mathbf{M}_A$ is a diagonal matrix of atomic masses. We note that while atoms in a given slab of constant l are coupled to only neighboring cells $l - 1$ and $l + 1$, the coupling in the m and n coordinate directions is not limited to nearest neighbors. This approximation can be relaxed simply by increasing the interaction range of the $\mathbf{K}$ matrices.

The derivation begins by taking a Laplace transform (LT) and discrete Fourier transform (DFT) of (2.34)

$$s^2\hat{\mathbf{U}}'(p,q,r,s) - s\mathbf{u}'(p,q,r,0) - \dot{\mathbf{u}}'(p,q,r,0) = \hat{\mathbf{A}}(p,q,r)\hat{\mathbf{U}}'(p,q,r,s) + \mathbf{M}_A^{-1}\hat{\mathbf{F}}_0^{ext}(p,q,r,s) \quad (2.35)$$

where p, q and r correspond to spatial indices l, m and n and the hatted notation indicates the discrete Fourier transform. $\hat{\mathbf{A}}(p,q,r)$ is the Fourier transform of $\mathbf{M}_A^{-1}\mathbf{K}_{l,m,n}$, and Laplace transformed variables are indicated by the transformed variable s. The goal of this process will be to eliminate the atoms in the $l > 0$ cells by solving for them in terms of the $l \leq 0$ degrees of freedom and resubstituting that expression into (2.34). In this manner, we will avoid the explicit solution for the $l > 0$ degrees of freedom while implicitly including their effects into the remaining system dynamics.

The key step in removing the unwanted $l > 0$ fine scale degrees of freedom is in realizing that the motion of the boundary ($l = 0$) atoms can be caused either by the displacements of the atoms to be kept, or by an external force acting upon the boundary atoms. Therefore, it will be assumed that the motion of the boundary atoms is in fact caused by the external force which acts only at $l = 0$,

$$\mathbf{f}_{l,m,n}^{ext}(t) = \delta_{l,0}\mathbf{f}_{0,m,n}^{ext}(t) \tag{2.36}$$

(2.35) can be solved to give the Laplace transformed/discrete Fourier transformed displacements in terms of the external force

$$\begin{aligned}\hat{\mathbf{U}}'(p,q,r,s) &= \hat{\mathbf{G}}(p,q,r,s)\\ &\quad\times\left(\mathbf{M}_A^{-1}\hat{\mathbf{F}}_0^{ext}(p,q,r,s) + s\mathbf{u}'(p,q,r,0) + \dot{\mathbf{u}}'(p,q,r,0)\right)\end{aligned} \tag{2.37}$$

where

$$\hat{\mathbf{G}}(p,q,r,s) = \left(s^2\mathbf{I} - \hat{\mathbf{A}}(p,q,r)\right)^{-1} \tag{2.38}$$

Taking the inverse Fourier transform of (2.37) in the x direction gives the displacement in the x direction at atomic position l

$$\tilde{\mathbf{U}}_l'(q,r,s) = \mathbf{M}_A^{-1}\tilde{\mathbf{G}}_l(q,r,s)\hat{\mathbf{F}}_0^{ext}(q,r,s) + \tilde{\mathbf{R}}_l^d(q,r,s) \tag{2.39}$$

where

$$\tilde{\mathbf{R}}_l^d(q,r,s) = \sum_{l'=-L/2}^{L/2-1}\tilde{\mathbf{G}}_{l-l'}(q,r,s)\left(s\mathbf{U}_{l'}'(q,r,0) + \dot{\mathbf{U}}_{l'}'(q,r,0)\right) \tag{2.40}$$

where L is the total number of unit cells in the x-direction. By writing (2.39) for both $l = 0$ and $l = 1$, we can obtain the displacements $\tilde{\mathbf{U}}_1'$ in terms of $\tilde{\mathbf{U}}_0'$, thereby eliminating $\hat{\mathbf{F}}_0^{ext}$ and obtaining

$$\tilde{\mathbf{U}}_1'(q,r,s) = \tilde{\mathbf{Q}}(q,r,s)\left(\tilde{\mathbf{U}}_0'(q,r,s) - \tilde{\mathbf{R}}_0^d(q,r,s)\right) + \tilde{\mathbf{R}}_1^d(q,r,s) \tag{2.41}$$

where

$$\tilde{\mathbf{Q}}(q,r,s) = \tilde{\mathbf{G}}_1(q,r,s)\tilde{\mathbf{G}}_0^{-1}(q,r,s) \tag{2.42}$$

By inverting the Fourier transform of (2.41) and using the convolution property of the DFT, we obtain

$$\mathbf{U}'_{1,m,n}(s) = \sum_{m'=-M/2}^{M/2-1} \sum_{n'=-N/2}^{N/2-1} \mathbf{Q}_{m-m',n-n'}(s) \left(\mathbf{U}'_{0,m',n'}(s) - \mathbf{R}^d_{0,m',n'}(s)\right) + \mathbf{R}^d_{1,m,n}(s) \quad (2.43)$$

where the random component of the displacement acting upon plane l can be written as

$$\mathbf{R}^d_{l,m,n}(s) = \sum_{l'=-L/2}^{L/2-1} \sum_{m'=-M/2}^{M/2-1} \sum_{n'=-N/2}^{N/2-1} \mathbf{G}_{l-l',m-m',n-n'}(s) \times \left(s\mathbf{U}'_{l',m',n'}(0) + \dot{\mathbf{U}}'_{l',m',n'}(0)\right) \quad (2.44)$$

Inversion of the Laplace transform for the above equation gives a time-dependent vector of random displacements,

$$\mathbf{R}^d_{l,m,n}(t) = \sum_{l'=-L/2}^{L/2-1} \sum_{m'=-M/2}^{M/2-1} \sum_{n'=-N/2}^{N/2-1} (\dot{\mathbf{g}}_{l-l',m-m',n-n'}(t)\mathbf{u}'_{l',m',n'}(0) + \mathbf{g}_{l-l',m-m',n-n'}(t)\dot{\mathbf{u}}'_{l',m',n'}(0)) \quad (2.45)$$

where $\mathbf{u}'_{l,m,n}(0)$, $\dot{\mathbf{u}}'_{l,m,n}(0)$ are the fine scale initial conditions determined by thermodynamic properties of the system, and

$$\mathbf{g}_{l,m,n}(t) = \mathcal{L}^{-1}\left(\mathbf{G}_{l,m,n}(s)\right) \quad (2.46)$$

$$\dot{\mathbf{g}}_{l,m,n}(t) = \mathcal{L}^{-1}\{s\mathbf{G}_{l,m,n}(s)\} \quad (2.47)$$

Since the fine scale portion of the coarse grain atom displacements $\mathbf{q}_{1,m,n}$ are known by (2.43), a complete coupled set of equations could be written as

$$\mathbf{M}_A\ddot{\mathbf{q}} = \mathbf{f}(\mathbf{q}) \quad (2.48)$$

$$\mathbf{M}\ddot{\mathbf{d}} = \mathbf{N}^T\mathbf{f}(\mathbf{u}) \quad (2.49)$$

$$\mathbf{q}_{1,m,n}(t) = \int_0^t \mathbf{Q}_{m-m',n-n'}(t-\tau) \times \left(\mathbf{q}_{0,m',n'}(\tau) - \bar{\mathbf{u}}_{0,m',n'}(\tau) - \mathbf{R}^d_{0,m',n'}(\tau)\right) d\tau + \bar{\mathbf{u}}_{1,m,n}(t) + \mathbf{R}^d_{1,m,n}(t) \quad (2.50)$$

In this formulation, the coarse grain atom displacements $\mathbf{q}_{1,m,n}(t)$ are semi-analytically controlled as a function of the fine scale portion of the coarse grain atom displacements (the first term on the right hand side of (2.50)) combined with a contribution of the coarse scale displacements interpolated to the coarse grain atom position. In the impedance force formulation derived later in this section, the coarse grain atom motion is completely prescribed by the finite element shape functions; in this displacement formulation, the dependence on the shape functions is reduced, but not completely eliminated.

We note the similarity between the coarse grain atoms and so-called "ghost" atoms, which are necessary in MD simulations such that spurious relaxation of the system due to surface effects does not occur. In the bridging scale formulation, because all atoms are initially present, the coarse grain atoms have not been artificially created; however, like ghost atoms, their motion is not explicitly solved for in the MD equation of motion.

Previous approaches to deriving non-reflecting boundary conditions, e.g. [11, 12] have utilized a force boundary condition, so we bow to convention and continue with the derivation. Equation (2.43) becomes useful, if the linearized forces acting on $l = 0$ slab of atoms due to the $l = 1$ slab of atoms is rewritten as, recalling (2.34)

$$\mathbf{F}^{1\to 0}_{m,n}(s) = \sum_{m'=m-\mu}^{m+\mu} \sum_{n'=n-\nu}^{n+\nu} \mathbf{K}_{-1,m-m',n-n'} \mathbf{U}'_{1,m',n'}(s) \tag{2.51}$$

Substituting (2.43) into (2.51) and taking the inverse Laplace transform, the *impedance force boundary condition* of the layer $l = 1$ slab of atoms acting upon the $l = 0$ slab of atoms becomes

$$\begin{aligned} \mathbf{f}^{imp}_{m,n}(t) = \boldsymbol{\mathcal{L}}^{-1}\{\mathbf{F}^{1\to 0}_{m,n}(s)\} = \sum_{m'=-M/2}^{M/2-1} \sum_{n'=-N/2}^{N/2-1} \int_0^t \boldsymbol{\theta}_{m-m',n-n'}(t-\tau) \\ \times \left(\mathbf{u}'_{0,m',n'}(\tau) - \mathbf{R}^d_{0,m',n'}(\tau)\right) d\tau + \mathbf{R}^f_{0,m,n}(t) \end{aligned} \tag{2.52}$$

where the time history kernel $\boldsymbol{\theta}(t)$ is defined to be

$$\boldsymbol{\theta}_{m,n}(t) = \boldsymbol{\mathcal{L}}^{-1}(\boldsymbol{\Theta}_{m,n}(s)) \tag{2.53}$$

$$\boldsymbol{\Theta}_{m,n}(s) = \sum_{m'=m-\mu}^{m+\mu} \sum_{n'=n-\nu}^{n+\nu} \mathbf{K}_{-1,m-m',n-n'} \mathbf{Q}_{m',n'}(s) \tag{2.54}$$

and the random force $\mathbf{R}^f_{0,m,n}(t)$ is given by

$$\mathbf{R}^f_{0,m,n}(t) = \sum_{l'=-L/2}^{L/2-1} \sum_{m'=-M/2}^{M/2-1} \sum_{n'=-N/2}^{N/2-1} \mathbf{K}_{-1,m-m',n-n'} \mathbf{R}^d_{1,m,n}(t) \tag{2.55}$$

Here, the random displacement vector $\mathbf{R}^d$ is given by (2.45).

As can be seen, the exact evaluation of the second term on the right hand side of (2.52) requires a summation over all other unit cells along the boundary. Clearly, it would be computationally inefficient to actually perform the exact summation in practice, particularly if the lattice is large. Therefore, we rewrite (2.52) as

$$\mathbf{f}^{imp}_{m,n}(t) = \sum_{m'=-n_c}^{n_c} \sum_{n'=-n_c}^{n_c} \int_0^t \boldsymbol{\theta}_{m-m',n-n'}(t-\tau) \times \left(\mathbf{u}'_{0,m',n'}(\tau) - \mathbf{R}^d_{0,m',n'}(\tau)\right) d\tau + \mathbf{R}^f_{0,m,n}(t) \quad (2.56)$$

where n_c refers to a maximum number of atomic neighbors which will be used to compute the impedance force. Now, the fine scale equation of motion (2.34) for the boundary $l = 0$ atoms can be rewritten as

$$\ddot{\mathbf{u}}'_{0,m,n} = \mathbf{A}_{0,m,n}\mathbf{u}'_{0,m,n} + \mathbf{M}_A^{-1} \sum_{m'=-n_c}^{n_c} \sum_{n'=-n_c}^{n_c} \int_0^t \boldsymbol{\theta}_{m-m',n-n'}(t-\tau) \left(\mathbf{u}'_{0,m',n'}(\tau) - \mathbf{R}^d_{0,m',n'}(\tau)\right) d\tau + \mathbf{M}_A^{-1}\mathbf{R}^f_{0,m,n}(t) \quad (2.57)$$

Note that the second term on the right hand side of (2.57) represents the implicit effects of the $l > 0$ cells which were mathematically eliminated. Adding (2.57) and (2.48) and noting that

$$\mathbf{M}_{A0}^{-1}\mathbf{f}_{0,m,n}(\bar{\mathbf{u}}) + \mathbf{A}_{0,m,n}\mathbf{u}'_{0,m,n} = \mathbf{M}_{A0}^{-1}\mathbf{f}_{0,m,n} \quad (2.58)$$

where $\mathbf{M}_{A0}^{-1}$ is the matrix of atomic masses for the boundary plane atoms, we obtain the modified equation of motion for the boundary atoms which does not involve any unknown degrees of freedom of the cells (l, m, n) with $l > 0$

$$\mathbf{M}_{A0}\ddot{\mathbf{q}}_{0,m,n}(t) = \mathbf{f}_{0,m,n}(t) + \sum_{m'=-n_c}^{n_c} \sum_{n'=-n_c}^{n_c} \int_0^t \boldsymbol{\theta}_{m-m',n-n'}(t-\tau) \left(\mathbf{u}'_{0,m',n'}(\tau) - \mathbf{R}^d_{0,m',n'}(\tau)\right) d\tau + \mathbf{R}^f_{0,m,n}(t) \quad (2.59)$$

The final step to writing the MD equations of motion for the boundary atoms is to note that the fine scale component of the MD displacements can be written as

$$\mathbf{u}'_{0,m',n'}(\tau) = \mathbf{q}_{0,m',n'}(\tau) - \bar{\mathbf{u}}_{0,m',n'}(\tau) \quad (2.60)$$

The final form for the coupled MD and FE equations of motion thus can be written as

$$\mathbf{M}_A\ddot{\mathbf{q}} = \mathbf{f}(\mathbf{q}) + \mathbf{f}^{imp}_{0,m,n}(t) + \mathbf{R}^f_{0,m,n}(t) \quad (2.61)$$

$$\mathbf{M}\ddot{\mathbf{d}} = \mathbf{N}^T\mathbf{f}(\mathbf{u}) \quad (2.62)$$

where $\mathbf{R}^f$ is a stochastic thermal force (2.55), and $\mathbf{f}^{imp}$ is the impedance force,

$$\mathbf{f}^{imp}_{0,m,n}(t) = \sum_{m'=-n_c}^{n_c} \sum_{n'=-n_c}^{n_c} \int_0^t \boldsymbol{\theta}_{m-m',n-n'}(t-\tau) \times \left(\mathbf{q}_{0,m',n'}(\tau) - \bar{\mathbf{u}}_{0,m',n'}(\tau) - \mathbf{R}^d_{0,m',n'}(\tau)\right) d\tau \quad (2.63)$$

Equations (2.61)-(2.63) represent the major results of this section.

3 Cauchy-Born Rule

In the coarse scale-only region, where the MD force is unavailable due to the elimination of those atomistic degrees of freedom, an approximation to the right hand side of (2.62), the $\mathbf{N}^T\mathbf{f}(\mathbf{u})$ term, must be made. The goal is to utilize the MD potential in the expression for the coarse scale force despite the absence of the MD degrees of freedom. We shall discuss two methods for applying this force, the Cauchy-Born rule, and the virtual atom cluster (VAC) model.

The Cauchy-Born rule is a homogenization method by which continuum stress and stiffness measures can be obtained directly from an interatomic potential; the link between the atomistic and continuum is achieved by way of the deformation gradient $\mathbf{F}$. As the Cauchy-Born rule is a homogenization method, the major assumption invoked when using the approach is that the lattice underlying any continuum point is required to deform homogeneously according to the continuum deformation gradient $\mathbf{F}$. Further expositions on the Cauchy-Born rule, and its application to solid mechanics can be found in [13], [14], and [15].

To utilize the Cauchy-Born rule in the bridging scale context, we first assume that the potential energy $U(\mathbf{u})$ for the system can be decomposed as

$$U(\mathbf{u}) = \sum_{\alpha} W_{\alpha}(\mathbf{u}) \Delta V_{\alpha} \tag{3.64}$$

where W_{α} is the potential energy *density* centered at atom α. Comparing the right hand sides of (1.19) and (1.22), we find the relation

$$(\mathbf{N}^T\mathbf{f})_I = -\frac{\partial U(\mathbf{u})}{\partial \mathbf{d}_I} \tag{3.65}$$

Substituting (3.64) into (3.65), we obtain

$$(\mathbf{N}^T\mathbf{f})_I = -\sum_{\alpha} \frac{\partial W_{\alpha}(\mathbf{u})}{\partial \mathbf{d}_I} \Delta V_{\alpha} \tag{3.66}$$

In order to use the Cauchy-Born rule, we use a chain rule on (3.66) to obtain

$$(\mathbf{N}^T\mathbf{f})_I = -\sum_{\alpha} \frac{\partial W_{\alpha}}{\partial \mathbf{F}_{\alpha}^T} \frac{\partial \mathbf{F}_{\alpha}^T}{\partial \mathbf{d}_I} \Delta V_{\alpha} \tag{3.67}$$

Simplying further,

$$(\mathbf{N}^T\mathbf{f})_I = -\sum_{\alpha} \frac{\partial N_I}{\partial \mathbf{X}}|_{\mathbf{X}=\mathbf{X}_{\alpha}} \frac{\partial W_{\alpha}}{\partial \mathbf{F}_{\alpha}^T} \Delta V_{\alpha} \tag{3.68}$$

Noting that the derivative of the energy density W with respect to $\mathbf{F}^T$ gives the first Piola-Kirchoff stress $\boldsymbol{\mathcal{P}}$, the summation in (3.68) can be approximated by a discrete summation as

$$(\mathbf{N}^T\mathbf{f})_I = -\sum_q \frac{\partial N_I}{\partial \mathbf{X}}(\mathbf{X}_q)\boldsymbol{\mathcal{P}}(\mathbf{X}_q)w_q \tag{3.69}$$

where w_q is the integration weight associated with point $\mathbf{X}_q$.

4 Atomistic Model

The three-dimensional MD/bridging scale calculations presented in this section utilize the Lennard-Jones potential (4.70).

$$\Phi(r_{ij}) = 4\epsilon\left((\frac{\sigma}{r_{ij}})^{12} - (\frac{\sigma}{r_{ij}})^{6}\right) \tag{4.70}$$

where r is the distance between two atoms i & j. The examples were run with parameter values $\sigma = \epsilon = 1$ considering nearest-neighbor interactions only, while all atomic masses were chosen as $m = 1$. The examples utilize the equilibrium interatomic distance, $r_{eq} = 2^{1/6}\sigma$, corresponding to the minimum of the potential (4.70).

An FCC lattice was considered for the three-dimensional simulations, the atoms were initially in an equilibrium configuration. Because of the symmetry between the top and bottom layers of the lattice, the time history kernels for the top and bottom layers can be related by

$$\boldsymbol{\theta}_n^{top}(t) = \boldsymbol{\theta}_{-n}^{bottom}(t) \tag{4.71}$$

and the storage requirements for the time history kernels can be reduced by one half. All simulations were performed with the random force $\mathbf{R}(t)$ in (2.61) set equal to zero, indicating an MD region at zero temperature.

For the regions which satisfy a coarse scale-only description, the Cauchy-Born rule was utilized to calculate the coarse scale internal force. The LJ 6-12 potential was used to describe the continuum strain energy density, such that the coarse scale internal force could be derived from the same potential that was used for the MD force calculations. 8-node trilinear hexagonal (brick) finite elements were utilized for all numerical examples.

All units related to atomistic simulations in this section, such as velocity, position and time, are given in reduced units. It should be noted that because of the choices of mass, σ and ϵ as unity, all normalization factors end up as unity. Finally, all numerical examples shown in this work were performed using the general purpose simulation code Tahoe, which was developed at Sandia National Laboratories ([16]).

5 Dynamic Crack Propagation in Three Dimensions

In this section, we validate the method on dynamic crack propagation examples within an FCC crystal. For the MD simulation, we utilize the

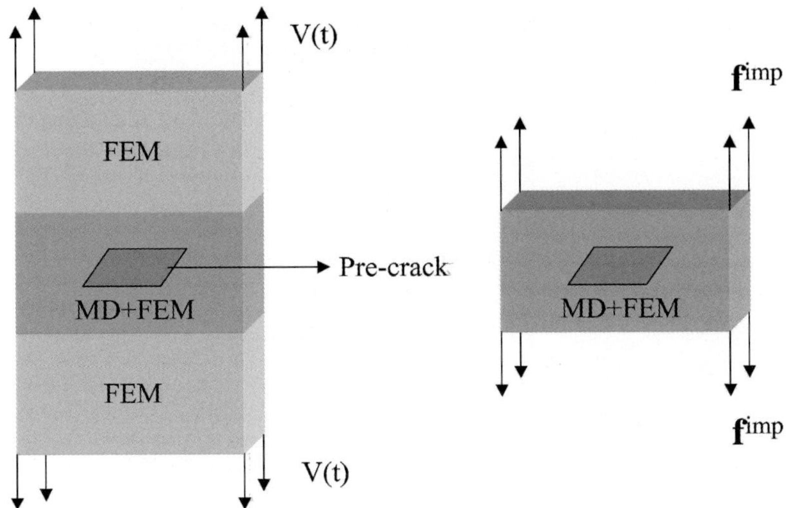

Figure 2. Left: Schematic of 3D bridging scale crack propagation example. Right: Application of MD impedances forces (2.63) to top and bottom (001) planes of reduced MD domain.

Lennard-Jones (LJ) 6-12 potential, though with a slightly altered form. The LJ potential we utilize contains a smooth cut-off, such that the revised potential takes the form

$$\Phi(r_{ij}) = \Phi_{LJ}(r_{ij}) - \Phi_{LJ}(r_c) - (r_{ij} - r_c)\Phi'_{LJ}(r_c) \tag{5.72}$$

where $\Phi_{LJ}(r_{ij})$ is the standard, unshifted LJ 6-12 potential (4.70), which is a function of the distance r_{ij} between two atoms i and j, and the shifted distance r_c is defined to be

$$r_c = \alpha\sigma \tag{5.73}$$

The shifted LJ potential described in (5.72) has a smooth cut-off in both the force and energy at the value $r_{ij} = \alpha$. For the new shifted potential, we utilize parameter values $\sigma = \epsilon = 1$, $\alpha = 1.50$; all atomic masses are taken to be unity.

The dynamic fracture problem schematic is shown in Figure (2). As is shown, the specimen under consideration is covered by finite elements everywhere, while the atomistic region is confined to the central region of the domain. A pre-crack is specified in the atomistic region by prescribing that two adjacent planes of atoms do not interact, and the crack opens naturally in a mode-I type failure under the ramp velocity loading. The normalized velocity applied to the top and bottom (001) surfaces was taken to be $V_{max} = .035$.

We note that the pre-crack is initially fully contained within the interior of the MD domain; the interactions between the crack and the surfaces shown later are a result of the propagation of the crack. A visual image of the pre-crack is given in Figure 3. For visualization purposes, the images in this section

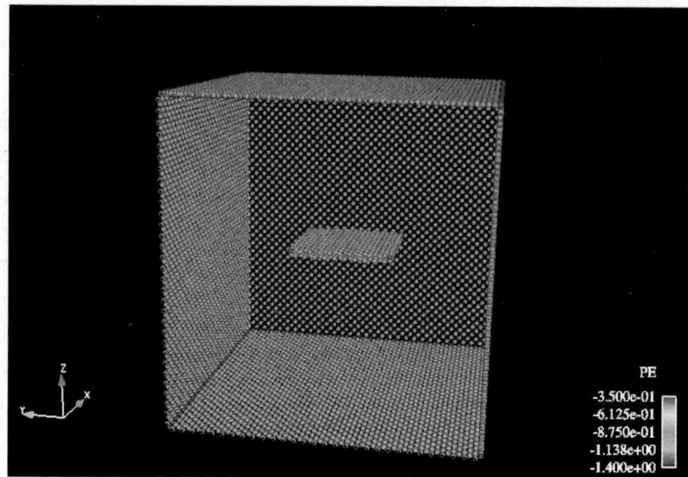

Figure 3. Initial pre-crack for dynamic crack propagation example. Contours of potential energy shown. Only those atoms with potential energy greater than ninety percent of the equilibrium value are shown. See Color Plate 1 on page 301.

only display those atoms whose potential energy is greater than ninety percent of the equilibrium value; this technique is useful for highlighting the defective parts of the lattice which may be of interest in fracture and failure simulations.

The simulations were run using nearest neighbor interactions only. The bridging scale simulation employed 1024 8-node hexahedral finite elements and 117121 atoms, while the full atomistic simulation was comprised of 225121 atoms. Different time steps are used for both simulations; the MD time step was taken to be $\Delta t_{md} = .014$, and 20 MD time steps were run for each FEM time step. The MD impedance force (2.63) is applied to the top and bottom planes of the reduced MD region as shown in Figure (2). All other boundary faces of the reduced MD region are taken to be free surfaces.

The time history kernel $\boldsymbol{\theta}(t)$ was numerically calculated for an FCC lattice structure and the shifted LJ potential for the (001) plane of atoms. The 3×3 $\boldsymbol{\theta}(t)$ matrices in (2.63) for the top and bottom surfaces of the reduced MD region in Figure (2) can be found to be related as

$$\begin{pmatrix} \boldsymbol{\theta}_{11}^{top}(t) & \boldsymbol{\theta}_{12}^{top}(t) & \boldsymbol{\theta}_{13}^{top}(t) \\ \boldsymbol{\theta}_{21}^{top}(t) & \boldsymbol{\theta}_{22}^{top}(t) & \boldsymbol{\theta}_{23}^{top}(t) \\ \boldsymbol{\theta}_{31}^{top}(t) & \boldsymbol{\theta}_{32}^{top}(t) & \boldsymbol{\theta}_{33}^{top}(t) \end{pmatrix} = \begin{pmatrix} \boldsymbol{\theta}_{11}^{bot}(t) & \boldsymbol{\theta}_{12}^{bot}(t) & -\boldsymbol{\theta}_{13}^{bot}(t) \\ \boldsymbol{\theta}_{21}^{bot}(t) & \boldsymbol{\theta}_{22}^{bot}(t) & -\boldsymbol{\theta}_{23}^{bot}(t) \\ -\boldsymbol{\theta}_{31}^{bot}(t) & -\boldsymbol{\theta}_{32}^{bot}(t) & \boldsymbol{\theta}_{33}^{bot}(t) \end{pmatrix} \tag{5.74}$$

Similar relationships relating the $\boldsymbol{\theta}(t)$ matrices for opposite faces of the FCC cube can also be determined.

A comparison between the bridging scale simulation and the full atomistic simulation is given in four distinct snapshots, which chronicle the time history of the crack propagation. In the first snapshot seen in Figure (4), the initially

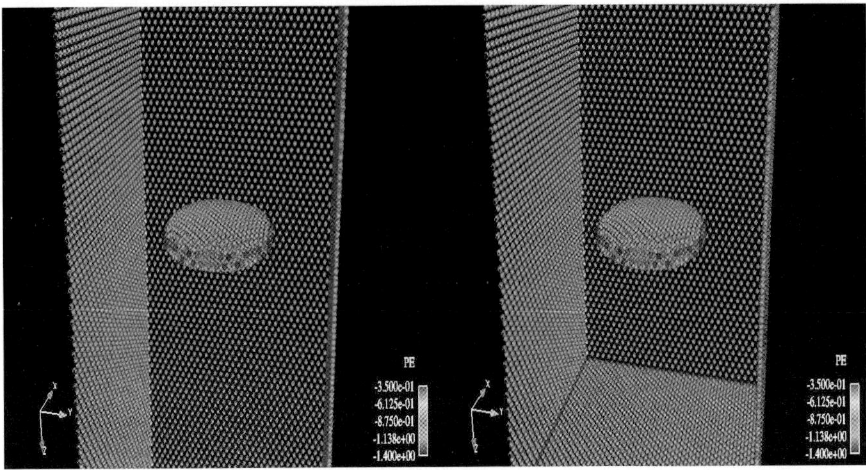

Figure 4. Onset of crack growth for (left) full MD simulation and (right) bridging scale simulation. See Color Plate 2 on page 301.

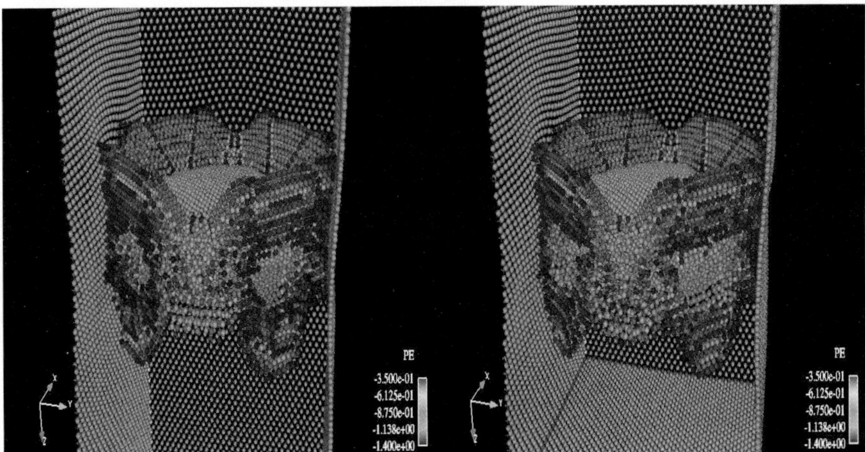

Figure 5. Out of plane crack branching for (left) full MD simulation and (right) bridging scale simulation. See Color Plate 3 on page 302.

square pre-crack has evolved under loading to resemble a penny-shaped crack. Figure (4) shows the crack just before propagation initiates; the figure and all subsequent figures show contours of potential energy, while the crack opening displacement has been magnified by a factor of five for easy viewing.

After the crack propagation has initiated and the crack nears the surfaces of the cube, the onset and subsequent branching of the crack is seen in Figure (5). In this simulation, it important to note that the branching is

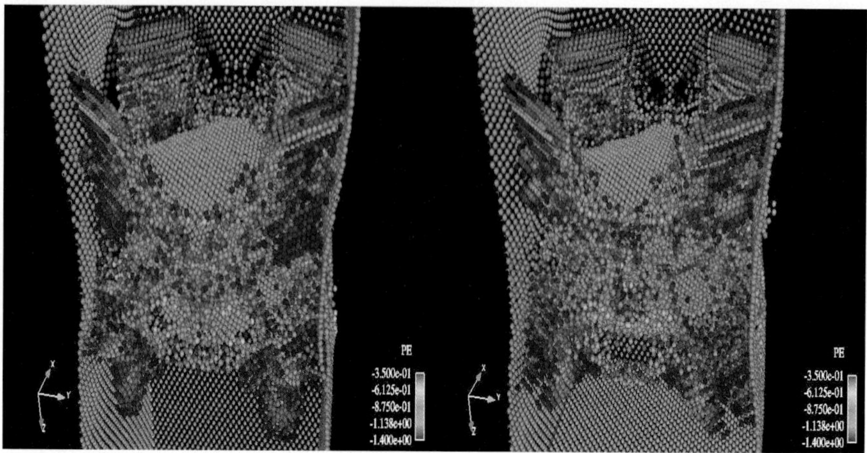

Figure 6. Final configuration in (left) full MD simulation and (right) bridging scale simulation. See Color Plate 4 on page 302.

a surface effect, i.e. the branching is caused when the crack front approaches the free surface. The branching is caused by a lack of constraint as the crack approaches the surface; in essence, a truly mode-I type loading does not exist near the surface, hence the crack is not constrained to propagate directly through the surface, and instead branches to cause material failure.

The final configurations of the bridging scale simulation and the full MD simulation are shown in Figure (6). Again, the bridging scale simulation matches the full MD simulation well, including the final configuration of the crack branches, and potential energy. Another interesting fact is that the crack branches have reached the MD boundary in the bridging scale simulations; this fact violates the postulates made in deriving the MD boundary condition, and is an example of a case in which some sort of adaptive insertion of new atoms or "remeshing" in the MD region would be necessary to continue such a large deformation failure simulation.

We note that the crack initiation time and progress during propagation appear to match correctly between both the full MD and bridging scale simulations, as measured visually. Unlike the two-dimensional multiple scale fracture simulations considered in [2], the crack tip position has not been detailed as a function of time, due to the inherent difficulty in tracking a planar crack front in three dimensions.

Finally, Figures (4-6) powerfully demonstrate the utility of multiple scale methods. In these images, it is clearly seen that only a small percentage of the atoms are largely perturbed from their equilibrium positions, even during catastrophic material failure processes such as fracture. Due to the fact that only a small percentage of the atoms play a critical role in describing the onset and subsequent propagation of cracks and defects, it seems clear that multiple scale methods are well-suited to eliminate the unnecessary atomistic degrees of freedom in favor of a continuum representation, as is done here.

6 Discussion on the Bridging Scale Method

Now that the bridging scale theory and applications have been demosntrated, a comparison between the bridging scale and the other multiscale methods is in order. The first point of comparison stems from the fact that the bridging scale is an approach inherently geared for finite temperature, dynamic problems. An approach in this direction has been undertaken in [17], in which the projection property of the bridging scale was utilized to derive a continuum temperature equation directly from the underlying atomistic motion. While it can be easily applied to static problems as in [8], the current formulation and applications have been for dynamic simulations. This lies in comparison to the quasicontinuum or CADD methods, which to date have been used only for quasi-static problems.

In comparing the bridging scale to the other dynamic multiscale methods (CGMD, MAAD), one clear advantage for the bridging scale is that the coarse scale is represented everywhere, and is *not* meshed down to the atomic scale. Furthermore, the coupled MD and FE equations of motion arise naturally from a Lagrangian formulation, which combines the kinetic and potential energies of the coupled system. The effect of this is that the coarse scale time step is not restricted by the smallest, atomic-sized elements in the mesh. More importantly, this allows the use of a staggered time integration scheme, as was detailed in [4]. Thus, the coarse scale variables can evolve on an appropriate time scale, while the fine scale variables can evolve (appropriately) on a much smaller time scale.

A major advantage of the bridging scale is that the high frequency MD waves which typically cannot be represented by the continuum are eliminated in a physically and mathematically sound manner by way of lattice mechanics arguments. The reduced MD lattice behaves as if part of a larger lattice due to the impedance force (2.63), which allows the high frequency information typically found in the MD region to naturally dissipate into the continuum. Furthermore, the time history kernel $\boldsymbol{\theta}(t)$ can be found by an automated numerical procedure which requires only standard Laplace and Fourier transform techniques. Moreover, the resulting size of $\boldsymbol{\theta}(t)$ is that of the minimum number of spatial degrees of freedom; this stands in contrast to previously derived damping kernels, e.g. [12, 18] whose size depended on the remaining number of degrees of freedom in the reduced MD region, or required that the geometry of the lattice be explicitly included in the formulation. The resulting formulation can be applied to both static and dynamics problems; examples include non-reflecting MD boundary conditons, as in [19] and [20], quasi-static nanoindentation, as in [21], multiple scale analysis of dynamic fracture in multiple dimensions, as in [2, 3], and analysis of dynamic strain localization in granular materials, as in [9].

The numerical examples shown in this review have been compared to full MD benchmark simulations. In all cases, the bridging scale simulations agree extremely well with the full MD benchmarks, even for highly nonlinear problems such as lattice fracture. These comparisons are extremely important in multiple scale simulations, as confidence in the ability of a method to faithfully reproduce the correct physics normally seen within a full MD simulation must be established.

References

1. H. S. Park, W. K. Liu, Introduction and tutorial on multiple scale analysis in solids, Computer Methods in Applied Mechanics and Engineering 193 (2004) 1733–1772.
2. H. S. Park, E. G. Karpov, W. K. Liu, P. A. Klein, The bridging scale for two-dimensional atomistic/continuum coupling, Philosophical Magazine 85 (1) (2005) 79–113.
3. H. S. Park, E. G. Karpov, P. A. Klein, W. K. Liu, Three-dimensional bridging scale analysis of dynamic fracture, Journal of Computational Physics 207 (2005) 588–609.
4. G. J. Wagner, W. K. Liu, Coupling of atomistic and continuum simulations using a bridging scale decomposition, Journal of Computational Physics 190 (2003) 249–274.
5. W. K. Liu, R. Uras, Y. Chen, Enrichment of the finite element method with the reproducing kernel particle method, Journal of Applied Mechanics 64 (1997) 861–870.
6. G. J. Wagner, W. K. Liu, Hierarchical enrichment for bridging scales and meshfree boundary conditions, International Journal for Numerical Methods in Engineering 50 (2001) 507–524.
7. L. T. Zhang, G. J. Wagner, W. K. Liu, A parallel meshfree method with boundary enrichment for large-scale cfd, Journal of Computational Physics 176 (2002) 483–506.
8. D. Qian, G. J. Wagner, W. K. Liu, A multiscale projection method for the analysis of carbon nanotubes, Computer Methods in Applied Mechanics and Engineering 193 (2004) 1603–1632.
9. H. Kadowaki, W. K. Liu, Bridging multi-scale method for localization problems, Computer Methods in Applied Mechanics and Engineering 193 (2004) 3267–3302.
10. W. T. Weeks, Numerical inversion of laplace tarnsforms using Laguerre functions, Journal of the Association for Computing Machinery 13 (3) (1966) 419–429.
11. S. A. Adelman, J. D. Doll, Generalized langevin equation approach for atom/solid-surface scattering: General formulation for classical scattering off harmonic solids, Journal of Chemical Physics 64 (1976) 2375–2388.
12. W. Cai, M. DeKoning, V. V. Bulatov, S. Yip, Minimizing boundary reflections in coupled-domain simulations, Physical Review Letters 85 (2000) 3213–3216.

13. E. Tadmor, M. Ortiz, R. Phillips, Quasicontinuum analysis of defects in solids, Philosophical Magazine A 73 (1996) 1529–1563.
14. P. A. Klein, A virtual internal bond approach to modeling crack nucleation and growth, Ph.D. Thesis (1999) Stanford University.
15. M. Arroyo, T. Belytschko, An atomistic-based finite deformation membrane for single layer crystalline films, Journal of the Mechanics and Physics of Solids 50 (2002) 1941–1977.
16. Tahoe, http://tahoe.ca.sandia.gov .
17. H. S. Park, E. G. Karpov, W. K. Liu, A temperature equation for coupled atomistic/continuum simulations, Computer Methods in Applied Mechanics and Engineering 193 (2004) 1713–1732.
18. W. E, Z. Y. Huang, A dynamic atomistic-continuum method for the simulation of crystalline materials, Journal of Computational Physics 182 (1) (2002) 234–261.
19. G. J. Wagner, E. G. Karpov, W. K. Liu, Molecular dynamics boundary conditions for regular crystal lattices, Computer Methods in Applied Mechanics and Engineering 193 (2004) 1579–1601.
20. E. G. Karpov, G. J. Wagner, W. K. Liu, A Green's function approach to deriving non-reflecting boundary conditions in molecular dynamics simulations, International Journal for Numerical Methods in Engineering 62 (9) (2005) 1250–1262.
21. E. G. Karpov, H. Yu, H. S. Park, W. K. Liu, Q. J. Wang, D. Qian, Multiscale boundary conditions in crystalline solids: Theory and application to nanoindentation, International Journal of Solids and Structures Accepted.

A New Stabilized Nodal Integration Approach

Michael Anthony Puso[1], Edward Zywicz[1], and J. S. Chen[3]

[1] Lawrence Livermore National Laboratory, Livermore CA 94550, USA
{puso, zywicz}@llnl.gov
[2] University of California at Los Angeles, Los Angeles CA 90095, USA
jschen@seas.ucla.edu

Summary. A new stabilized nodal integration scheme is proposed and implemented. In this work, focus is on the natural neighbor meshless interpolation schemes. The approach is a modification of the stabilized conforming nodal integration (SCNI) scheme and is shown to perform well in several benchmark problems.

Key words: Nodal integration, natural neighbor, particle method

1 Introduction

Nodal integration of the Galerkin weak form of the equations of motion is often desirable due to its efficiency and applicability in large deformation problems. For example, in the large deformation case, the stress and material history will move with the nodes and so no re-mapping of the quantities is required. Numerous formulations and types of meshless shape functions are available, here, we will focus on Galerkin formulations where natural neighbor (NN) interpolation is used; often referred to as the natural element method (NEM) [7]. Different nodal integration approaches have been developed such as stabilized conforming nodal integration (SCNI) [3, 10, 5] and a least squares stabilization approach [1]. The latter approach was applied using an EFG formulation and suggested the use of quadratic basis functions due to the second derivatives in the least squares term. Furthermore, the least squares formulation in [1] is not directly applicable to natural neighbor schemes since they are not smooth at the nodes. Recently we have found that the SCNI method applied to NN shape functions produces spurious low energy (not zero energy) modes in some problems and these modes did not appear to vanish with refinement. In this work, a modification to the SCNI method is made that appears to provide stable results as demonstrated in eigenvalue and large deformation examples provided here.

Natural neighbor (NN) shape functions have some distinct advantages over many meshless schemes in that they interpolate the data making it simpler to implement essential boundary conditions and the nodal adjacency only includes near neighbors. The NN/NEM approach has been implemented in the context of a meshless method by exploiting the concept of *alpha shapes* [4, 6] to determine/treat the free surface of cloud of points. In this work, the integration method in [10, 5] is modified by applying an additional stabilization term. In what follows: Section 2 introduces the global weak form, the NN interpolation scheme and the SCNI approach with added stabilization, Section 3 presents results demonstrating the necessity of the added stabilization and effectiveness of the proposed approach.

2 Formulation

The formulation will be introduced in the context of linear elasticity and the straightforward extensions to the nonlinear regime will be given at the end of the section.

2.1 Galerkin Method

Considering a body occupying the domain $\Omega \subset \mathbb{R}^3$, the governing equations of motion are given

$$\rho \ddot{\boldsymbol{u}} = \nabla \boldsymbol{\sigma} + \boldsymbol{b} \tag{2.1}$$

where $\boldsymbol{u}$ is the displacement field, $\boldsymbol{b}$ is the body force and $\boldsymbol{\sigma} = \boldsymbol{\sigma}(\varepsilon)$ is the Cauchy stress tensor in terms of strain

$$\varepsilon(\boldsymbol{u}) = 1/2(\nabla \boldsymbol{u} + \nabla \boldsymbol{u}^T). \tag{2.2}$$

Employing test function $\boldsymbol{v}$, the weak form of (2.1) can then be written

$$\int_\Omega \rho\, \boldsymbol{v} \cdot \ddot{\boldsymbol{u}}\, d\Omega + \int_\Omega \varepsilon(\boldsymbol{v}) : \boldsymbol{\sigma}(\varepsilon(\boldsymbol{v}))\, d\Omega = \int_\Omega \boldsymbol{v} \cdot \boldsymbol{b}\, d\Omega + \int_{\Gamma_t} \boldsymbol{v} \cdot \boldsymbol{t}\, d\Gamma \tag{2.3}$$

where applied tractions $\boldsymbol{t}$ are specified on the boundary $\boldsymbol{\Gamma}_t$.

2.2 Discrete Form

Following standard procedure, the discrete displacement field is defined

$$\boldsymbol{u}_h = \sum_{I=1}^{N} \phi_I(\boldsymbol{x}) \boldsymbol{u}_I \qquad \boldsymbol{x} \in \Omega_h \tag{2.4}$$

in terms of shape functions $\boldsymbol{\phi}_I$ and nodal displacements $\boldsymbol{u}_I$ over all nodes $I = 1, ..., N$ on the discretized domain Ω_h. Unlike finite elements, the definition of the domain Ω_h in meshless methods is not straightforward and is defined in what follows. Substituting (2.4) into the weak form (2.3) and applying nodal integration, the following discrete form results

$$\sum_{I=1}^{N} m_I \boldsymbol{v}_I \cdot \ddot{\boldsymbol{u}}_I + V_I \boldsymbol{\varepsilon}_I(\boldsymbol{v}_h) : \boldsymbol{\sigma}(\boldsymbol{\varepsilon}_I(\boldsymbol{u}_h)) - \boldsymbol{v}_I \cdot \boldsymbol{f}_I = 0 \tag{2.5}$$

where m_I is the nodal mass, f_I is a collocated form of the applied loads, $\boldsymbol{\varepsilon}_I$ is the nodal strain and V_I is the nodal volume. Since the NN shape functions interpolate the data, the nodal mass is simply $m_I = \rho V_I$. The body force contribution to $\boldsymbol{f}_I$ is computed similarly. Here, the traction force contribution to $\boldsymbol{f}_I$ is computed using the surface mesh but other techniques are outlined in [1]. The $\boldsymbol{\varepsilon}_I$ could be the symmetric gradient of the displacement at $\boldsymbol{x}_I$ or computed using SCNI as shown in what follows. In this work, V_I represents the volume of the Voronoi cell Ω_I about node I (Fig. 1(a)). The Voronoi diagram of a set of sites $P := \{\boldsymbol{x}_1, \boldsymbol{x}_2, \ldots \boldsymbol{x}_N\}$ is a partition such that all points $\boldsymbol{x}$ within a Voronoi cell Ω_I are closer to the generating site I than to any other site in P. Considering three dimensional space, this can be mathematically stated

$$\Omega_I = \{\boldsymbol{x} \in \mathbb{R}^3 : d(\boldsymbol{x}, \boldsymbol{x}_I) < d(\boldsymbol{x}, \boldsymbol{x}_J) \ \forall \ J \neq I\} \tag{2.6}$$

To handle arbitrary non-convex geometries, the *alpha* − *shapes* approach is used to compute a surface mesh from the cloud of points. The Voronoi diagram is then "clipped" by the surface in an approach similar to that in [4] and [6] thus producing the discretized domain Ω_h. Figure 1(b) illustrates this process for two dimensions and our implementation performs the equivalent in three dimensions. Of course, this approach may be to computationally intensive for some applications and one may resort to simpler approaches to compute nodal volumes (c.f. [1]).

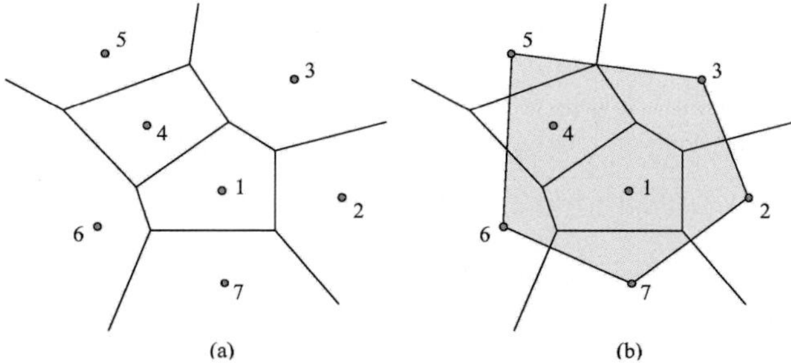

Figure 1. (a) Voronoi diagram about cloud of points. (b) Surface mesh and clipped Voronoi cells.

2.3 Natural Neighbor Interpolation

The Voronoi cell for point I is formed by polygonal (usually triangular) cell walls and the points with common cell walls are the set of natural (or nearest) neighbors $\mathcal{N}(\boldsymbol{x}_I)$ for point I. The Laplace (or non-Sibson) form of NN interpolation [8] computes interpolation functions at a point $\boldsymbol{x}$ in the following way. Consider the Voronoi diagram shown in black in Fig. 2, the interpolation function ϕ at a point $\boldsymbol{x}$ is computed by doing a *point insertion* where $\boldsymbol{x}$ is treated as an additional vertex in the set P and a new Voronoi cell is formed (shown in blue in Fig. 2(a)). For the two dimensional case, the heights h and

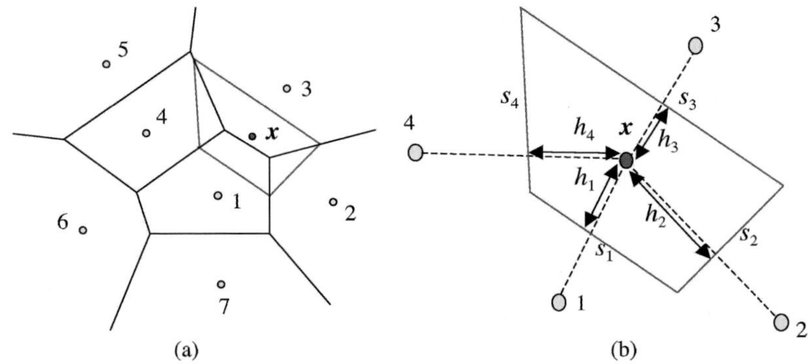

Figure 2. (a) Point insertion at x. (b) Geometry of Voronoi cell about $\boldsymbol{x}$

cell wall lengths s are used to compute the shape functions evaluated at point $\boldsymbol{x}$

$$\phi_J(\boldsymbol{x}) = \frac{s_J/h_J}{\sum_{K\in\mathcal{N}(\boldsymbol{x})} s_K/h_K} \qquad \forall J \in \mathcal{N}(\boldsymbol{x}). \tag{2.7}$$

In three dimensions, the quantity s is a cell wall area. The NN interpolation functions are both linear exact and interpolative.

2.4 Stabilized Conforming Nodal Integration (SCNI)

If smooth shape functions are used, one could merely take derivatives at the nodes to compute the nodal strain $\boldsymbol{\varepsilon}_I$ (although patch test satisfaction may be violated). Since NN shape functions are not smooth, a different approach is required. Here, the nodal strain is computed using the SCNI approach which is well described in [3, 10, 5]. In short, $\boldsymbol{\varepsilon}_I$ is computed from the volume average of the strain $\boldsymbol{\varepsilon}$ over the Voronoi cell domain Ω_I and using divergence theorem results in the surface integral

$$\boldsymbol{\varepsilon}_I(\boldsymbol{u}_h) = \frac{1}{V_I}\int_{\partial\Omega_I} (\boldsymbol{u}_h \otimes \boldsymbol{n} + \boldsymbol{n} \otimes \boldsymbol{u}_h)/2\, d\Gamma \tag{2.8}$$

where $\boldsymbol{n}$ is the unit normal along the contour surface of cell I. Numerical integration is employed in computing (2.8) such that one integration point at the centroid of each Voronoi facet is used. This approach exactly satisfies patch test and can be used without any additional stabilization in many situations as demonstrated in [3, 10, 5]. On the other hand, it is shown here that problems arise such that spurious low energy modes arise in some applications of SCNI motivating the following approach.

2.5 Modified SCNI

As mentioned, the SCNI formulation produces spurious modes as seen in Fig. 4(b). Although the SCNI approach does not have zero energy modes, a straightforward analysis can demonstrate that the method is not V_1 coercive. Referring to (2.3) and (2.5), the discrete bilinear form for linear elasticity is written

$$a(\boldsymbol{v}_h, \boldsymbol{u}_h) = \sum_{I=1}^{N} \boldsymbol{\varepsilon}_I(\boldsymbol{v}_h) : \boldsymbol{C}\boldsymbol{\varepsilon}_I(\boldsymbol{u}_h) V_I \tag{2.9}$$

where $\boldsymbol{C}$ is the elasticity tensor and the discrete strain energy results when $\boldsymbol{v}_h \doteq \boldsymbol{u}_h$. Considering a one dimensional BVP with evenly spaced nodes, the appropriate saw tooth displacement field would cause the nodal strain defined by (2.8) to be zero everywhere except at the boundaries. This would produce a zero energy mode for an infinite domain and a spurious low energy mode for a finite domain. In order to eliminate these modes the Voronoi cells at node I are further decomposed into sub-cells Ω_I^c as shown for the two dimensional case in Fig. 3. Here each sub-cell is a triangle/polyhedron formed from the generating node and a Voronoi edge/facet in two/three dimensions respectively. In practice, we can merge adjacent triangle/polyhedron into reduced set of sub-cells. Examples in this work use eight sub-cells by performing this merge process although some nodes may actually have less than eight sub-cells since they have less than eight Voronoi facets. The same SCNI approach is used to compute the strain over each sub-cell Ω_I^c

$$\boldsymbol{\varepsilon}_I^c(\boldsymbol{u}_h) = \frac{1}{V_I^c} \int_{\partial \Omega_I^c} (\boldsymbol{u}_h \otimes \boldsymbol{n} + \boldsymbol{n} \otimes \boldsymbol{u}_h)/2 \, d\Gamma \quad c = 1, ..., N_I^c \tag{2.10}$$

where N_I^c is the number of sub-cells used at node I and V_I^c is the volume of the sub-cell. The modified SCNI weak form is now proposed

$$\begin{aligned} a(\boldsymbol{v}_h, \boldsymbol{u}_h) = & \sum_{I=1}^{N} [\boldsymbol{\varepsilon}_I(\boldsymbol{v}_h) : \boldsymbol{C}\boldsymbol{\varepsilon}_I(\boldsymbol{u}_h) V_I \\ & + \sum_{c=1}^{N_I^c} \alpha_I^c (\boldsymbol{\varepsilon}_I(\boldsymbol{v}_h) - \boldsymbol{\varepsilon}_I^c(\boldsymbol{v}_h)) : \tilde{\boldsymbol{C}} (\boldsymbol{\varepsilon}_I(\boldsymbol{u}_h) - \boldsymbol{\varepsilon}_I^c(\boldsymbol{u}_h)) V^c] \end{aligned} \tag{2.11}$$

where α_I^c is a penalty parameter and $\tilde{C}$ is the elasticity tensor or some modified version. In the proceeding examples, the sub-cell penalty will be uniform i.e. $\alpha_I^c = 1$ amounting to what is sometimes referred to as "physical stabilization" [2, 9]. Here, the stiffness $\tilde{\boldsymbol{C}}$ is chosen in the following way: Isotropic elastic material: Lamé parameters μ and λ:

$$\tilde{\mu} := \mu \qquad \tilde{\lambda} := \max(\lambda, 25\tilde{\mu}) \tag{2.12}$$

Isotropic plastic material: Lamé parameter λ and plastic modulus E_T:

$$\tilde{\mu} := E_T/2 \qquad \tilde{\lambda} := \max(\lambda, 25\tilde{\mu}) \tag{2.13}$$

For non-isotropic materials, the choice for $\tilde{\mu}$ and $\tilde{\lambda}$ would be less obvious but still tractable. Using a maximum in (2.12) and (2.13) mitigates volumetric locking. The nodal deformation gradients $\boldsymbol{F}_I, \boldsymbol{F}_I^c$ are defined

$$\boldsymbol{F}_I(\boldsymbol{u}_h) = \frac{1}{V_I}\int_{\partial\Omega_I} \boldsymbol{u}_h \otimes \boldsymbol{n}\, d\Gamma \quad \boldsymbol{F}_I^c(\boldsymbol{u}_h) = \frac{1}{V_I^c}\int_{\partial\Omega_I^c} \boldsymbol{u}_h \otimes \boldsymbol{n}\, d\Gamma \quad e = 1 : N_I^c \tag{2.14}$$

and for large deformation kinematics, the internal virtual work (2.11) is redefined

$$\begin{aligned} a(\boldsymbol{v}_h, \boldsymbol{u}_h) = &\sum_{I=1}^{N} [\boldsymbol{F}_I(\boldsymbol{v}_h) : \boldsymbol{P}(\boldsymbol{F}_I(\boldsymbol{u}_h))V_I \\ &+ \sum_{e=1}^{N_I^c} \alpha_I^c(\boldsymbol{F}_I(\boldsymbol{v}_h) - \boldsymbol{F}_I^c(\boldsymbol{v}_h)) : \tilde{\boldsymbol{C}}(\boldsymbol{F}_I(\boldsymbol{u}_h) - \boldsymbol{F}_I^c(\boldsymbol{u}_h))V^c] \end{aligned} \tag{2.15}$$

where $\boldsymbol{P}$ is the first Piola Kirchoff stress and V_I and V_I^c are determined in the initial reference configuration. The numerical implementation replaces the internal virtual work (2.9) with (2.11) or (2.15) in the weak form of the equations of motion (2.3).

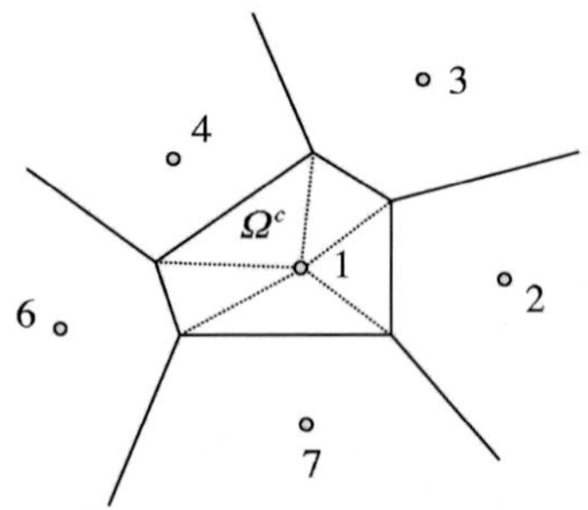

Figure 3. Subcells in the Voronoi cell formed from $\boldsymbol{x}_I$.

2.6 Implementation Details

In our implementation, the strain displacement matrices were computed at initialization and then stored to avoid re-computation of the Voronoi diagram at every time step. This also applies to the nonlinear kinematics case where the total Lagrangian approach (2.15) was employed.

The SCNI and the stabilization approach require the computation of shape functions at the Voronoi cell walls and polygons separating adjacent sub-cells for each node I. To expedite this calculation and to reduce connectivity, the smallest set of nodes were employed in evaluating shape functions:

(i) Only the set $\mathcal{N}(\boldsymbol{x}_I) \cup \mathcal{N}(\boldsymbol{x}_J)$ was used for Voronoi cell wall shape function evaluations where the cell wall was common to the Voronoi cells of node $\boldsymbol{x}_I$ and $\boldsymbol{x}_J$.
(ii) Only the set $\mathcal{N}(\boldsymbol{x}_I)$ was used for the interior sub-cell shape function evaluations.

If the above approach didn't work (i.e. the cell wasn't *closed*), more nodes were added to the set based on adjacency. Unlike changing the smoothing length of the weight functions with RKPM or EFG to get the appropriate amount of cover, adding nodes to the set to compute shape functions does not affect any previous computations.

The results here used two dimensional triangular finite element interpolation everywhere on the exterior boundary surface mesh to do shape function evaluations along the boundary. We have recently implemented two dimensional Laplace shape functions but have yet to compare the difference.

3 Results

The following results demonstrate the necessity and effectiveness of the additional stabilization. Furthermore, the proposed approach satisfies patch tests to machine precision and does not appear to lock in nearly incompressible situations. The constant pressure Q1P0 finite element is used for comparison since it performs reasonably well for the nearly incompressible case.

Eigenanalysis

This example illustrates the spurious modes using the standard SCNI approach and the improved behavior using additional stabilization. An eigenanalysis of a nearly incompressible $1 \times 1 \times 1$ block ($E = 1$, $\nu = 0.499$, $\rho = 1$) was performed and the first eigenmodes using the Q1P0 brick element, the SCNI NN approach and the proposed approach are plotted in Fig. 4. The frequency versus eigenmode is plotted in Fig. 5 and it is seen that results from the new approach are nearly identical to that of the Q1P0 for the first 100 modes. With SCNI, the frequencies for the first one hundred modes are low due to the presence of spurious modes.

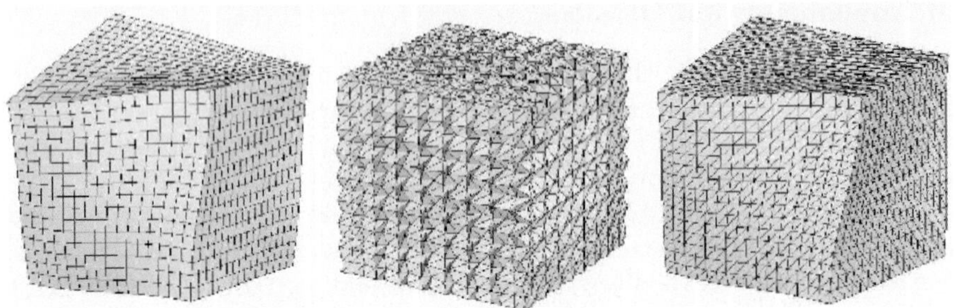

Figure 4. First eigenmode using: (a) brick elements, (b) SCNI (c) modified SCNI.

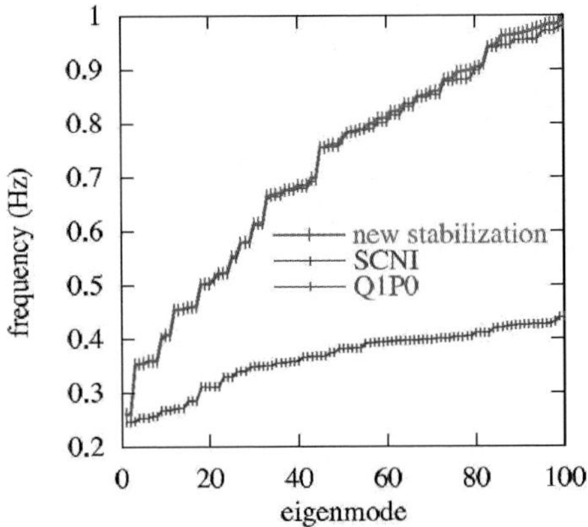

Figure 5. Frequency versus eigenmode for three approaches.

Rubber Billet

In this example, the performance on a large deformation problem is evaluated. A rubber billet 50.8 in length and diameter (Fig. 6(a)) with Neo Hookean material parameters ($E = 1$, $\nu = 0.49$) was compressed to 70% strain and only the bottom section was modeled due to symmetry. In the model, a unilateral contact surface was used to vertically constrain the outer surface as it expands (bulges) horizontally. The sequence of deformations from the proposed approach are shown in Fig. 6(a-c). The SCNI analysis terminated at 41% strain due to large oscillations (Fig. 6(d)) whereas the stress and deformation match nicely for the proposed approach and the Q1P0 hex element results at this strain threshold (Fig. 6(b,e)). Oscillations do become apparent in the last state shown in Fig. 6(c) compared to the Q1P0 results Fig. 6(e). The total force applied to the billet versus strain is shown plotted in Fig. 7 and it seen that the proposed approach matches the Q1P0 quite well.

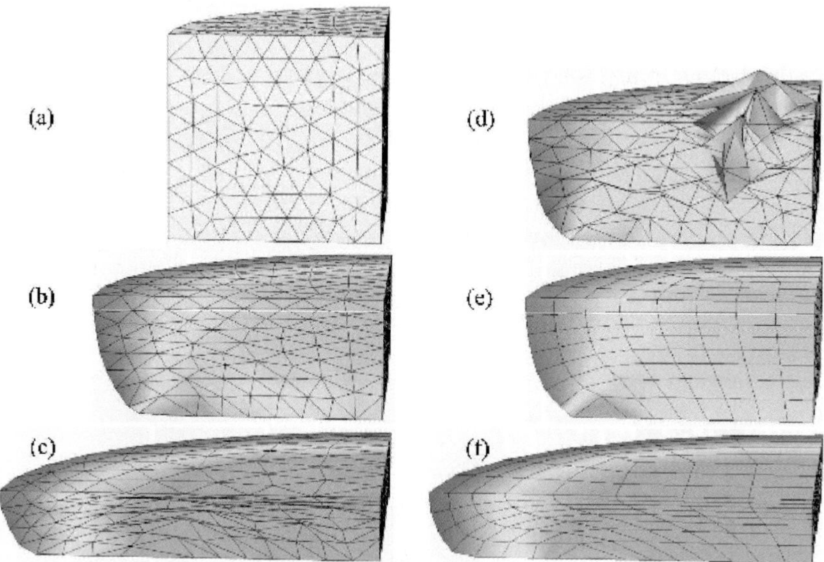

Figure 6. Rubber billet: (a-c) sequence of deformation of stabilized NEM (d) unstabilized NEM (e-f) deformed configurations of Q1P0. The vertical σ_z stress is indicated in (b-f).

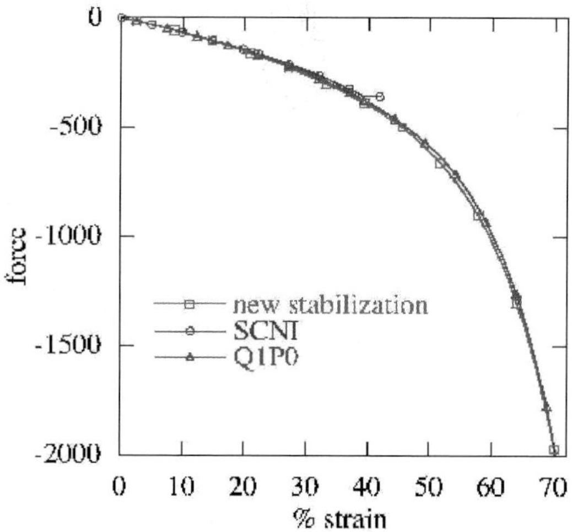

Figure 7. Force versus percent strain for rubber billet.

Taylor Bar

A standard benchmark for plasticity is the Taylor bar impact problem. Here, a copper bar impacts a rigid wall at high velocity. The following elastoplastic material properties were used: $E = 117$ GPa, $\nu = 0.35$, $\sigma_y = 0.4$ GPa, linear hardening modulus $E_T = 0.1$ GPa and density $\rho = 8930\,\mathrm{kg/m^3}$. The initial bar length was 32.4 mm, the initial radius was 3.2 mm and the initial velocity was 227 m/s. The final deformed states are shown in Fig. 8 where the color indicates displacement magnitude. Again, the new stabilization compares well with the Q1P0 finite element result but the SCNI shows oscillations.

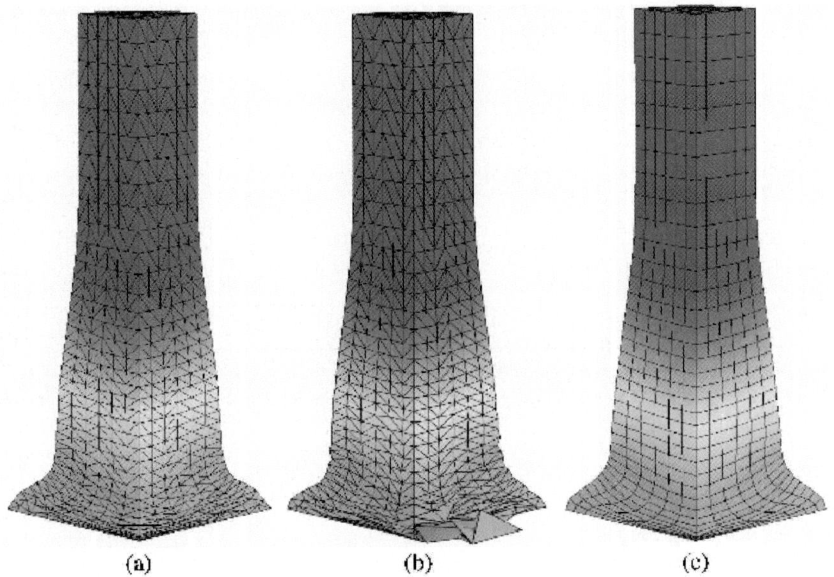

Figure 8. Final configuration and plot of displacement magnitude from (a) new stabilized (b) SCNI and (c) Q1P0 approaches.

4 Discussion

A new stabilized nodal integration approach was developed and applied to the natural neighbor meshless method. The approach performed well in eigenvalue and nonlinear benchmark problems. Here, a total Lagrangian approach was used for large deformation. Extension of the method using an "Eulerian" type kernel for extremely large deformations occurring in penetration will be investigated. The method appeared to perform well for the nearly incompressible case but would require modifications to treat the fully incompressible case.

References

1. S. Beissel and T. Belytschko, *Nodal integration of the element-free Galerkin method*, Computer Methods in Applied Mechanics and Engineering **139** (1996), 49–74.
2. T. Belytschko and Bindeman L.P., *Assumed strain stabilization of the eight node hexahedral*, Computer Methods in Applied Mechanics and Engineering **105** (1993), 225–260.
3. J.S. Chen, C.T. Wu, S. Yoon, and Y. You, *A stabilized conforming nodal integration for Galerkin mesh-free methods*, International Journal for Numerical Methods in Engineering **50** (2001), 435–466.
4. E. Cueto, B. Calvo, and M. Doblare, *Modeling three-dimensional piece-wise homogenous domains using the α-shape based Natural Element Method*, International Journal for Numerical Methods in Engineering **54** (2002), 871–897.
5. D. Gonzalez, E. Cueto, M.A. Martinez, and M. Doblare, *Numerical integration in Natural neighbor Galerkin methods*, International Journal for Numerical Methods in Engineering **60** (2003), 2077–2104.
6. S. Idelsohn, E. Onate, N. Calvo, and F. Del Pin, *The meshless finite element method*, International Journal for Numerical Methods in Engineering **58** (2003), 893–912.
7. Sukumar N., Moran B., and Belytschko T., *The Natural element method in solid mechanics*, International Journal for Numerical Methods in Engineering **43** (1998), 839–887.
8. Sukumar N., Moran B., Yu Semenov Y., and Belikov V.V., *Natural neighbor Galerkin methods*, International Journal for Numerical Methods in Engineering **50** (1998), 1–27.
9. M.A. Puso, *A highly efficient enhanced assumed strain physically stabilized hexahedral element*, International Journal for Numerical Methods in Engineering **49** (2003), 1024–1064.
10. J.W. Yoo, B. Moran, and J.S. Chen, *Stabilized conforming nodal integration in the Natural Element Method*, International Journal for Numerical Methods in Engineering **60** (2004), 861–890.

Multigrid and M-Matrices in the Finite Pointset Method for Incompressible Flows

Benjamin Seibold

Technische Universität Kaiserslautern, Fachbereich Mathematik, D-67663 Kaiserslautern, Germany
seibold@mathematik.uni-kl.de

Summary. The Finite Pointset Method (FPM) is a meshfree Lagrangian particle method for flow problems. We focus on incompressible, viscous flow equations, which are solved using the Chorin projection method. In the classical FPM second order derivatives are approximated by a least squares approximation. In general this approach yields stencils with both positive and negative entries. We present how optimization routines can force the stencils to have only positive weights aside from the central point, and investigate conditions on the particle geometry. This approach yields an M-matrix structure, which is of interest for various linear solvers, for instance multigrid. We solve the arising linear systems using algebraic multigrid.

Key words: Finite Pointset Method, meshfree finite difference method, optimization, M-matrix, algebraic multigrid.

1 Introduction

The *Finite Pointset Method* (FPM) was introduced by Kuhnert [8] in 1999 as *General Smoothed Particle Hydrodynamics*. While in the classical SPH [14, 15] space derivatives are approximated using symmetric kernels, the FPM approximates them using the *moving least squares* method, as done by Dilts [3]. Furthermore, in FPM, particles have no mass, they are merely interpolation points for the field information, such as density, velocity, pressure, energy, etc. The FPM was applied successfully to problems with complex, time-dependent geometries [10] and free surface flows [11].

Tiwari and Manservisi [13] applied the FPM to incompressible flows using the Chorin projection method [1]. Such computations require one or several (if solved implicitly) Poisson problems to be solved in each time step. Hence, efficient methods are required to solve these. We seek Poisson solvers that do not use or generate a mesh. Two aspects are of importance:

(i) **The discretization of the Poisson equation**, to obtain a linear system. We use finite differences to approximate the Laplace operator. The classical FPM approach makes use of all particles inside a given *smoothing length*. We present a new approach which takes the FPM points as candidates and selects out of them a minimum number of points, which are positioned such that the Laplace stencils are positive.

(ii) **Solving the obtained system efficiently.** We investigate a BICG solver as well as an algebraic multigrid (AMG) [18] solver. The finite difference discretization yields in general non-symmetric matrices, thus various nice properties known from symmetric matrices are not present here.

In Sect. 2 we briefly outline the FPM for incompressible flows. In Sect. 3 the finite difference approximations are given and linear minimization with sign constraints is motivated. In Sect. 4 we derive conditions on the point geometry, such that positive stencils exist. Sect. 5 deals with optimization routines for the stencil computation, and Sect. 6 concludes how to construct the M-matrix. In Sect. 7 we explain the application of multigrid to our problem, and Sect. 8 provides numerical investigations and examples.

2 FPM for the Incompressible Navier-Stokes Equation

We consider the instationary incompressible Navier-Stokes equations in Lagrangian form

$$\begin{aligned} \frac{D\boldsymbol{u}}{Dt} &= -\nabla p + \nu \Delta \boldsymbol{u} + \boldsymbol{g} \\ \nabla \cdot \boldsymbol{u} &= 0 \end{aligned} \tag{2.1}$$

with appropriate initial and boundary conditions. Here $\boldsymbol{u}$ is the velocity, ν is the kinematic viscosity, p is pressure per density, and $\boldsymbol{g}$ represents the acceleration due to external forces. Applications may involve temperature evolution equations and time-dependent boundary conditions.

We solve the above system using the *Chorin projection method* [1], as described in [9]. In each time step we move the particles and solve (2.1) as if the flow was compressible

$$\begin{aligned} \boldsymbol{x}^{n+1} &= \boldsymbol{x}^n + \Delta t\ \boldsymbol{v}^n \\ \tilde{\boldsymbol{u}}^{n+1} &= \boldsymbol{u}^n + \Delta t\ \nu\ \Delta \boldsymbol{v}^n + \Delta t\ \boldsymbol{g}^n \end{aligned} \tag{2.2}$$

then project $\tilde{\boldsymbol{u}}^{n+1}$ onto the subspace of divergence free velocity fields

$$\boldsymbol{u}^{n+1} = \tilde{\boldsymbol{u}}^{n+1} + \nabla p^{n+1},$$

where p^{n+1} solves the Poisson equation

$$-\Delta p^{n+1} = \nabla \cdot \tilde{\boldsymbol{u}}^{n+1}. \tag{2.3}$$

with possibly mixed boundary conditions Dirichlet and Neumann. If the viscosity ν is large, step (2.2) should be computed implicitly, which leads to additional Poisson problems to be solved in each time step. In such cases, the main effort can lie in solving equations of the type (3.4) on the point cloud of the particles.

3 Finite Difference Approximation

Consider the Poisson equation to be solved inside a suitable domain $\Omega \subset \mathbb{R}^d$

$$\begin{cases} \Delta u = f & \text{in } \Omega \\ u = g & \text{on } \Gamma_D \\ \frac{\partial u}{\partial n} = h & \text{on } \Gamma_N \end{cases} \tag{3.4}$$

where $\Gamma = \Gamma_D \cup \Gamma_N = \partial\Omega$. Let an arbitrary point cloud $X = \{\boldsymbol{\xi}_1, \ldots, \boldsymbol{\xi}_n\} \subset \overline{\Omega}$ be given, which consists of interior points $X_i = \{\boldsymbol{\xi}_1, \ldots, \boldsymbol{\xi}_{n_i}\} \subset \Omega$ and boundary points $X_b = \{\boldsymbol{\xi}_{n_i+1}, \ldots, \boldsymbol{\xi}_n\} \subset \partial\Omega$. A great variety of finite difference methods (as described in Chap. III in [7]) exists to convert this problem into a finite dimensional linear system

$$A\hat{\boldsymbol{u}} = \hat{\boldsymbol{f}}, \tag{3.5}$$

where $\hat{\boldsymbol{u}}$ contains approximations to $u(\boldsymbol{\xi}_i)$. The moving weighted least squares (MWLS) method [4, 12] computes coefficients, such that the solution of (3.4) is best approximated in some polynomial basis. Other approaches in the context of generalized finite difference methods (GFDM) approximate the Laplace operator by a stencil, which is computed by guaranteeing exactness for polynomials up to a given order. Nearly all methods in this context have two aspects in common: Firstly, a quadratic functional is minimized or a problem is solved in a least-squares sense, secondly, for each interior point $\boldsymbol{\xi}_i$ the Laplace stencil is obtained by solving a small linear system. In the following we slightly deviate from this path by formulating a linear minimization problem and imposing sign constraints on the stencil values.

3.1 Constraints by Taylor Expansion

Consider an arbitrary interior point of the point cloud, denoted by $\boldsymbol{\xi}_0$. A neighborhood $U(\boldsymbol{\xi}_0) = B(\boldsymbol{\xi}_0, r) = \{\boldsymbol{\xi} \in \overline{\Omega} : \|\boldsymbol{\xi} - \boldsymbol{\xi}_0\| < r\}$ is considered (due to its SPH-background, in the FPM the length r is called *smoothing length*). Let the points inside the neighborhood $U(\boldsymbol{\xi}_0)$ be numbered $(\boldsymbol{\xi}_0, \boldsymbol{\xi}_1, \ldots, \boldsymbol{\xi}_m)$. Assume r is chosen large enough, such that $m \geq k$, where $k = \frac{d(d+3)}{2}$ (the number of constraints (3.10) and (3.11)). Define the distance vectors $\boldsymbol{x}_i = \boldsymbol{\xi}_i - \boldsymbol{\xi}_0 \; \forall i = 0, \ldots, m$. Assume $u \in C^2(\Omega, \mathbb{R})$. Inside the neighborhood $B(\boldsymbol{\xi}_0, r)$ the function value at each neighboring point $u(\boldsymbol{\xi}_i)$ can be expressed by a Taylor expansion

$$u(\boldsymbol{\xi}_i) = u(\boldsymbol{\xi}_0) + \nabla u(\boldsymbol{\xi}_0) \cdot \boldsymbol{x}_i + \frac{1}{2}\nabla^2 u(\boldsymbol{\xi}_0) : \left(\boldsymbol{x}_i \cdot \boldsymbol{x}_i^T\right) + e_i \tag{3.6}$$

Here the colon denotes the matrix scalar product $A : B = \sum_{i,j} A_{ij} B_{ij}$, and e_i is the error in the expansion. A linear combination with coefficients $(\alpha_0, \ldots, \alpha_m)$ equals

$$\begin{aligned}\sum_{i=0}^{m} \alpha_i u(\boldsymbol{\xi}_i) = u(\boldsymbol{\xi}_0) \left(\sum_{i=0}^{m} \alpha_i\right) + \nabla u(\boldsymbol{\xi}_0) \cdot \left(\sum_{i=1}^{m} \alpha_i \boldsymbol{x}_i\right) \\ + \nabla^2 u(\boldsymbol{\xi}_0) : \left(\frac{1}{2}\sum_{i=1}^{m} \alpha_i \left(\boldsymbol{x}_i \cdot \boldsymbol{x}_i^T\right)\right) + \left(\sum_{i=1}^{m} \alpha_i e_i\right)\end{aligned} \tag{3.7}$$

Expression (3.7) shall approximate the Laplacian at the central point

$$\sum_{i=0}^{m} \alpha_i u(\boldsymbol{\xi}_i) = \Delta u(\boldsymbol{\xi}_0) + O(r^3) \tag{3.8}$$

This is satisfied, if the following constraints hold

$$\sum_{i=0}^{m} \alpha_i = 0 \tag{3.9}$$

$$\sum_{i=1}^{m} \boldsymbol{x}_i \alpha_i = 0 \tag{3.10}$$

$$\sum_{i=1}^{m} \left(\boldsymbol{x}_i \cdot \boldsymbol{x}_i^T\right) \alpha_i = 2I \tag{3.11}$$

Here (3.11) follows, since $\nabla^2 u(\boldsymbol{\xi}_0) : I = \Delta u(\boldsymbol{\xi}_0)$. Constraints (3.10) and (3.11) can be formulated as a linear system of equations

$$V \cdot \boldsymbol{\alpha} = \boldsymbol{b} \tag{3.12}$$

where $V \in \mathbb{R}^{k \times m}$ is the Vandermonde matrix given by $\boldsymbol{x}_1, \ldots, \boldsymbol{x}_m$, and $\boldsymbol{\alpha} \in \mathbb{R}^m$ is the sought vector of weights. In 2d the system reads as

$$V = \begin{pmatrix} x_1 & \ldots & x_m \\ y_1 & \ldots & y_m \\ x_1 y_1 & \ldots & x_m y_m \\ x_1^2 & \ldots & x_m^2 \\ y_1^2 & \ldots & y_m^2 \end{pmatrix}, \quad \boldsymbol{b} = \begin{pmatrix} 0 \\ 0 \\ 0 \\ 2 \\ 2 \end{pmatrix}$$

The central weight α_0 is then given by (3.9). In general one chooses r large enough such that $m > k$, i.e. system (3.12) is underdetermined. A minimization problem is formulated to single out a unique stencil $\boldsymbol{\alpha}$.

3.2 Quadratic Minimization

The weighted least squares approach leads to minimizing the functional

$$J_0 = \sum_{i=0}^{m} \frac{\alpha_i^2}{w_i} \quad \text{or} \quad J_1 = \sum_{i=1}^{m} \frac{\alpha_i^2}{w_i}, \tag{3.13}$$

depending on the formulation. Here $w_i = w(\|\boldsymbol{x_i}\|)$. The distance-weight function $w(d)$ is chosen to decay with d increasing, such that points closer to the central point have a larger influence. In [7] various distance-weight functions are discussed. The quadratic minimization (QM) problem (3.13) with the constraints $V \cdot \boldsymbol{\alpha} = \boldsymbol{b}$ can be solved using Lagrange multipliers

$$\boldsymbol{\alpha} = (WV^T)(VWV^T)^{-1} \cdot \boldsymbol{b} \tag{3.14}$$

Here W is a diagonal matrix containing the weights w_i. The Vandermonde matrix V is given by (3.10) and (3.11) when minimizing J_1 and additionally contains (3.9) when minimizing J_0. In the FPM the same formula (3.14) is derived [17] by minimizing $\sum_{i=0}^{m} w_i e_i^2$, where e_i are the errors given by (3.6).

For each central point, computing the stencil via (3.14) requires to solve a small $k \times k$ system, an effort that often must not be neglected. For the matrix VWV^T to be regular, some conditions on the geometry of the points have to be imposed, e.g. the points must not lie on one line. In [7] examples of such failures are provided. However, these conditions are rather relaxed compared to the conditions the existence of a positive stencil imposes (see Sect. 4).

3.3 Linear Minimization with Sign Constraints

As an alterative to QM we suggest obtaining the stencil $\boldsymbol{\alpha}$ via the linear minimization (LM) problem

$$\min \sum_{i=1}^{m} \frac{\alpha_i}{w_i}, \text{ s.t. } V \cdot \boldsymbol{\alpha} = \boldsymbol{b},\ \boldsymbol{\alpha} \geq 0, \tag{3.15}$$

Three points speak for this approach:

- QM can yield negative weights. As an example consider the central point $\boldsymbol{\xi}_0 = (0,0)$ and 6 points on the unit circle $\boldsymbol{\xi}_i = (\cos(\frac{\pi}{2}\varphi_i), \sin(\frac{\pi}{2}\varphi_i))$, where $(\varphi_1, \ldots, \varphi_6) = (0,1,2,3,0.1,0.2)$. As the neighboring points lie on the unit circle, the distance-weight function does not influence the result. QM of J_1 in (3.13) yields the stencil $\boldsymbol{\alpha}_{\text{quad}} = (0.846, 1.005, 0.998, 1.003, 0.312, -0.164)$. However, there is a solution $\boldsymbol{\alpha} \geq 0$ satisfying $V \cdot \boldsymbol{\alpha} = b$, namely $\boldsymbol{\alpha}_{\text{lin}} = (1,1,1,1,0,0)$. Indeed, this solution is obtained by formulation (3.15).
  If a positive stencil $\boldsymbol{\alpha} \geq 0$ can be obtained for every point, the system matrix in (3.5) is an M-matrix (see Sect 6 on boundary conditions). The desirability of this structure has been pointed out in [2, 4].

- Approximation (3.8) is the better, the smaller the total error term $e = \sum_{i=1}^{m} \alpha_i e_i$. It becomes particularly large, if the error terms accumulate, i.e. for $\boldsymbol{\alpha} \geq 0$ w.l.o.g. $e_i \geq 0\ \forall i$. Assumed that due to the Taylor expansion $e_i = c\|\boldsymbol{x_i}\|^3$, minimizing e yields formulation (3.15) with $w(d) = |d|^{-3}$.
- Problem (3.15) is a linear program in standard form. Assumed the constraints admit a solution (see Sect. 4), then due to the *fundamental theorem of linear programming* [6] there is a basic solution, in which only k of the m stencil entries α_i are different from zero. This yields significantly fewer matrix entries as in QM. Hence, applying the sparse system matrix in (3.5) will be significantly faster.

On the other hand, QM as presented in Sect. 3.2 has advantages. The stencil is given by a solution formula (3.14), various interpretations (MWLS, GFDM, FPM) allow to adapt desired properties, the approximating function has a good smoothness (if the distance-weight function is smooth enough [12]), etc.

4 Conditions for the Existence of Positive Stencils

Let $(\boldsymbol{x}_1, \ldots, \boldsymbol{x}_m) \subset \mathbb{R}^d$ be the point positions distributed around a central point in the origin, at which the Laplace operator should be approximated. Let V and $\boldsymbol{b}$ be defined as in Sect. 3.1. We investigate under which conditions on the points' geometry in $\mathbb{R}^d$ the underdetermined system $V \cdot \boldsymbol{\alpha} = \boldsymbol{b}$ has a solution $\boldsymbol{\alpha} \geq 0$ (which is a problem in $\mathbb{R}^k$).

Due to *Farka's Lemma* [6], $V \cdot \boldsymbol{\alpha} = \boldsymbol{b}$ has no solution $\boldsymbol{\alpha} \geq 0$, if and only if system $V^T \cdot \boldsymbol{w} \geq 0$ has a solution satisfying $\boldsymbol{b}^T \cdot \boldsymbol{w} < 0$. The i-th component of $V^T \cdot \boldsymbol{w}$ can be written as

$$\left(V^T \cdot \boldsymbol{w}\right)_i = \boldsymbol{a}^T \cdot \boldsymbol{x}_i + \boldsymbol{x}_i^T \cdot A \cdot \boldsymbol{x}_i ,$$

where $\boldsymbol{a} = (w_1, \ldots, w_d)^T$ and A is the symmetric matrix

$$A = \begin{pmatrix} w_4 & w_3 \\ w_3 & w_5 \end{pmatrix} \text{ (2d)} \quad \text{resp.} \quad A = \begin{pmatrix} w_7 & w_4 & w_5 \\ w_4 & w_8 & w_6 \\ w_5 & w_6 & w_9 \end{pmatrix} \text{ (3d)}$$

Given $\boldsymbol{w}$ (resp. $\boldsymbol{a}$ and A), we consider the function

$$f(\boldsymbol{x}) = \boldsymbol{a}^T \cdot \boldsymbol{x} + \boldsymbol{x}^T \cdot A \cdot \boldsymbol{x}$$

Since A is symmetric, an orthogonal matrix $S \in O(d)$ exists, such that $S^T A S = D$, where $D = diag\,(\lambda_1, \ldots, \lambda_d)$. In the new coordinates we define

$$g(\boldsymbol{x}) = f(S\boldsymbol{x}) = \boldsymbol{d}^T \cdot \boldsymbol{x} + \boldsymbol{x}^T \cdot D \cdot \boldsymbol{x}$$

where $\boldsymbol{d} = S^T \boldsymbol{a}$. Assumed that all eigenvalues $\lambda_i \neq 0$, we can write

$$g(\boldsymbol{x}) = (\boldsymbol{x} - \boldsymbol{c})^T \cdot D \cdot (\boldsymbol{x} - \boldsymbol{c}) - \boldsymbol{c}^T \cdot D \cdot \boldsymbol{c}$$

where $c = -\frac{1}{2}D^{-1}\boldsymbol{d}$. One can show that choosing $\boldsymbol{w} \in \mathbb{R}^k$ arbitrarily is equivalent to choosing $S \in O(d)$, $\boldsymbol{\lambda} = (\lambda_1, \ldots, \lambda_d) \in \mathbb{R}^d$ and $\boldsymbol{c} \in \mathbb{R}^d$ arbitrarily. For any $\boldsymbol{\lambda}, \boldsymbol{c} \in \mathbb{R}^d$ define the domain

$$H_{\boldsymbol{\lambda},\boldsymbol{c}} = \{\boldsymbol{x} \in \mathbb{R}^d : g(\boldsymbol{x}) \geq 0\}$$

For a set of points $X = \{\boldsymbol{x}_1, \ldots, \boldsymbol{x}_m\}$ define $SX = \{S\boldsymbol{x}_1, \ldots, S\boldsymbol{x}_m\}$. Due to *Farka's Lemma*, system $V \cdot \boldsymbol{\alpha} = \boldsymbol{b}$ has no solution $\boldsymbol{\alpha} \geq 0$, if and only if $S \in O(d)$, $\boldsymbol{c}, \boldsymbol{\lambda} \in \mathbb{R}^d$ with $\sum_{i=1}^d \lambda_i < 0$ exist, such that

$$SX \subset H_{\boldsymbol{\lambda},\boldsymbol{c}}. \tag{4.16}$$

Criterion (4.16) means that the set of points X can be transformed (via $S \in O(d)$), such that it is contained in the set $H_{\boldsymbol{\lambda},\boldsymbol{c}}$ for some $\boldsymbol{c}, \boldsymbol{\lambda} \in \mathbb{R}^d$ with $\sum_{i=1}^d \lambda_i < 0$. It is exactly equivalent to the non-existence of a positive Laplace stencil. However, due to the nonlinearity in g, it is difficult to translate condition (4.16) into geometric means. Instead, we derive a necessary (but not sufficient) as well as a sufficient (but not necessary) criterion on the point geometry for the existence of a positive Laplace stencil. To our knowledge the latter has not been given yet.

4.1 A Necessary Criterion for Positive Stencils

If $V \cdot \boldsymbol{\alpha} = \boldsymbol{b}$ has a solution $\boldsymbol{\alpha} \geq 0$, then for any $S \in O(d)$, $\boldsymbol{c}, \boldsymbol{\lambda} \in \mathbb{R}^d$ with $\sum_{i=1}^d \lambda_i < 0$, there is a point $\boldsymbol{x}_i$ with $S\boldsymbol{x}_i \notin H_{\boldsymbol{\lambda},\boldsymbol{c}}$. For the particular choice $\lambda_1 = -1$, $\lambda_i = 0 \ \forall i > 1$ and $c_1 \gg \max_i \|\boldsymbol{x}_i\|$ it follows that for any $S \in O(d)$ at least one point must satisfy $x_1 < 0$. This yields the following

Theorem 1. *If a set of points $X \subset \mathbb{R}^d$ around the origin admits a positive Laplace stencil, then they must not lie in one and the same halfspace (with respect to an arbritrary hyperplane through the origin).*

This result is well known and has been presented e.g. in [4]. Obviously, due to the particular choice of $\boldsymbol{\lambda}$ above, this criterion is very crude, but very easy to formulate in geometric means. A more careful estimate of (4.16) may yield stricter necessary criteria.

4.2 A Sufficient Criterion for Positive Stencils

For any $\boldsymbol{c}, \boldsymbol{\lambda} \in \mathbb{R}^d$ with $\sum_{i=1}^d \lambda_i < 0$ we construct a domain $G_{\boldsymbol{\lambda},\boldsymbol{c}} \supset H_{\boldsymbol{\lambda},\boldsymbol{c}}$. The domain $G_{\boldsymbol{\lambda},\boldsymbol{c}}$ will be the whole $\mathbb{R}^d$ aside from a cone centered at the origin. If we can ensure that for any $\boldsymbol{c}, \boldsymbol{\lambda} \in \mathbb{R}^d$, $S \in O(d)$ there is at least one point $S\boldsymbol{x}_i \notin G_{\boldsymbol{\lambda},\boldsymbol{c}}$, then $S\boldsymbol{x}_i \notin H_{\boldsymbol{\lambda},\boldsymbol{c}}$. Hence, a positive Laplace stencil exists.

Theorem 2 (Domain in 2d). *Let $\mathbf{c}, \boldsymbol{\lambda} \in \mathbb{R}^2$ with $\lambda_1 + \lambda_2 < 0$. There exists always a cone $C_{\boldsymbol{v}}$ defined by*

$$\boldsymbol{v} \cdot \boldsymbol{x} > \frac{1}{\sqrt{1+\beta^2}} \|\boldsymbol{x}\|,$$

where $\beta = \sqrt{2} - 1$ (a cone with total opening angle 45°, the vector $\boldsymbol{v}$ depends on $\boldsymbol{\lambda}$ and $\mathbf{c}$), such that $G_{\boldsymbol{\lambda},\mathbf{c}} = \mathbb{R}^d \setminus C_{\boldsymbol{v}}$ satisfies $H_{\boldsymbol{\lambda},\mathbf{c}} \subset G_{\boldsymbol{\lambda},\mathbf{c}}$.

Proof. We show that $H_{\boldsymbol{\lambda},\mathbf{c}}$ and $C_{\boldsymbol{v}}$ do not intersect. Since the problem is invariant under interchanging coordinates, only two cases need to be considered:

- **Case $(--)$:** $\lambda_1 < 0$, $\lambda_2 < 0$
  The set $H_{\boldsymbol{\lambda},\mathbf{c}}$ is the interior of an ellipse centered at $\mathbf{c}$ with $0 \in \partial H_{\boldsymbol{\lambda},\mathbf{c}}$. The vector $\boldsymbol{v} = -(\frac{\lambda_2}{\lambda_1} c_1, \frac{\lambda_1}{\lambda_2} c_2)$ is the outer normal vector to the ellipse. Obviously the cone $C_{\boldsymbol{v}}$ touches the ellipse only at the origin.
- **Case $(+-)$:** $\lambda_1 > 0$, $\lambda_2 < 0$
  Define $\mu_1 = \frac{|\lambda_1|}{|\lambda_2|} < 1$. The domain $H_{\boldsymbol{\lambda},\mathbf{c}}$ is defined by

  $$\tilde{g}(x_1, x_2) = \mu_1(x_1^2 - 2c_1 x_1) - (x_2^2 - 2c_2 x_2) \geq 0$$

  Due to symmetry we can assume $c_1, c_2 \geq 0$. For all $\boldsymbol{x} \in B$, where $B = \{(x_1, x_2) | x_1 > 0, x_2 < 0, |x_1| < |x_2|\}$, the function $\tilde{g}$ satisfies

  $$\begin{aligned} \tilde{g}(\boldsymbol{x}) &= \mu_1(|x_1|^2 - 2c_1|x_1|) - (|x_2|^2 + 2c_2|x_2|) \\ &< (\mu_1 - 1)|x_2|^2 - 2(\mu_1 c_1 |x_1| + c_2|x_2|) < 0, \end{aligned}$$

  hence $H_{\boldsymbol{\lambda},\mathbf{c}} \cap C_{\boldsymbol{v}} = \emptyset$. The domain B is a 2d cone with opening angle 45°, where $\boldsymbol{v} = (\frac{1}{2}\sqrt{2-\sqrt{2}}, \frac{1}{2}\sqrt{2+\sqrt{2}})$, which proves the claim.

The case $\lambda_1 = 0$ reduces to the trivial 1d case.

Theorem 3 (Domain in 3d). *Let $\mathbf{c}, \boldsymbol{\lambda} \in \mathbb{R}^3$ with $\lambda_1 + \lambda_2 + \lambda_3 < 0$. There exists always a cone $C_{\boldsymbol{v}}$ defined by*

$$\boldsymbol{v} \cdot \boldsymbol{x} > \frac{1}{\sqrt{1+\beta^2}} \|\boldsymbol{x}\|,$$

where $\beta = \sqrt{\frac{1}{6}(3-\sqrt{6})}$ (a cone with total opening angle 33.7°), such that $G_{\boldsymbol{\lambda},\mathbf{c}} = \mathbb{R}^d \setminus C_{\boldsymbol{v}}$ satisfies $H_{\boldsymbol{\lambda},\mathbf{c}} \subset G_{\boldsymbol{\lambda},\mathbf{c}}$.

Proof. The following cases need to be considered:

- **Case $(---)$:** $\lambda_1 < 0$, $\lambda_2 < 0$, $\lambda_3 < 0$
  This case is analogous to the 2d case.

- **Case** $(++-)$**:** $\lambda_1 > 0,\ \lambda_2 > 0,\ \lambda_3 < 0$
Define $\mu_1 = \frac{|\lambda_1|}{|\lambda_3|} < 1$, and $\mu_2 = \frac{|\lambda_2|}{|\lambda_3|} < 1$. The domain $H_{\boldsymbol{\lambda},\boldsymbol{c}}$ is defined by

$$\tilde{g}(x_1, x_2, x_3) = \mu_1(x_1^2 - 2c_1x_1) + \mu_2(x_2^2 - 2c_2x_2) - (x_3^2 - 2c_3x_3) \geq 0$$

Due to symmetry we can assume $c_1, c_2, c_3 \geq 0$. For all $\boldsymbol{x} \in B$, where $B = \{(x_1, x_2, x_3)|x_1, x_2 > 0, x_3 < 0, |x_1|, |x_2| < \sqrt{\frac{1}{2}}|x_3|\}$, the function $\tilde{g}$ satisfies

$$\begin{aligned}\tilde{g}(\boldsymbol{x}) &= \mu_1(|x_1|^2 - 2c_1|x_1|) + \mu_2(|x_2|^2 - 2c_2|x_2|) - (|x_3|^2 + 2c_3|x_3|) \\ &< (\tfrac{1}{2}(\mu_1 + \mu_2) - 1)|x_3|^2 - 2(\mu_1c_1|x_1| + \mu_2c_2|x_2| + c_3|x_3|) < 0.\end{aligned}$$

Note that B is not a cone. However, a 3d cone can always be contained inside B. Some geometric considerations yield that the cone with maximum opening angle contained inside B is given by $\beta = \sqrt{\frac{1}{6}(3 - \sqrt{6})}$ and $\boldsymbol{v} = \frac{1}{\sqrt{41-16\sqrt{6}}}\left(2(\sqrt{3} - \sqrt{2}), 2(\sqrt{3} - \sqrt{2}), 1\right)$.
- **Case** $(+--)$**:** $\lambda_1 > 0,\ \lambda_2 < 0,\ \lambda_3 < 0$
Define $\mu_2 = \frac{|\lambda_2|}{|\lambda_1|}$ and $\mu_3 = \frac{|\lambda_3|}{|\lambda_1|}$. Since $\mu_2 + \mu_3 > 1$, we assume w.l.o.g. $\mu_3 \geq \frac{1}{2}$. The domain $H_{\boldsymbol{\lambda},\boldsymbol{c}}$ is defined by

$$\tilde{g}(x_1, x_2, x_3) = (x_1^2 - 2c_1x_1) - \mu_2(x_2^2 - 2c_2x_2) - \mu_3(x_3^2 - 2c_3x_3) \geq 0$$

Due to symmetry we can assume $c_1, c_2, c_3 \geq 0$. For all $\boldsymbol{x} \in B$, where $B = \{(x_1, x_2, x_3)|x_1 > 0, x_2, x_3 < 0, |x_1|, |x_2| < \sqrt{\frac{1}{2}}|x_3|\}$ the function $\tilde{g}$ satisfies

$$\begin{aligned}\tilde{g}(\boldsymbol{x}) &= (|x_1|^2 - 2c_1|x_1|) - \mu_2(|x_2|^2 + 2c_2|x_2|) - \mu_3(|x_3|^2 + 2c_3|x_3|) \\ &= \underbrace{(|x_1|^2 - \mu_2|x_2|^2 - \mu_3|x_3|^2)}_{<(\frac{1}{2}-\mu_3)|x_3|^2 \leq 0} - 2(\mu_1c_1|x_1| + \mu_2c_2|x_2| + c_3|x_3|) < 0.\end{aligned}$$

Here the domain B is the same as in case $(++-)$, merely reflected at the x_1, x_3 plane. Hence, a 3d cone can be placed in the same way.

The cases $\lambda_1 = 0$ or $\lambda_2 = 0$ reduce to the 2d case.

Theorem 4. *Let the point cloud have a mesh size $h_{\max}$ such that $\forall \boldsymbol{x} \in \Omega \exists \boldsymbol{x}_i \in X : \|\boldsymbol{x}_i - \boldsymbol{x}\| < h_{\max}$ (the point cloud has no holes larger than $h_{\max}$, see [12] for an equivalent definition). Let γ be the opening angle of the cone derived in Thm. 2 and Thm. 3. If the smoothing length satisfies*

$$r > \frac{1}{\sin(\gamma/2)} h_{\max}$$

then for every interior point there exists a positive stencil.

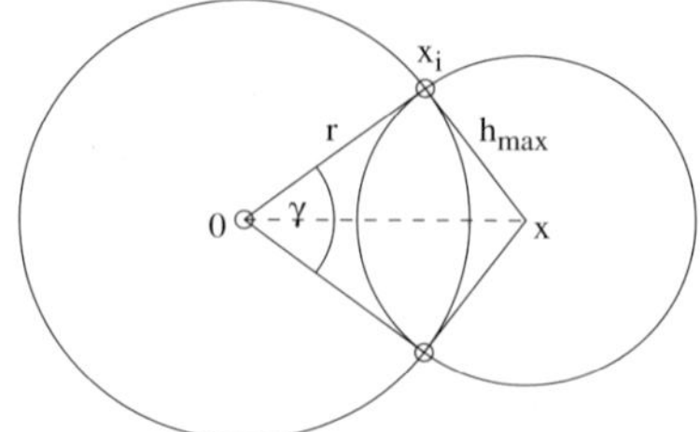

Figure 1. Relation between *smoothing length* and mesh size

Proof. Fig. 1 shows a ball with radius r around the central point and a cone with opening angle γ. Assumed, the cone contains no point, then there must be a ball of radius $h_{\max}$ which contains no points. The claim follows by considering the triangle $(0, \boldsymbol{x}, \boldsymbol{x}_i)$.

The condition on the mesh size given in Thm. 4 is guaranteed in the FPM by constant inserting of points. Hence, one can enforce the Laplace stencils for every interior point to be positive by choosing the *smoothing length* r large enough. The ratios given by Thm. 4 are

$$\frac{r}{h_{\max}} > \sqrt{1 + \frac{1}{\beta^2}} = \begin{cases} \sqrt{4 + 2\sqrt{2}} & = 2.61 \quad \text{in 2d} \\ \sqrt{7 + 2\sqrt{6}} & = 3.45 \quad \text{in 3d} \end{cases}$$

Note that the estimates in Thms. 2 and 3 are not sharpest possible. Indeed, one can derive smaller ratios. However, the algebra becomes significantly more complex. Additionally, as a sufficient criterion only, positive stencils can very well be obtained for significantly smaller ratios, e.g. if the point geometry is particularly nice. We shall provide sharper estimates in future work.

5 Optimization for Interior Points

For every interior point we consider points in the neighborhood, set up the linear program (3.15) and solve using the two stage simplex algorithm [6]. It is efficient to first choose few candidate points. If the simplex yields no feasible solution, then the *smoothing length* is increased to the value given by Thm. 4.

Our optimization problem is rather atypical, since we are confronted with lots of small, non-sparse systems. The effort of the optimization routines has to compare with the effort for the matrix multiplications and Gauß elimination for computing the stencils via (3.14). Assumed that the number of simplex steps roughly equals the number of constraints , one can estimate that the numerical effort should be of the same order. In Sect. 8.1 we show computational results.

The simplex algorithm is implemented into the FPM Fortran-code. We compare the FPM-simplex the with different routines in the optimization

Table 1. Optimization Run Times

	fpm simplex	primal simplex	dual simplex	barrier method
presolve on		17s	14s	38s
presolve off	12s	23s	20s	39s

software CPLEX. The run times for 42000 interior points are shown in Table 1. The FPM simplex performs well in comparison with the CPLEX routines. We assume that initialization and checks in CPLEX are not negligible for the small problems. Optimization heuristics (CPLEX presolve) do not improve the results. Finally, communication requires extra time, which is not considered here. Interior point methods (here barrier methods) perform worse than the simplex method for problems of the type (3.15).

6 Setting Up the System Matrix

In the previous sections we have shown that for interior points positive stencils can be guaranteed. Points with Dirichlet boundary conditions are also no problem. For each such point, an identity row in included into the system matrix. This way the matrix size equals the number of points. One can alternatively eliminate the Dirichlet points from the system matrix.

Neumann boundary conditions are included in a similar manner as interior points. At the boundary point, $\frac{\partial u}{\partial n}$ is approximated using neighboring points. In many applications, a second order accurate approximation is required. The stencil must satisfy

$$\begin{pmatrix} x_1 & \cdots & x_m \\ y_1 & \cdots & y_m \\ x_1 y_1 & \cdots & x_m y_m \\ x_1^2 & \cdots & x_m^2 \\ y_1^2 & \cdots & y_m^2 \end{pmatrix} \cdot \boldsymbol{\alpha} = \begin{pmatrix} n_x \\ n_y \\ 0 \\ 0 \\ 0 \end{pmatrix} \qquad \text{(2d)}$$

which is incompatible with a positive stencil, since then the quadratic conditions cannot be satisfied. A first order approximation does not consider the quadratic rows, thus an $\boldsymbol{\alpha} \geq 0$ can always be chosen.

In principle, problem (3.4) can always be approximated with an M-matrix. In applications, however, one often approximates Neumann boundary conditions with second order accuracy, sacrificing the M-matrix property. We do so in the applications presented in Sect. 8.2, and observe that this does not significantly worsen the performance when solving system (3.5). On the contrary, choosing too few points in the Neumann stencils sometimes leads to instabilities.

Figure 2 shows the sparse matrix structure for QM, Fig. 3 shows the structure for LM. Obviously the LM-matrix is significantly sparser. Note that

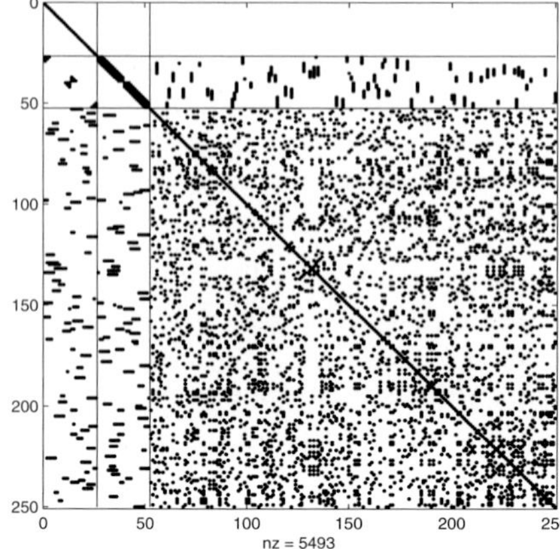

Figure 2. Quadratic minimization

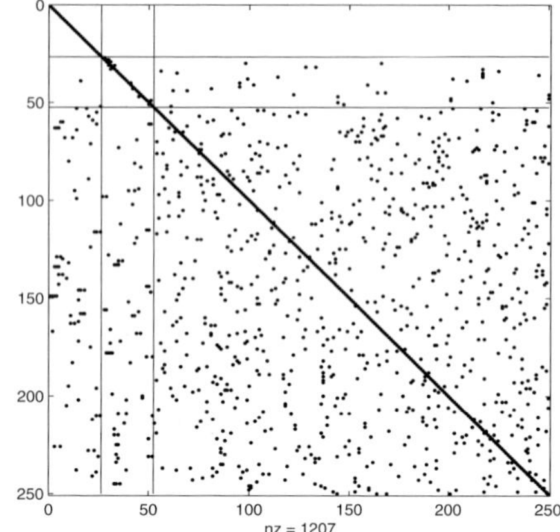

Figure 3. Linear minimization

the QM-matrix is symmetrically structured ($a_{ij} = 0 \Leftrightarrow a_{ji} = 0$). The LM-matrix loses this property. Apart from the numerical results in Sect. 8.1 we do not know whether this property influences the performance of solvers.

7 Algebraic Multigrid

As in many methods for incompressible flows, multigrid is of interest for the FPM. Especially for large problems, the optimal effort $O(n)$ is desirable. We consider multigrid as a solver for the problems (2.3) and linear problems due to implicit time steps. Various approaches exist on how to apply multigrid to such linear problems [18]. Since in the FPM the point geometry is given by the fluid flow, working directly geometrically is complicated. Instead, we consider Algebraic Multigrid (AMG), as presented in App. A in [18]. Advantages are

- Discretizing (2.3) and solving (3.5) can be treated (almost) independently of each other. AMG can even be used as a black box solver.
- Points cloud coarsening and appropriate restriction and prolongation are done automatically. One does not have to consider the geometry.

Of course, these advantages can be disadvantages. With geometric multigrid one has more control on the geometry, more flexibility, and coarsening is easier to interpret. Additionally, setting up the graph and selecting the C/F-variables in AMG is an initial overhead.

We use the SAMG code of the *Fraunhofer Institut für Algorithmen und Wissenschaftliches Rechnen* (SCAI). The code is powerful and flexible. Various AMG versions described in [16], such as aggregation-type AMG are available.

Some good results and estimates have been given for AMG applied to the class of symmetric M-matrices [16, 18]. Unfortunately, the matrices obtained in the FPM are non-symmetric, which no AMG theory is available for. As shown in Sect. 6, we can enforce an M-matrix structure, but the matrices remain non-symmetric. On the other hand, we approximate the Laplace operator, so we hope that some of the good properties for symmetric M-matrices carry over. The numerical investigations in Sect. 8 give some insight.

8 Numerical Results

We apply the derived methods to a Poisson model problem in order to compare run times and performance, as well as to industrial applications in order to investigate whether the LM approximation performs stable and robust.

8.1 Poisson Test Problem

We consider a 3d cylinder with its top deformed into a tip. Typical FPM point clouds are generated. We impose Dirichlet boundary conditions and a right hand side, which arises from solving a projection step (2.3). We generate the linear system (3.5) first by QM as described in Sect. 3.2, second by LM as described in Sect. 6. The latter yields an M-matrix, the former does not. Fig. 4 shows the time required for setting up the linear system, depending on the number of points. The linear optimization as described in Sect. 5 is by a factor of 1.2 slower than computing (3.14). Both types of linear systems are solved by the stabilized BICG method with ILU preconditioning, as well as by the AMG code described in Sect. 7. The corresponding run times are shown in Fig. 6 and Fig. 7. Two observations can be made:

- The AMG solver is significantly faster than the BICG solver. Additionally, the dependence of run time on problem size is approximately linear for AMG, and slightly superlinear for BICG. While the latter meets our expectations, the former surprises, as a right hand in a projection step should not have too large smooth error components [5]. AMG seems to fit the problem well. Algebraic smoothness need not coincide with geometric smoothness [18]. Aggregation [16] also speeds up solving.
- The M-matrix property does not improve the performance of the solver. The matrices obtained by LM are solved significantly faster due to the matrices' higher sparsity. In Fig. 6 and 7 we show a solve with the same LM matrices, but structured as the QM matrices (i.e. zeros are saved). The run times equal the QM run times. Obviously, the QM matrices are rather "well behaved", although they are no M-matrices. Still, LM yields a significant speedup and a much better memory performance (see Fig. 5).

8.2 Industrial Applications

We consider two industrial simulations of incompressible viscous flows, which the FPM was successfully applied to, and replace the QM by our LM.

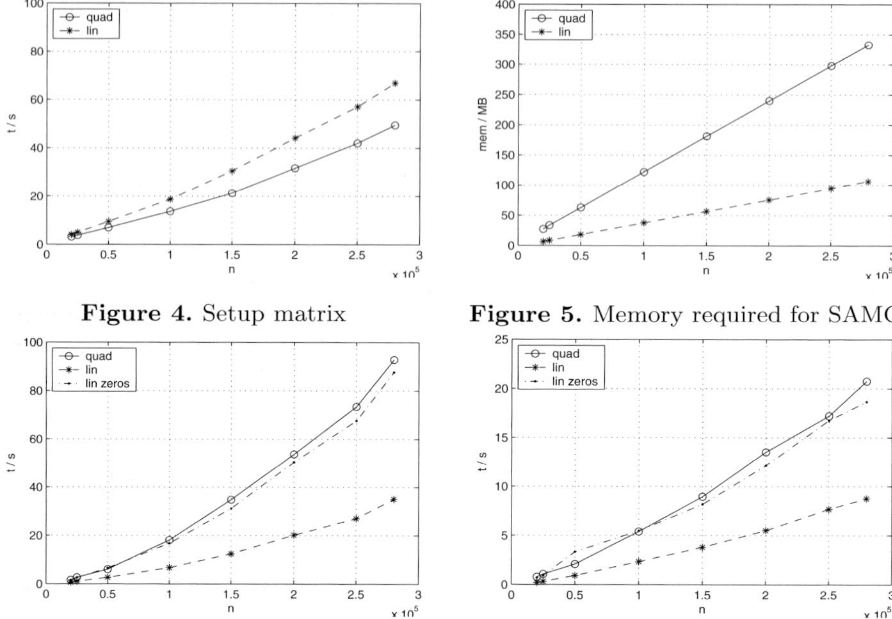

Figure 4. Setup matrix

Figure 5. Memory required for SAMG

Figure 6. Solve with BICGstab

Figure 7. Solve with SAMG

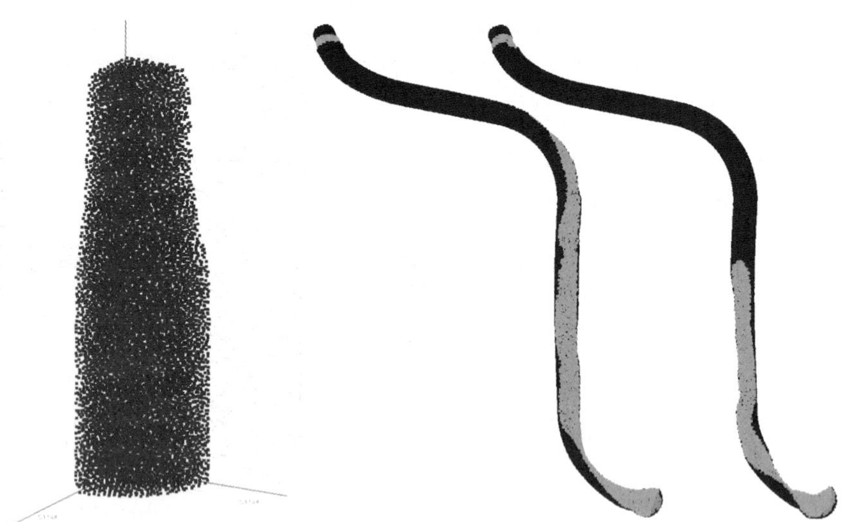

Figure 8. Glass melting

Figure 9. Tank filling

- **Melting of a glass cylinder.** Figure 8 shows the point cloud of a melting glass cylinder. The geometry is rather well behaved, there are many interior points. Indeed, the FPM works well with LM. The code runs stable and there is no visible difference between the LM and QM results.
- **Filling of a gasoline tank.** Figure 9 shows the same setup, on the left computed with QM, on the right computed with LM. The geometry is complex, many particles are boundary points, many free surfaces are present. The FPM with LM runs stable. However, the results deviate significantly. Since no analytical solution is available, we cannot prefer one to the other solution. However, one can observe problems with LM when two free surfaces are close to each other compared to the *smoothing length*. Since LM selects a minimal stencil, not enough boundary points might be selected. We will investigate this problem further.

9 Conclusions and Outlook

We have presented an alternative approach for a finite difference approximation of the Poisson equation, based on linear optimization. It guarantees an M-matrix which is significantly sparser than matrices obtained by the classical least-squares approximation. We proved that under reasonable assumptions on the point cloud, the M-matrix structure can always be achieved. Numerical tests showed that the linear optimization approach speeds up solving Poisson problems significantly. We investigated algebraic multigrid as a solver for linear systems which arise in the FPM for incompressible flows, discretized both by least-squares and by linear optimization. The SAMG solver could solve the given linear systems significantly faster than a BICGstab solver, and showed to be work well with the FPM.

We will further investigate the linear optimization approach. Especially the "moving" approach known from least-squares is not clear here. An important issue is the performance of AMG on the arising non-symmetric M-matrices. Neumann boundary conditions still cause problems. An interesting question (also for interior points) is how to construct good positive stencils, which have slightly more points than the minimum number obtained with linear optimization. Another question to investigate is which geometric operations AMG actually performs on the particles, and whether geometric multigrid approaches can compare with or beat AMG here. We will investigate linear optimization and AMG in a time dependent problem with complex geometries.

Acknowledgement. We would like to thank Axel Klar, Sven Krumke, Jörg Kuhnert, Helmut Neunzert, Klaus Stüben and Sudarshan Tiwari for helpful discussions and comments and an unknown referee for helpful comments. This work was supported by the DFG Graduiertenkolleg "Mathematik und Praxis".

References

1. A.J. Chorin, *Numerical solution of the Navier-Stokes equations*, Math. Comput. **22** (1968), 745–762.
2. L. Demkowicz, A. Karafiat, and T. Liszka, *On some convergence results for FDM with irregular mesh*, J. Comp. Meth. Appl. Mech. **42** (1984), 343–355.
3. G.A. Dilts, *Moving-least-squares-particle hydrodynamics – I. consistency and stability*, Int. J. Numer. Meth. Eng. **44** (1999), 1115–1155.
4. L. Duarte, T. Liszka, and W. Tworzyako, *H-p meshless cloud method*, J. Comp. Meth. Appl. Mech. **139** (1996), no. 1–7, 263–288.
5. M. Griebel, T. Dornseifer, and T. Neunhoeffer, *Numerical simulation in fluid dynamics: a practical introduction*, Society for Industrial and Applied Mathematics, 1998.
6. H.W. Hamacher and K. Klamroth, *Linear and network optimization*, Vieweg, 2000.
7. M. Kleiber, *Handbook of computational solid mechanics*, Survey and Comparison of Contemporary Methods, Springer, 1998.
8. J. Kuhnert, *General smoothed particle hydrodynamics*, Dissertation, Department of Mathematics, University of Kaiserslautern, 1999.
9. J. Kuhnert and S. Tiwari, *Finite pointset method based on the projection method for simulations of the incompressible Navier-Stokes equations*, Meshfree methods for Partial Differential Equations (M. Griebel and M.A. Schweitzer, eds.), vol. 26, Springer LNCSE, 2002, pp. 373–388.
10. ______, *A meshfree method for incompressible fluid flows with incorporated surface tension*, Revue aurope'enne des elements finis **11/7–8** (2002), 965–987.
11. ______, *Particle method for simulation of free surface flows*, Proceedings of the Ninth International Conference on Hyperbolic Problems (T. Hou and E. Tadmor, eds.), Springer, 2003, pp. 889–898.
12. D. Levin, *The approximation power of moving least-squares*, Math. Comput. **67** (1998), 1517–1531.
13. S. Manservisi and S. Tiwari, *Modeling incompressible flows by least-squares approximation*, The Nepali Math. Sc. Report **20, 1&2** (2002), 1–23.
14. J.J. Monaghan, *An introduction to SPH*, Comput. Phys. Commun. **48** (1988), 89–96.
15. J.J. Monaghan, *Smoothed particle hydrodynamics*, Ann. Rev. Astron. Astrophys. **30** (1992), 543–574.
16. K. Stüben, *A review of algebraic multigrid*, Journal of Computational and Applied Mathematics **128** (2001), 281–309.
17. S. Tiwari, *A LSQ-SPH approach for solving compressible viscous flows*, Intern. series of num. math. **141** (2000).
18. U. Trottenberg, C.W. Oosterlee, and A. Schüller, *Multigrid*, Academic Press, 2001.

Assessment of Generalized Finite Elements in Nonlinear Analysis

Yuichi Tadano and Hirohisa Noguchi

Keio University, 3-14-1, Hiyoshi, Kouhoku-ku, Yokohama 223-8522, Japan
{tadano, noguchi}@mech.keio.ac.jp

Summary. This paper presents two different formulations for in-plane generalized finite elements for geometrical non-linear analysis. The results from analyses which employ the proposed elements are also presented. One of the proposed elements has one additional degree of freedom at each node and shows good performance for analysis in which bending deformation is dominant. The other can reproduce quadratic deformation mode with only corner nodes and it has no linear dependency, which is a well known problem of generalized finite elements. The formulation is based on the rate form of the virtual work principle and obtained by a simple extension of standard FEM. The convergence of analytical solutions and the robustness against element distortion are investigated and the results are compared with those of standard displacement based first and second order elements. In most cases, the proposed elements provide good solution convergence which is similar to, if not better than, those of conventional second order elements. Additionally, it is also shown that high-precision solutions can be obtained even if the mesh is strongly distorted.

Key words: Geometrical Nonlinear Analysis, Compatible Element, Element Distortion, Locking Free, Generalized Finite Element Method.

1 Introduction

To improve the accuracy of the finite element method (FEM), many kinds of approaches have been proposed from past to present and one practical and well-known way is the generalized finite element method (GFEM). GFEM was introduced by several researches in the 80's [1, 2]. The main concept of GFEM is that the precision of solution can be drastically enhanced by introducing additional nodal degrees of freedom.

In the recent decade, the enhancement of GFEM based on the partition of unity method (PUM) which was first introduced by Babuška [3] has been actively worked on [4, 5, 6, 7, 8]. The idea of the partition of unity approach has good compatibility with the GFEM framework although it was originally used in a kind of meshfree method [9]. It has been found that GFEM with PUM can drastically improve the accuracy of FEM by simple extension of the conventional finite element formulation. It is also known that the conventional FEM is a special case in the GFEM framework [10]. On the other hand, it also has several numerical shortcomings, for example, the computational cost is expensive due to increased degrees of freedom and the stiffness matrix becomes singular because of the linear dependency.

In the mean time, as another approach to enhance the performance of FEM, finite element with drilling freedom was first proposed by Allman [11], which is an extension of a triangular element and many modification is presented after that [12, 13]. In Allman's element, each node of elements has an independently additional degree of freedom which describes in-plane rotation and the accuracy of solution is enhanced in several problems, especially bending deformation. Recently, Sekiguchi and Kikuchi reviewed and reexamined elements with drilling freedom and proposed a new element formulation including drilling freedom in the framework of the standard displacement method [14]. This element can be interpreted as an element based on the GFEM framework.

Despite the outstanding performance of GFEM, most studies focus only on linear problems and only a few nonlinear analyses are reported. For example, Barros et al. incorporated GFEM into material nonlinear analysis [15] and Terada et al. analyzed discontinuous surface and debonding problems with the finite cover method, which is a kind of GFEM [16]. However, to date, there have been no studies taking into consideration geometrical nonlinearity in GFEM.

In this paper, we propose two kinds of in-plane finite elements in the framework of GFEM and discuss geometrically nonlinear analyses conducted with these. We also discuss investigations on the convergence of analysis solution and the robustness against element distortion through comparison with first and second order elements based on the standard displacement method.

2 Formulation

2.1 Governing Equations

In this study, the formulation is based on the rate form of equilibrium equation. Therefore, the governing equation is the rate form of Cauchy's first law of motion;

$$\nabla \cdot \dot{\boldsymbol{\pi}} + \rho \dot{\mathbf{g}} = \mathbf{0} \tag{2.1}$$

In the above equation, the inertia force is neglected because only static condition is considered in the present analysis. $\boldsymbol{\pi}$, ρ, and $\mathbf{g}$ are the first Piola-Kirchhoff stress tensor, density, and body force vector respectively. They are functions of the coordinates. ∇ is the gradient operator and superposed dots denote material time derivative.

The rate form of the virtual work principle based on (2.1), in which volumetric force is neglected, is given as;

$$\int_v \dot{\boldsymbol{\pi}} : \delta \mathbf{L} = \int_s \dot{\mathbf{t}} \cdot \delta \dot{\mathbf{u}} \mathrm{d}s \tag{2.2}$$

where $\mathbf{L}$, $\mathbf{t}$, and $\mathbf{u}$ are the velocity gradient tensor, traction vector, and displacement vector respectively. δ means virtual quantities and v and s are current volume and surface. As for constitutive equation, material objectivity should be considered because finite rotation occurs in geometrical nonlinear analysis. Therefore, in the present formulation, the following constitutive relation between the deformation rate tensor (i.e. the symmetric part of velocity gradient) $\mathbf{D}$ and Jaumann rate of the Cauchy stress $\boldsymbol{\sigma}$ is assumed.

$$\overset{\circ}{\boldsymbol{\sigma}} = \mathbf{C} : \mathbf{D} \tag{2.3}$$

$$\overset{\circ}{\boldsymbol{\sigma}} = \dot{\boldsymbol{\sigma}} - \mathbf{W}\boldsymbol{\sigma} + \boldsymbol{\sigma}\mathbf{W} \tag{2.4}$$

$\mathbf{C}$ is the fourth order constitutive tensor and $\mathbf{W}$ is the asymmetric part of velocity gradient tensor. There is the following relation between the first Piola-Kirchhoff stress rate and the Cauchy stress rate.

$$\dot{\boldsymbol{\pi}} = \dot{\boldsymbol{\sigma}} - \mathbf{L}\boldsymbol{\sigma} + (\mathrm{tr}\mathbf{L})\,\boldsymbol{\sigma} \tag{2.5}$$

By substituting (2.3) and (2.4) into the above, the following equation can be obtained.

$$\dot{\boldsymbol{\pi}} = \mathbf{C} : \mathbf{D} - \mathbf{D}\boldsymbol{\sigma} - \boldsymbol{\sigma}\mathbf{D} + \boldsymbol{\sigma}\mathbf{L}^{\mathrm{T}} + (\mathrm{tr}\mathbf{L})\,\boldsymbol{\sigma} \tag{2.6}$$

where the relation $\mathbf{W}\boldsymbol{\sigma} - \boldsymbol{\sigma}\mathbf{W} - \mathbf{L}\boldsymbol{\sigma} = -\mathbf{D}\boldsymbol{\sigma} - \boldsymbol{\sigma}\mathbf{D} + \boldsymbol{\sigma}\mathbf{L}^{\mathrm{T}}$ is used. By substituting (2.6) into (2.2) and separating terms with respect to the deformation rate and velocity gradient, the following equation can be obtained.

$$\int_v \left[(\mathbf{C} : \mathbf{D} - \mathbf{D}\boldsymbol{\sigma} - \boldsymbol{\sigma}\mathbf{D}) : \delta\mathbf{D} + \left\{ \boldsymbol{\sigma}\mathbf{L}^{\mathrm{T}} + (\mathrm{tr}\mathbf{L})\,\boldsymbol{\sigma} \right\} : \delta\mathbf{L} \right] \mathrm{d}v = \int_s \dot{\mathbf{t}} \cdot \delta\dot{\mathbf{u}} \mathrm{d}s \tag{2.7}$$

2.2 Interpolation Functions

The general expression of interpolation function in GFEM is given as

$$u = \sum_{n=1}^{m} \left[\sum_j L_j a_j^n \right] N^n \tag{2.8}$$

Here, L_j and N^n are polynomial terms and shape function respectively. a_j^n is an unknown coefficient. m is the number of nodes in the element and superscript n denotes the values at node. Equation (2.8) is reduced to the conventional FEM formulation when $L_j = \{1\}$ is used. In the nonlinear analysis based on rate form, (2.8) should be applied to interpolate the velocity field instead of the displacement field. Therefore, the displacement terms in the equation are replaced by the displacement rates;

$$\dot{u} = \sum_{n=1}^{m} \left[\sum_{j} L_j a_j^n \right] N^n \tag{2.9}$$

In this paper, two kinds of generalized in-plane element are proposed in this framework. Each element has only corner nodes with additional degree of freedom. The first element presented is the element with an additional freedom like drilling motion. The other one can describe quadratic mode with four additional degrees of freedom at corner node. These elements are called QD4G1 and QD4G2 in the following section. The concepts of QD4G1 and QD4G2 are illustrated in Fig. 1.

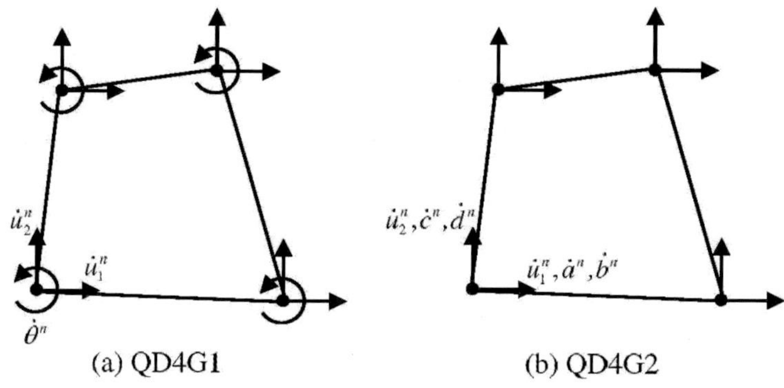

Figure 1. Generalized elements in present study

QD4G1: Element with Drilling Freedom

In QD4G1, the following set of polynomial terms is adopted.

$$\begin{cases} L_j = \{1,\ x_2\} & \text{for } \dot{u}_1 \\ L_j = \{1,\ x_1\} & \text{for } \dot{u}_2 \end{cases} \tag{2.10}$$

A subscript denotes the degree of freedom and it takes 1 or 2 in two-dimensional analysis. With (2.10), the interpolation functions of displacement rate are given as;

$$\dot{u}_1 = \sum_{n=1}^{4} \left[\dot{u}_1^n - (x_2 - x_2^n)\,\dot{\theta}^n \right] N^m \tag{2.11}$$

$$\dot{u}_2 = \sum_{n=1}^{4} \left[\dot{u}_2^n - (x_1 - x_1^n)\,\dot{\theta}^n \right] N^m \tag{2.12}$$

where $\dot{\theta}^n$ is an additional nodal degree of freedom. x_1 and x_2 are defined not in the local coordinates of element but in the global coordinates. The bi-linear shape function for the conventional 4 node element is used as N^n. Equations (2.11) and (2.12) give a compatibility of the displacement velocity field.

The original form of (2.11) and (2.12) is proposed by Sekiguchi and Kikuchi [14] for linear analysis. In the case of linear analysis, the additional degree of freedom corresponds to rotational motion. However, in the present case, $\dot{\theta}^n$ has no obvious physical meaning. It is noted that additional boundary condition corresponding to $\dot{\theta}^n$ has to be imposed to avoid unphysical deformation because QD4G1 has linear dependency. Additionally, in the case of distributed loads, unphysical external force must be applied to $\dot{\theta}^n$, even for uniform tensile loading.

QD4G2: Element with Enriched Quadratic Terms

In QD4G2, the reduced quadratic polynomial terms, which is motivated by Wilson's non-conforming element [17, 18], are used.

$$L_j = \{1,\ (x_1)^2,\ (x_2)^2\} \tag{2.13}$$

Using (2.13), the interpolation functions are written as;

$$\dot{u}_1 = \sum_{n=1}^{4} \left[\dot{u}_1^n + \left\{ (x_1)^2 - (x_1^n)^2 \right\} \dot{a}^n + \left\{ (x_2)^2 - (x_2^n)^2 \right\} \dot{b}^n \right] N^m \tag{2.14}$$

$$\dot{u}_2 = \sum_{n=1}^{4} \left[\dot{u}_2^n + \left\{ (x_1)^2 - (x_1^n)^2 \right\} \dot{c}^n + \left\{ (x_2)^2 - (x_2^n)^2 \right\} \dot{d}^n \right] N^m \tag{2.15}$$

Here, $\dot{a}^n$, $\dot{b}^n$, $\dot{c}^n$ and $\dot{d}^n$ are additional degrees of freedom. x_1 and x_2 are defined in the global coordinate and N^n is the same bi-linear shape function as QD4G1. QD4G2 is also a compatible element. Due to the additional degrees of freedom, a complete perfect cubic polynomial function can be reproduced for no-distorted rectangular element. One of the important features of QD4G2 is that no linear dependency problem occurs because it can represent a displacement rate field without any duplicated term. As for treatment of boundary condition, in case of distributed loads, unphysical load must be applied to additional degrees of freedom.

2.3 Finite Element Discretization

The finite element discretization form of (2.7) is given as;

$$\left\{\delta\underline{\dot{\mathbf{U}}}\right\}[\underline{\mathbf{K}}]\left\{\underline{\dot{\mathbf{U}}}\right\}=\left\{\delta\underline{\dot{\mathbf{U}}}\right\}\left\{\underline{\dot{\mathbf{F}}}\right\} \tag{2.16}$$

$\left\{\delta\underline{\dot{\mathbf{U}}}\right\}$, $\left\{\underline{\dot{\mathbf{U}}}\right\}$ and $\left\{\delta\underline{\dot{\mathbf{F}}}\right\}$ are discretized virtual displacement rate, displacement rate and external force rate respectively, and $[\underline{\mathbf{K}}]$ is the tangent stiffness matrix. Matrix or vector with underbar is a quantity corresponding to the total system equation. Due to the arbitrary property of virtual displacement rate, the relation between displacement rate $\left\{\dot{\mathbf{U}}\right\}$ and external force rate $\left\{\dot{\mathbf{F}}\right\}$ can be expressed as;

$$[\mathbf{K}]\left\{\dot{\mathbf{U}}\right\}=\left\{\dot{\mathbf{F}}\right\} \tag{2.17}$$

$$\left\{\dot{\mathbf{U}}\right\}\equiv\{\dot{\mathbf{U}}^1\ \dot{\mathbf{U}}^2\ \dot{\mathbf{U}}^3\ \dot{\mathbf{U}}^4\} \tag{2.18}$$

$$\left\{\dot{\mathbf{F}}\right\}\equiv\{\dot{\mathbf{F}}^1\ \dot{\mathbf{F}}^2\ \dot{\mathbf{F}}^3\ \dot{\mathbf{F}}^4\} \tag{2.19}$$

The tangent stiffness matrix $[\underline{\mathbf{K}}]$ consists of terms corresponding to the deformation rate $\mathbf{D}$ and the velocity gradient $\mathbf{L}$. Therefore, it can be rewritten as;

$$[\mathbf{K}]=[\mathbf{K}_{\mathrm{D}}]+[\mathbf{K}_{\mathrm{L}}] \tag{2.20}$$

$$[\mathbf{K}_{\mathrm{D}}]\equiv\int_{v^e}[\mathbf{B}_{\mathrm{D}}]^{\mathrm{T}}[\mathbf{D}_{\mathrm{D}}][\mathbf{B}_{\mathrm{D}}]\mathrm{d}v^e \tag{2.21}$$

$$[\mathbf{K}_{\mathrm{L}}]\equiv\int_{v^e}[\mathbf{B}_{\mathrm{L}}]^{\mathrm{T}}[\mathbf{D}_{\mathrm{L}}][\mathbf{B}_{\mathrm{L}}]\mathrm{d}v^e \tag{2.22}$$

where v^e is the volume of element. $[\mathbf{B}_{\mathrm{D}}]$ and $[\mathbf{B}_{\mathrm{L}}]$ are matrices which translate the displacement rate vector into the deformation rate and the velocity gradient. $[\mathbf{D}_{\mathrm{D}}]$ and $[\mathbf{D}_{\mathrm{L}}]$ are parts of the constitutive relation matrix in terms of deformation rate and velocity gradient. Each component can be easily obtained in the same manner as that of the general nonlinear finite element formulation based on the updated Lagrangian method.

$\dot{\mathbf{U}}^n$ is a vector including unknown coefficients at each node and is written as;

$$\begin{aligned}\dot{\mathbf{U}}^n &\equiv \{\dot{u}_1^n\ \ \dot{u}_2^n\ \ \dot{\theta}^n\} && \text{for QD4G1}\\ \dot{\mathbf{U}}^n &\equiv \{\dot{u}_1^n\ \ \dot{u}_2^n\ \ \dot{a}^n\ \ \dot{b}^n\ \ \dot{c}^n\ \ \dot{d}^n\} && \text{for QD4G2}\end{aligned} \tag{2.23}$$

$\dot{\mathbf{F}}^n$ is an external force rate vector whose components correspond to those in $\dot{\mathbf{U}}^n$. By discretizing the virtual work by surface loading, the concrete form of $\dot{\mathbf{F}}^n$ can be obtained. The virtual work caused by surface loading $\delta\dot{W}$ is equivalent to the right-hand side of (2.2).

$$\begin{aligned}\delta\dot{W} &= \int_s \dot{\mathbf{t}}\cdot\delta\dot{\mathbf{u}}\mathrm{d}s\\ &= \int_s\left(\dot{t}_1\delta\dot{u}_1+\dot{t}_2\delta\dot{u}_2\right)\mathrm{d}s\end{aligned} \tag{2.24}$$

By substituting (2.11) and (2.12) for QD4G1 or (2.13) and (2.14) for QD4G2 into the above equation, the components of $\dot{\mathbf{F}}^n$ can be given.

2.4 Equilibrium Equation in Nonlinear Analysis

The governing equations in the section 2.1 just provide equilibrium of the rates of internal and external forces at one moment and do not provide an equality of internal and external forces. The discretized expression of the relation is;

$$\left\{\underline{\dot{\mathbf{Q}}}\right\} = \left\{\underline{\dot{\mathbf{F}}}\right\} \tag{2.25}$$

In the case of the quasi-static state, in which the inertia force can be neglected, by integrating both sides of (2.25) from initial time 0 to current time t, the following equations can be obtained.

$$\{\underline{\mathbf{Q}}\} = \{\underline{\mathbf{F}}\} \tag{2.26}$$

$$\{\underline{\mathbf{Q}}\} \equiv \int_0^t \left\{\underline{\dot{\mathbf{Q}}}\right\} \mathrm{d}t \tag{2.27}$$

$$\{\underline{\mathbf{F}}\} \equiv \int_0^t \left\{\underline{\dot{\mathbf{F}}}\right\} \mathrm{d}t \tag{2.28}$$

The nodal displacement vector is also given by the same way:

$$\{\underline{\mathbf{U}}\} \equiv \int_0^t \left\{\underline{\dot{\mathbf{U}}}\right\} \mathrm{d}t \tag{2.29}$$

Equation (2.26) is nonlinear equation in terms of displacement because the relation between displacement and internal force is generally nonlinear. For this reason, to solve the equation, convergent calculation by a nonlinear implicit method should be introduced. In the present formulation, the tangent stiffness matrix $[\underline{\mathbf{K}}]$ is not always consistent with internal force, therefore, the modified Newton method is utilized and the converged solution can be obtained in nonlinear analysis.

3 Numerical Examples

In this section, two numerical examples are demonstrated. As the first example, a rolling-up analysis of thin plate is conducted. The second analysis is on the bending of thick and quasi-incompressible beam. Both involve geometrical nonlinearity and are computed in the two-dimensional space. In the analyses, the phenomenon in which a stiffer solution is obtained (so-called the locking), may occur when the conventional finite elements are used.

In each analysis, the convergence of the solution with respect to a number of elements and the robustness against element distortion are investigated.

The results obtained by the conventional 4-, 8- and 9-node isoparametric finite elements (QD4, QD8 and QD9) are also shown. Full integration is adopted for all kinds of elements; i.e. the numbers of integration points are 2×2 for QD4 and QD4G1, and 3×3 for QD8, QD9 and QD4G2. It is assumed that the material has isotropic elasticity with Young's modulus E and Poisson's ratio ν.

3.1 Rolling-up of Plate

The first problem, rolling-up analysis of thin plate, is shown in Fig. 2. In the present case, the Poisson's ratio ν is set to zero. The cross section shape of the beam is rectangle whose area is A and the specimen is subjected to a moment M. Under the above condition, the rotation angle at the tip ϕ can theoretically be given as;

$$\phi = 12ML/EAt^2 \tag{3.30}$$

In this analysis, the tip of plate is subject to moment $M = \pi EAt^2/6L$ and the exact solution is $\phi = 2\pi$.

The number of elements along the thickness direction is always set to one and that along the length direction is changed. The definition of element distortion in this analysis is shown in Fig. 3. h is the characteristic length of element. Parameter β characterizes an element distortion and it takes the range of $0 \leq \beta < 1$. $\beta = 0$ corresponds to regular rectangle mesh and element degenerates into triangle shape with $\beta = 1$.

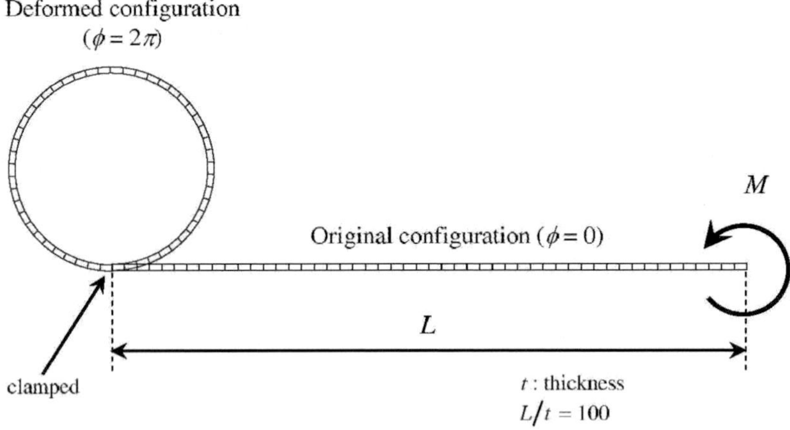

Figure 2. Rolling-up of plate

The convergence in tip rotation with respect to the number of elements is shown in Fig. 4. The element has no distortion or $\beta = 0$. All results in the following section are normalized by the theoretical one. The errors in tip

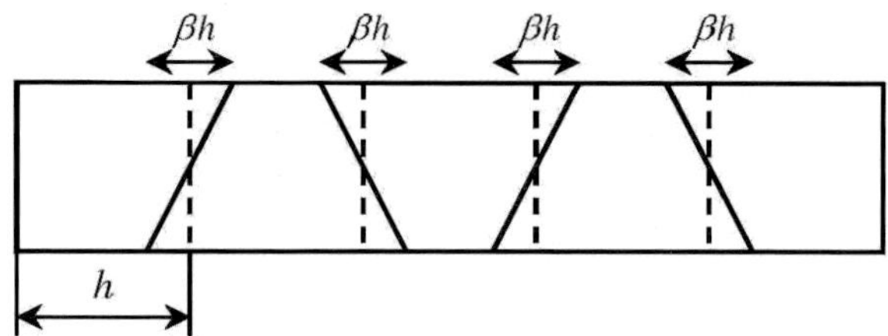

Figure 3. Definition of element distortion in rolling-up analysis

rotation with QD4G1 and QD4G2 are 4.7% and 8.1% when the number of elements is only 10. On the other hand, the conventional elements provide a stiffer solution especially with a smaller number of elements. For second order elements QD8 and QD9, this is caused by membrane locking, which results from unphysical membrane strain in curved element. In case of QD4, shear locking with unphysical strain in pure bending gives a descent of the convergence.

Figure 5 shows the results obtained by distorted meshes. The number of element is fixed at 100, which gives regular tetragon element. QD4G2 provides an almost exact solution even if $\beta \approx 1$ and the robustness against element distortion is similar to that of QD9. The accuracy of QD4G1 slightly decreases when β increases, however, there is no drastic degradation of the solution with $\beta \approx 1$.

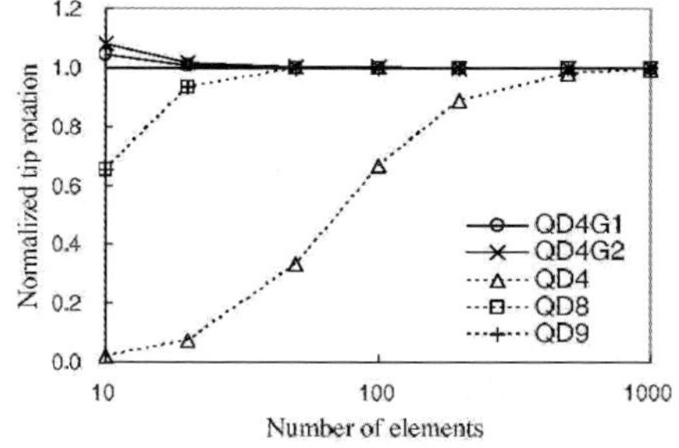

Figure 4. Rolling-up of plate: Normalized tip rotation vs. number of elements

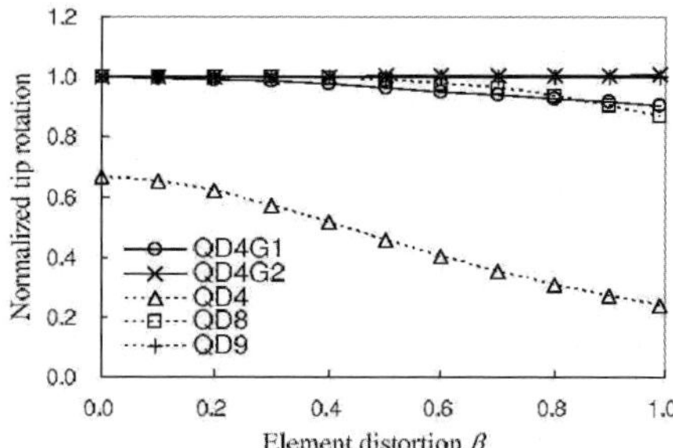

Figure 5. Rolling-up of plate: Normalized tip rotation vs. element distortion

Figures 6 and 7 indicate errors in tip rotation with respect to the number of elements with regular mesh and distorted mesh when $\beta = 0.99$. With regular mesh, QD4G1 and QD4G2 privide good results as well or better than those of QD8 or QD9. In the case with distorted mesh, QD4G2 has a good performance especially with a small number of elements although the rate of convergence is slightly slower than QD9.

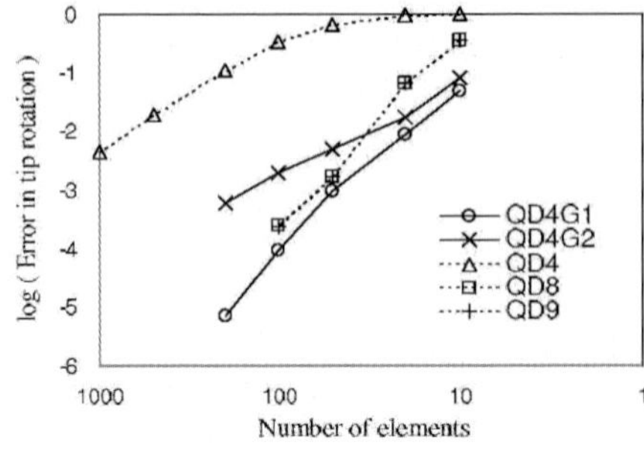

Figure 6. Rolling-up of plate with regular mesh: Error of tip rotation with respect to number of elements

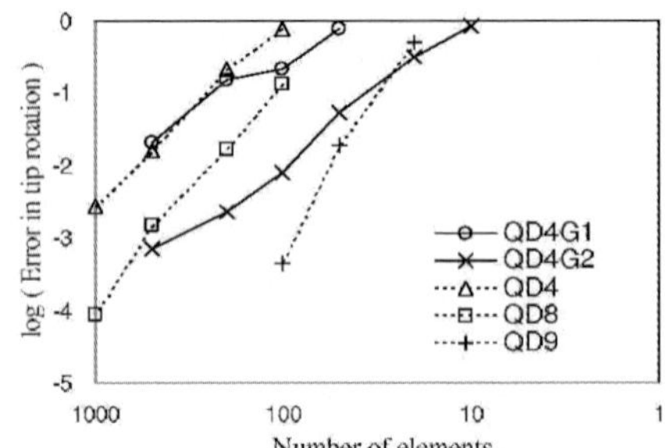

Figure 7. Rolling-up of plate with distorted mesh: Error of tip rotation with respect to number of elements

This analysis shows that both of QD4G1 and QD4G2 provide good results for the convergence of solution and robustness against element distortion and there is no shear or membrane locking in pure bending deformation.

3.2 Bending of Quasi-Incompressible and Thick Beam

As the second example, bending of quasi-incompressible and thick beam shown in Fig. 8 is analyzed. To reproduce the incompressible condition, Poisson's ratio ν is set to 0.499 and plane strain state is assumed. Under this condition, volumetric locking may occur in numerical analysis. It is subject to a point loading $P = 5 \times 10^{-4} EA$ at the tip of the beam and the tip deflection corresponding to it is about $0.1L$. Therefore, geometrical nonlinearity cannot be neglected. In this problem, the converged solution with extraordinary fine mesh is used as a reference because no theoretical solution is available. The results in the following section are normalized using this.

In this analysis, the element distortion is defined by the method shown in Fig. 9. Here, θ and r are given as random numbers of $0 \leq \theta < 2\pi$ and $0 \leq r < \gamma h/2$ with the parameter γ which characterizes the element distortion. The figure also shows examples of regular and distorted meshes.

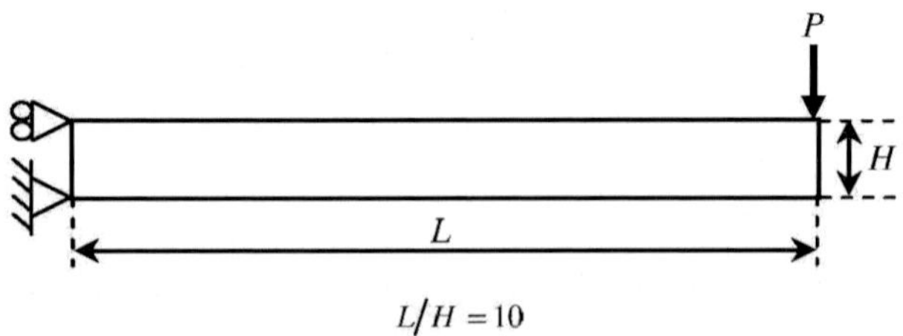

Figure 8. Bending of quasi-incompressible and thick beam

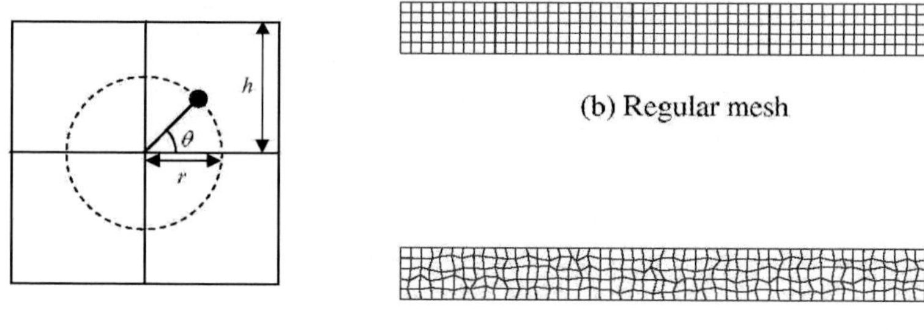

(a) Definition of element distortion (c) Distorted mesh with $\gamma = 0.80$

Figure 9. Definition of element distortion in bending of thick beam

The convergence in tip deflection with respect to the number of elements along the length direction is shown in Fig. 10. The element has no distortion or $\gamma = 0$ and the ratio of the length to the height of the element is always 2:1. The results with QD4G2, QD8 and QD9 are almost exact even if the number of elements along the thickness direction is 1 (total element number is 5). On the other hand, QD4G1 and QD4 give very stiff response because of volumetric locking which is caused by the constraint of (almost) incompressible deformation. These results indicate that volumetric locking does not occur with QD4G2 although it cannot be avoided with QD4G1.

Figure 11 shows the results with distorted meshes. The number of elements is fixed at 50×5. The solution obtained by QD4G2 is almost exact with $\gamma = 0.80$ although the element distortion affects the precision of the solution with all conventional elements, even for QD9.

Figures 12 and 13 show errors in tip deflection with respect to the number of elements obtained by regular mesh and distorted mesh with $\gamma = 0.80$. With regular mesh, the convergence rate of QD4G2 is more or less the same as QD8

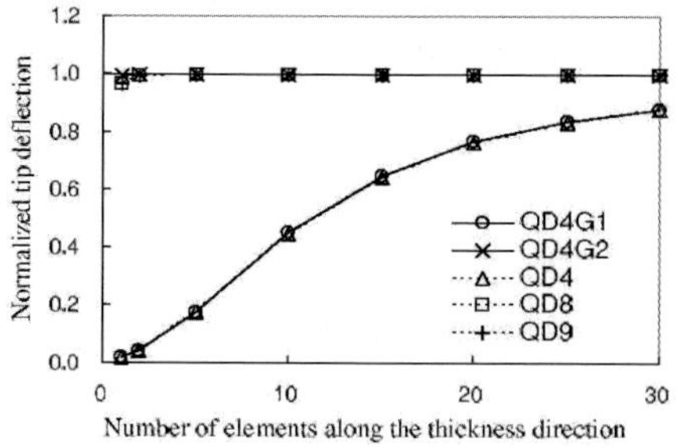

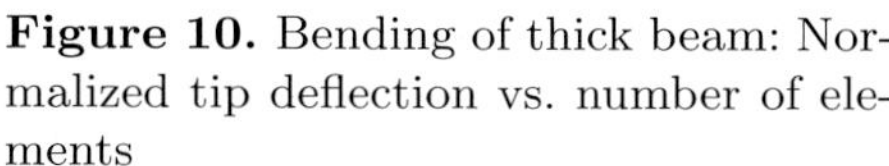
Figure 10. Bending of thick beam: Normalized tip deflection vs. number of elements

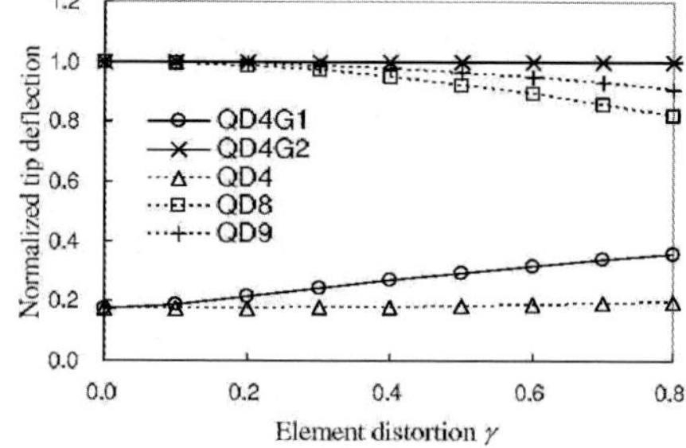

Figure 11. Bending of thick beam: Normalized tip deflection vs. element distortion

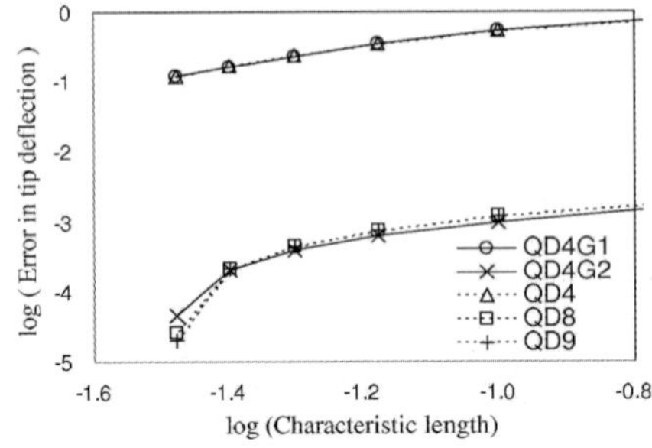

Figure 12. Bending of thick beam with regular mesh: Error of tip rotation with respect to number of elements

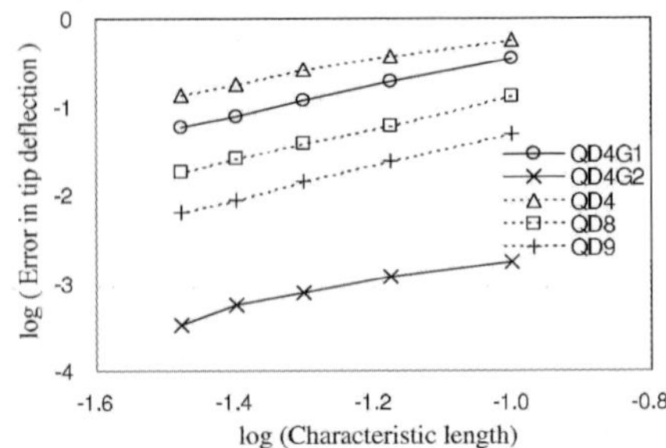

Figure 13. Bending of thick beam with distorted mesh: Error of tip rotation with respect to number of elements

and QD9. In the results with distorted mesh, QD4G2 presents very high performance relative to QD8 and QD9, suggesting that QD4G2 is highly robust against element distortion in the present problem compared with conventional second order elements.

These results prove that QD4G2 provides good convergence with the number of elements and is quite robust against element distortion for the problem where incompressible constraint is dominant.

4 Conclusion

In this paper, the formulation of two kinds of in-plane finite elements is presented in the framework of GFEM based on PUM and they are introduced into geometrical nonlinear analysis. For the element with drilling freedom, QD4G1, the solution convergence and robustness against element distortion are similar to, if not better than, conventional quadratic elements in the analysis with pure bending. On the other hand, it cannot avoid volumetric locking. With the second proposed element QD4G2, which can express quadratic deformation mode with only corner nodes, no shear locking, membrane locking and volumetric locking occurs. It is shown that QD4G2 is a highly reliable element in analyses where some locking may occur even if conventional second order elements are used.

References

1. I. Babuška and J. E. Osborn, *Generalized finite element methods: Their performance and their relation to mixed methods*, SIAM J. Num. Anal. **20** (1983), 510–536.
2. O. C. Zienkiewicz, *Generalized finite element method: State of the art and future directions*, Trans. ASME J. Appl. Mech. **50** (1983), 1210–1217.

3. I. Babuška, G. Caloz and J. Osborn, *Special finite element methods for a class of second order elliptic problems with rough coefficients*, SIAM J. Numer. Anal. **31** (1994), 945–981.
4. J. M. Melenk and I. Babuška, *The partition of unity finite element method: Theory and application*, Comput. Methods Appl. Mech. Engrg. **139** (1996), 289–314.
5. I. Babuška and J. M. Melenk, *The partition of unity finite element method*, Int. J. Numer. Methods Engng. **40** (1997), 727–758.
6. R. L. Taylor, O. C. Zienkiewicz and E. Oñate, *A hierarchical finite element method based on the partition of unity*, Comput. Methods Appl. Mech. Engrg. **152** (1998), 73–84.
7. I. Babuška, U. Banerjee and E. Osborn, *On principles for the selection of shape functions for the Generalized Finite Element Method*, Comput. Methods Appl. Mech. Engrg. **191** (2002), 5595–5629.
8. I. Babuška, U. Banerjee and J. Osborn, *Generalized finite element methods - main ideas, results and perspective*, Int. J. Comp. Methods **1** (2004), 67–103.
9. I. Babuška, U. Banerjee and J. Osborn, *Survey of meshless and generalized finite element methods*, Acta Numerica **12** (2003), 1–125.
10. T. Strouboulis, K. Copps and I. Babuška, *The generalized finite element method*, Comput. Methods Appl. Mech. Engrg. **190** (2001), 4081–4193.
11. D. J. Allman, *A compatible triangular element including vertex rotations for plane elasticity analysis*, Comput. Struct. **19** (1984), 1–8.
12. D. J. Allman, *A quadrilateral finite element including vertex rotations for plane elasticity analysis*, Int. J. Num. Methods Engrg. **26** (1988), 717–730.
13. T. J. R. Hughes, *On drilling degrees of freedom*, Comput. Methods Appl. Mech. Engrg. **72** (1989), 105–121.
14. M. Sekiguchi and N. Kikuchi, *Re-examination of membrane elements with drilling freedom*, Proc. of Fifth World Congress on Computational Mechanics (H. A. Mang F. G. Rammerstorfer and J. Eberhardsteiner, eds.), 2002.
15. F. B. Barros, S. P. B. Proenca and C. S. de Barcellos, *Generalized finite element method in structural nonlinear analysis - a p-adaptive strategy*, Comput. Mech. **33** (2004) 95–107.
16. K. Terada, M. Asai and M. Yamagishi, *Finite cover method for linear and non-linear analyses of heterogeneous solids*, Int. J. Num. Methods Engrg. **58** (2003), 1321–1346.
17. R. L. Taylor, P. J. Beresford and E. L. Wilson, *A non-conforming element for stress analysis*, Int. J. Numer. Methods Engng. **10** (1976), 1211–1219.
18. O. C. Zienkiewicz and R. L. Taylor, *The finite element method: Fifth edition Volume 1*, Butterworth Heinemann, 2000.

A Meshfree Method for Simulations of Interactions between Fluids and Flexible Structures

Sudarshan Tiwari, Sergey Antonov, Dietmar Hietel, Jörg Kuhnert, Ferdinand Olawsky and Raimund Wegener

Fraunhofer-Institut für Techno- und Wirtschaftmathematik (ITWM)
Gottlieb-Daimler-Straße, Geb. 49, D-67663 Kaiserslautern, Germany
tiwari@itwm.fraunhofer.de

Summary. We present the application of a meshfree method for simulations of interaction between fluids and flexible structures. As a flexible structure we consider a sheet of paper. In a two-dimensional framework this sheet can be modeled as curve by the dynamical Kirchhoff-Love theory. The external forces taken into account are gravitation and the pressure difference between upper and lower surface of the sheet. This pressure difference is computed using the Finite Pointset Method (FPM) for the incompressible Navier-Stokes equations. FPM is a meshfree, Lagrangian particle method. The dynamics of the sheet are computed by a finite difference method. We show the suitability of the meshfree method for simulations of fluid-structure interaction in several applications.

Key words: Meshfree Method, FPM, Fluid Structure Interaction, Sheet of Paper, Dynamical Coupling

1 Introduction

There exists a wide range of problems where the motion of flexible structures is driven by a fluid flow. Such kind of problems are in printing processes the prediction of the fluttering of a thin sheet [11] or in bio-mechanics the opening and closing behavior of aortic heart valves [2]. In general, the structure is considered as a moving wall and its dynamics are described in a Lagrangian framework. Therefore, it is suitable to use a Lagrangian formulation also for the fluid. Whenever the structure moves rapidly, mesh based methods like finite volume or finite element methods are limited due to grid adaption by re-meshing during the simulation. It requires the use of interpolation techniques to recover the fluid dynamical variables on the new mesh. This procedure not only introduces artificial diffusivity, but is also problematic with respect to accuracy, robustness and efficiency. To resolve this problem, we consider a meshfree method called Finite Pointset Method (FPM) [4, 10].

FPM is a Lagrangian particle method, where the fluid domain is represented by a finite number of arbitrarily distributed particles (pointset). These particles move with the fluid velocity and carry all fluid dynamical informations like pressure and velocity. In FPM we approximate the spatial derivatives at an arbitrary point from its surrounding neighbor particles by the weighted least squares method. Furthermore, the elliptic equations occurring in the treatment of incompressible flows are also solved in this framework [4, 5, 9, 10].

In this paper we consider a sheet of paper in an incompressible flow. For simplicity we restrict ourself to a two-dimensional setting. Therefore, the thin sheet can be represented and modeled as a curve in a two-dimensional flow domain using the dynamical Kirchhoff-Love theory. The external forces taken into account are gravitation and the pressure difference between upper and lower surface of the sheet. This pressure difference is computed using the Finite Pointset Method (FPM).

The paper is organized as follows. In section 2 we present the mathematical model for the fluid flow and the flexible structure. The numerical schemes for the corresponding equations are described in section 3. Here, specific aspects are the computation of the pressure difference acting on the sheet of paper and the coupling between fluid flow and the dynamics of the flexible structure. Finally, in section 4 three numerical examples are presented.

2 Governing Equations

2.1 Fluid

We consider a fluid flow in a two-dimensional bounded domain $\bar{\Omega} \subset \mathbb{R}^2$ with boundary Γ. Let $\boldsymbol{v}$ and p be the velocity and pressure fields representing the state variables. The incompressible Navier-Stokes equations in the Lagrangian form are given by

$$\frac{D\boldsymbol{x}}{Dt} = \boldsymbol{v}, \tag{2.1}$$

$$\frac{D\boldsymbol{v}}{Dt} = -\nabla p + \nu \, \Delta \, \boldsymbol{v} + \boldsymbol{g}, \tag{2.2}$$

$$\nabla \cdot \boldsymbol{v} = 0. \tag{2.3}$$

Here, $\boldsymbol{g}$ denotes the gravitation, ν the kinematic viscosity, D/Dt the material derivative and $\boldsymbol{x}$ the trajectories of material points.

These equations are to be solved with appropriate initial and boundary conditions. We use no slip conditions on solid walls and also on both surfaces of the sheet of paper with regard to its movement. At inflow and outflow we use Dirichlet and Neumann conditions for the velocity.

2.2 Flexible Structure

As mentioned the flexible structure under consideration is a sheet of paper. In general, the mathematical model of a sheet of paper is based on the bending stiffness of a thin plate. We restrict ourself on a two-dimensional setting. Since the thickness of the sheet is much smaller than its length the paper can be described by a curve of constant length. For the Kirchhoff-Love theory of elastic rods capable of large bending deformation see[1, 6]. Here, we resume the final equations of this theory.

Let $\boldsymbol{r} = \boldsymbol{r}(s,t) \in \mathbb{R}^2$ with $t \in \mathbb{R}^+$ be a curve parameterized by the arc length $s \in [0, L]$, where L is the length of the curve. The dynamics of the sheet of paper is given by balancing the acting forces with the acceleration

$$\omega\, \partial_{tt}\boldsymbol{r}(s,t) = \partial_s[T(s,t)\partial_s\boldsymbol{r}(s,t)] - B\, \partial_{ssss}\boldsymbol{r}(s,t) + [p]\,\boldsymbol{n} + \omega\,\boldsymbol{g}. \tag{2.4}$$

Here, the unknown tension T acts as a Lagrangian multiplier referencing to the additional constraint of in-extensibility

$$\|\partial_s\boldsymbol{r}(s,t)\|^2 = 1. \tag{2.5}$$

The base weight ω and the bending stiffness B are given material constants of the sheet of paper. $\boldsymbol{n}$ denotes the unit normal vector orthogonal to the curve. The first two terms on the right hand side of (2.4) are the internal forces due to tension and bending, whereas the last two terms are the external forces due to the pressure difference $[p]$ between both sides of the sheet and due to gravity $\boldsymbol{g}$.

We note that the system (2.4, 2.5) has some similarity to the incompressible Navier-Stokes equations, where the pressure is a Lagrangian parameter referencing to the incompressibility constraint.

The complete model for the sheet of paper needs appropriate initial and boundary conditions. For the boundaries we consider two different types of possible situations. Either an end is fixed with given position and direction by

$$\boldsymbol{r}(0,t) = \boldsymbol{r}^0(t), \quad \partial_s\boldsymbol{r}(0,t) = \boldsymbol{e}^0(t) \tag{2.6}$$

or it is free with Neumann boundary conditions given by

$$\partial_{ss}\boldsymbol{r}(L,t) = 0, \quad \partial_{sss}\boldsymbol{r}(L,t) = 0, \quad T(L,t) = 0. \tag{2.7}$$

Wall Contact

In some applications the sheet of paper can hit a wall. Such events are detected geometrically and are taken into account by adding the geometrical constraint of the wall. To handle this situation an additional force perpendicular to the wall is introduced in the dynamical equation (2.4). Its magnitude is another Lagrangian multiplier referencing to the geometrical constraint.

3 Numerical Schemes

3.1 Flexible Structure

Space Discretization

For the space discretization in arc length $s \in [0, L]$ we use an equidistant division with $s_i = i\Delta s, \quad i = 0, \ldots, N$. The equations are approximated in a conservative finite difference scheme at a curve point $\boldsymbol{r}_i$ at s_i by using

$$\partial_s(T\partial_s \boldsymbol{r})_i = \frac{1}{\Delta s}\left[(T\partial_s \boldsymbol{r})_{i+\frac{1}{2}} - (T\partial_s \boldsymbol{r})_{i-\frac{1}{2}}\right], \tag{3.8}$$

$$\partial_s(B\partial_{sss} \boldsymbol{r})_i = \frac{1}{\Delta s}\left[(B\partial_{sss} \boldsymbol{r})_{i+\frac{1}{2}} - (B\partial_{sss} \boldsymbol{r})_{i-\frac{1}{2}}\right]. \tag{3.9}$$

Here, the fluxes at the cell faces are approximated by

$$(\partial_s \boldsymbol{r})_{i+\frac{1}{2}} = \frac{1}{24\,\Delta s}\left[\boldsymbol{r}_{i-1} - 27\,\boldsymbol{r}_i + 27\,\boldsymbol{r}_{i+1} - \boldsymbol{r}_{i+2}\right], \tag{3.10}$$

$$(\partial_{sss} \boldsymbol{r})_{i+\frac{1}{2}} = \frac{1}{\Delta s^3}\left[\boldsymbol{r}_{i-1} - 3\,\boldsymbol{r}_i + 3\,\boldsymbol{r}_{i+1} - \boldsymbol{r}_{i+2}\right]. \tag{3.11}$$

Due to the higher order nature of the underlying equation it is important to use a higher order approximation of the tangential $\partial_s \boldsymbol{r}$ as given by (3.10).

Time Integration

We consider an implicit time integration with constant time step Δt. The resulting system for inner grid points $i = 1, \cdots, N-1$ is given by

$$\begin{aligned} \omega\,\frac{\boldsymbol{r}_i^{n+1} - 2\,\boldsymbol{r}_i^n - \boldsymbol{r}_i^{n-1}}{\Delta t^2} &= \frac{1}{\Delta s}\left[T_{i+\frac{1}{2}}^{n+1}(\partial_s \boldsymbol{r})_{i+\frac{1}{2}}^{n+1} - T_{i-\frac{1}{2}}^{n+1}(\partial_s \boldsymbol{r})_{i-\frac{1}{2}}^{n+1}\right] \\ &-\frac{B}{\Delta s}\left[(\partial_{sss} \boldsymbol{r})_{i+\frac{1}{2}}^{n+1} - (\partial_{sss} \boldsymbol{r})_{i-\frac{1}{2}}^{n+1}\right] + [p]_i^n\,\boldsymbol{n}_i^n + \omega\,\boldsymbol{g}. \end{aligned} \tag{3.12}$$

Here, the normal vector $\boldsymbol{n}_i$ is orthogonal to the tangential vector defined by the average of $(\partial_s \boldsymbol{r})_{i\pm\frac{1}{2}}$.

The boundary conditions at both ends are treated straight forward. The in-extensibility constraint is approximated by

$$(\partial_s \boldsymbol{r})_{i-\frac{1}{2}}^{n+1} \cdot (\partial_s \boldsymbol{r})_{i-\frac{1}{2}}^{n+1} = 1. \tag{3.13}$$

Hence, we get a system of nonlinear equations (3.12, 3.13). This is solved by means of an iterative method based on partial linearization.

3.2 Navier-Stokes Equations

The system (2.1–2.3) describes the incompressible Navier-Stokes equations in a Lagrangian form. We solve the equations with our particle method FPM (Finite Pointset Method) in combination with Chorin's projection method [4, 8, 10]. FPM is characterized by a weighted least squares method for approximation of spatial derivatives.

Therefore, this section is divided into three parts: the projection method, the general treatment of derivatives in FPM and the application of this technique for solving the resulting elliptic equation for the pressure.

Chorin's Projection Method

To handle the incompressibility we use Chorin's projection method [3]. This method consists of two fractional steps and is of first order accuracy in time. In the first step the new particle positions and intermediate velocities $\boldsymbol{v}^*$ are computed by

$$\boldsymbol{x}^{n+1} = \boldsymbol{x}^n + \Delta t\ \boldsymbol{v}^n, \tag{3.14}$$

$$\boldsymbol{v}^* = \boldsymbol{v}^n + \Delta t\ \nu \Delta \boldsymbol{v}^* + \Delta t\ \boldsymbol{g}. \tag{3.15}$$

Then, in the second step we correct $\boldsymbol{v}^*$ considering

$$\boldsymbol{v}^{n+1} = \boldsymbol{v}^* - \Delta t\ \nabla p^{n+1} \tag{3.16}$$

together with the incompressibility constraint

$$\nabla \cdot \boldsymbol{v}^{n+1} = 0. \tag{3.17}$$

By taking the divergence of equation (3.16) and by making use of (3.17) we finally obtain a Poisson equation for the pressure

$$\Delta p^{n+1} = \frac{\nabla \cdot \boldsymbol{v}^*}{\Delta t}. \tag{3.18}$$

The boundary condition for p is obtained by projecting equation (3.16) onto the normal vector $\boldsymbol{n}$ of the boundary Γ. Thus, we obtain a Neumann boundary condition

$$\left(\frac{\partial p}{\partial \boldsymbol{n}}\right)^{n+1} = -\frac{1}{\Delta t}(\boldsymbol{v}_\Gamma^{n+1} - \boldsymbol{v}_\Gamma^*) \cdot \boldsymbol{n}. \tag{3.19}$$

Assuming vanishing normal velocity components on Γ, we obtain

$$\left(\frac{\partial p}{\partial \boldsymbol{n}}\right)^{n+1} = 0. \tag{3.20}$$

We note that the particle positions change only in the first step. The intermediate velocity $\boldsymbol{v}^*$ is then obtained on these new particle positions. Finally, the pressure Poisson equation and the divergence free velocity vector are also computed on these particle positions.

Least Squares Approximation of Derivatives in FPM

Let $\psi : \bar{\Omega} \to \mathbb{R}$ be a scalar function and ψ_i its given values at the particle positions $\boldsymbol{x}_i$ for $i = 1, \ldots, N$. Consider the problem to approximate spatial derivatives of ψ at a particle position $\boldsymbol{x}$ based on the function values of its neighbor particles. In order to restrict the number of neighbors we introduce a weight function $w = w(\boldsymbol{x}_i - \boldsymbol{x}; h)$ with compact support of size h.

The weight function can be chosen quite arbitrary. However, here we choose a cutted Gaussian weight function of the form

$$w(\boldsymbol{x}_i - \boldsymbol{x}; h) = \begin{cases} \exp(-\alpha \frac{\|\boldsymbol{x}_i - \boldsymbol{x}\|^2}{h^2}), & \text{if } \frac{\|\boldsymbol{x}_i - \boldsymbol{x}\|}{h} \le 1 \\ 0, & \text{else.} \end{cases}$$

The positive constant α describes the exponential decay with a typical value $\alpha = 6$. In our implementation the weighting radius h can be chosen as a function in space and time. However, no adaptive method to modify h during the computation is realized yet. Working with a user defined h implies that new particles will have to be brought into play if the particle distribution becomes too sparse or, logically, particles will have to be removed from the computation as the distribution becomes too dense.

Let $P(\boldsymbol{x}, h) = \{\boldsymbol{x}_i : i = 1, 2, \ldots, m\}$ be the set of m neighbor points of $\boldsymbol{x} = (x, y)$ inside the radius h. We note that the central particle $\boldsymbol{x}$ is one element of the neighbor set $P(\boldsymbol{x}, h)$. To ensure consistency in two dimensions the number m of these particles should be at least 6 and they should neither be on the same line nor on the same circle.

Now, we approximate derivatives of a function by using its Taylor series expansion and the least squares approximation. Hence, consider the Taylor expansions around $\boldsymbol{x} = (x, y)$ at its m neighbors $\boldsymbol{x}_i$

$$\begin{aligned} \psi(\boldsymbol{x}_i) &= \psi(\boldsymbol{x}) + (x_i - x)\partial_x \psi(\boldsymbol{x}) + (y_i - y)\partial_y \psi(\boldsymbol{x}) + \frac{(x_i - x)^2}{2} \partial_{xx} \psi(\boldsymbol{x}) \\ &\quad + (x_i - x)(y_i - y)\partial_{xy} \psi(\boldsymbol{x}) + \frac{(y_i - y)^2}{2} \partial_{yy} \psi(\boldsymbol{x}) + \mathcal{O}(h^3). \end{aligned} \tag{3.21}$$

Using these expansions up to quadratic terms this leads to m equations for the unknown derivatives of ψ at position $\boldsymbol{x}$. Here, we note that $\boldsymbol{x}$ is one of the particle positions in $P(\boldsymbol{x}, h)$. Therefore, $\psi = \psi(\boldsymbol{x})$ is supposed to be known. The resulting system for the unknowns $\boldsymbol{a} = (\psi_x, \psi_y, \psi_{xx}, \psi_{xy}, \psi_{yy})^T$ is not under-determined for $m > 5$ but can in general only be solved as a least squares solution by minimizing the error $\boldsymbol{e}$ in the resulting system

$$\boldsymbol{e} = M\boldsymbol{a} - \boldsymbol{b} \tag{3.22}$$

where

$$M = \begin{pmatrix} dx_1 & dy_1 & \frac{1}{2}dx_1^2 & dx_1 dy_1 & \frac{1}{2}dy_1^2 \\ \vdots & \vdots & \vdots & \vdots & \vdots \\ dx_m & dy_m & \frac{1}{2}dx_m^2 & dx_m dy_m & \frac{1}{2}dy_m^2 \end{pmatrix},$$

$\boldsymbol{b} = (\psi_1 - \psi, \ldots, \psi_m - \psi)^T$, $\boldsymbol{e} = (e_1, \ldots, e_m)^T$, $dx_i = x_i - x$, $dy_i = y_i - y$.

Thus, the unknown vector $\boldsymbol{a}$ is computed by minimizing the weighted error over all neighboring points. That means we have to minimize the quadratic form

$$J = \sum_{i=1}^{m} w_i e_i^2 = (M\boldsymbol{a} - \boldsymbol{b})^T W (M\boldsymbol{a} - \boldsymbol{b}) \tag{3.23}$$

with $W = \operatorname{diag}(w_1, \ldots, w_m)$ and $w_i = w(\boldsymbol{x}_i - \boldsymbol{x}; h)$. The minimization of J with respect to the unknowns $\boldsymbol{a}$ formally yields (if $M^T W M$ is nonsingular)

$$\boldsymbol{a} = (M^T W M)^{-1} (M^T W) \boldsymbol{b}. \tag{3.24}$$

Solving Elliptic Equations with FPM

Now, we consider the following type of linear second order partial differential equations

$$A\ \psi + C\ (\psi_{xx} + \psi_{yy}) = f, \tag{3.25}$$

which represents all equations resulting from the discussed Chorin's projection scheme. Here, the coefficients A and C are real constants and $f = f(\boldsymbol{x})$ is a given real valued function. To determine the intermediate velocity, we have $A = 1$ and $C = -\Delta t\ \nu$ and in case of the Poisson equation for the pressure we have $A = 0$ and $C = 1$. At the boundary we use either Dirichlet conditions or Neumann conditions

$$\frac{\partial \psi}{\partial \boldsymbol{n}} = \phi \quad \text{on} \quad \Gamma \tag{3.26}$$

with prescribed normal derivative ϕ.

To our knowledge, there are two types of methods to solve elliptic equations in a given meshfree configuration. The first one is presented in [7], which can be directly derived from equation(3.24). The second one is presented in [9], where equation (3.25) and possibly Neumann conditions (3.26) are added as constraints in the least squares approximation. Both methods are compared in [5]. It is found that the method presented in [9] is more stable and that Neumann conditions can be easily included in the approximation. Therefore, this method is described shortly in the following.

Based on the set of neighbors $\boldsymbol{x}_i$, $i = 1, \ldots, m$ we use the Taylor expansions (3.21). Unlike the above approximation of derivatives we now have to consider also the value ψ at the central particle $\boldsymbol{x}$ to be unknown. For inner particles of the domain we just add the elliptic equation (3.25) in the least squares procedure. For a boundary particle we have to take into account the additional boundary conditions. To illustrate the method we describe the case of a boundary particle with Neumann condition (3.26). Here, the modified system $\tilde{\boldsymbol{e}} = \tilde{M}\tilde{\boldsymbol{a}} - \tilde{\boldsymbol{b}}$ is defined by

$$\tilde{M} = \begin{pmatrix} 1 & dx_1 & dy_1 & \frac{1}{2}dx_1^2 & dx_1 dy_1 & \frac{1}{2}dy_1^2 \\ \vdots & \vdots & \vdots & \vdots & \vdots & \vdots \\ 1 & dx_m & dy_m & \frac{1}{2}dx_m^2 & dx_m dy_m & \frac{1}{2}dy_m^2 \\ A & 0 & 0 & C & 0 & C \\ 0 & n_1 & n_2 & 0 & 0 & 0 \end{pmatrix},$$

$\tilde{\boldsymbol{a}} = (\psi, \psi_x, \psi_y, \psi_{xx}, \psi_{xy}, \psi_{yy})^T$, $\tilde{\boldsymbol{b}} = (\psi_1, \ldots, \psi_m, f, \phi)^T$, $\tilde{\boldsymbol{e}} = (e_1, \ldots, e_{m+2})^T$. The solution now minimizes the functional

$$\tilde{J} = \sum_{i=1}^{m+2} w_i e_i^2 = (\tilde{M}\tilde{\boldsymbol{a}} - \tilde{\boldsymbol{b}})^T \tilde{W} (\tilde{M}\tilde{\boldsymbol{a}} - \tilde{\boldsymbol{b}}) \tag{3.27}$$

with $\tilde{W} = \text{diag}\,(w_1, \ldots, w_m, 1, 1)$ and $w_i = w(\boldsymbol{x}_i - \boldsymbol{x}; h)$ for $i = 1, \ldots, m$ while $w_{m+1} = w_{m+2} = 1$. Formally, the solution of this minimization problem can be written as

$$\tilde{\boldsymbol{a}} = (\tilde{M}^T \tilde{W} \tilde{M})^{-1} (\tilde{M}^T \tilde{W}) \tilde{\boldsymbol{b}}. \tag{3.28}$$

The value of ψ is the only relevant component of the solution $\tilde{\boldsymbol{a}}$. By introducing $\beta_1, \ldots, \beta_6$ as the first row of $(\tilde{M}^T \tilde{W} \tilde{M})^{-1}$ and using the explicit structure of $(\tilde{M}^T \tilde{W})\tilde{\boldsymbol{b}}$, we finally obtain for this component the equation

$$\psi - \sum_{i=1}^{m} w_i \left(\beta_1 + \beta_2 dx_i + \beta_3 dy_i + \beta_4 \frac{dx_i^2}{2} + \beta_5 dx_i dy_i + \beta_6 \frac{dy_i^2}{2} \right) \psi_i$$
$$= (\beta_1 A + \beta_4 C + \beta_5 C)\, f + (\beta_1 n_1 + \beta_2 n_2)\, \phi.$$

Hence, if we consider this procedure for all inner and boundary particles we get a sparse linear system for the unknowns ψ_j, $j = 1, \ldots, N$ at all particle positions. The discretization is of second order [5, 9, 10] the system can be solved by standard iteration methods. In our projection method for the incompressible Navier-Stokes equations, we have one linear system for the velocities and one for the pressure values. As an initial guess for the iterative solvers at time level $n + 1$, we use the corresponding values at time level n.

3.3 Computation of Pressure Difference at the Sheet of Paper

We consider the sheet of paper as a moving solid boundary. In our two-dimensional framework it is simply a curve. A particle located directly above the curve can have a neighbor located under the curve. But with respect to the fluid behavior they should not be coupled. Therefore, we introduce for particle position on the curve two boundary particles with opposite normal directions and indicate their orientation with $+1$ and -1.

Then all inner particles are marked with the orientation $+1$ if they are in the neighborhood of a boundary particle of orientation $+1$ and if they lie on the half plane defined by the normal of this boundary particle. The same procedure is done for the orientation -1. Finally, all particles have either an

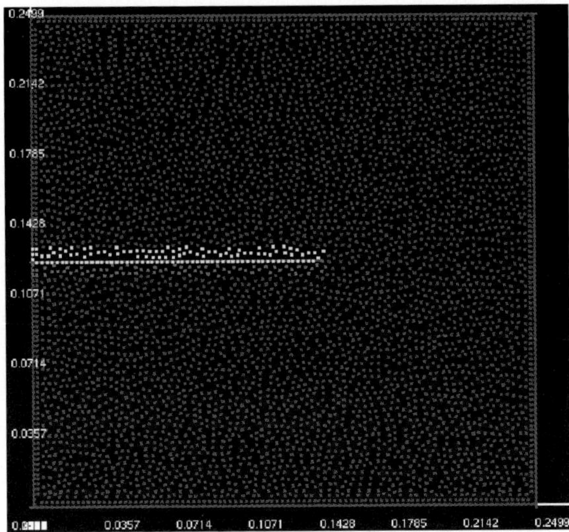

Figure 1. Orientations of the particles around a sheet of paper, blue indicates orientation 0, red +1 and yellow −1. See Color Plate 5 on page 303.

orientation 1, −1 or are marked by 0. Figure 1 shows an illustrative example for the orientations.

These orientations are used to organize the neighborhood relations. For particles of orientation 0 all other particles are allowed as neighbors. For particles of orientation +1 or −1 all other particles in the neighborhood with different sign of orientation are removed from the list of neighbors. Similar to the situation at boundaries the restriction to the one sided subgroup of particles reduces the typical number of neighbors. However, in order to obtain the accuracy of the weighted least squares approximation the scheme guarantees everywhere a minimum number of neighbors.

In general, it is not necessary that the positions on the sheet of paper used for its dynamics are also boundary particles in the fluid dynamical computation. In this case the pressure difference acting on the sheet of paper has to be computed by interpolation. For each side we use the already introduced weighted least squares approximation. Here, the neighboring particles taken into account are selected with the help of the orientations defined above.

3.4 Coupling Algorithm

We consider a simple numerical approach to simulate the interaction between fluid and flexible structure by decoupling the motion of the fluid and the structure at each time step. This is an fully explicit coupling scheme and is described as follows:

1. Initialization: $t_0 = 0$ and $n = 0$, where n is index of the time step
   (i) Initialize the curve $\boldsymbol{r}^0(s)$ by N segments with equal arc length Δs
   (ii) Define boundary particles on each arc segment with both orientations
   (iii) Initialize the particles in the whole computational domain (position, velocity, pressure)
2. Do $n = 0, 1, 2, \ldots, M$
   (i) Use the pressure p^n and the curve $\boldsymbol{r}^n$ describing the sheet of paper to calculate the new curve $\boldsymbol{r}^{n+1}$
   (ii) Use the curve $\boldsymbol{r}^{n+1}$ to compute the new positions $\boldsymbol{x}^{n+1}$ and velocities $\boldsymbol{v}^{n+1}$ of the boundary particles lying on the sheet of paper
   (iii) Use the new positions and velocities of boundary particles of the prior step to solve the Navier-Stokes equations and obtain new positions $\boldsymbol{x}^{n+1}$ and flow fields $\boldsymbol{v}^{n+1}$ and p^{n+1}
   (iv) End do

4 Numerical Examples

In this section we present some numerical applications. Since the sheet of paper has very small bending stiffness and low weight, the grid size as well as the time step for the simulation of its dynamics should be very small. However, the time step for the Navier-Stokes solver need not to be so small. Therefore, for both solvers we choose the time step of the sheet of paper. In the following applications, we consider the weight of paper $\omega = 0.1\,\text{kg/m}^2$, the bending stiffness $B = 5 \cdot 10^{-4}\,\text{kg}\,\text{m}^2/\text{s}^2$ and the gravity force $g = 9.81\,\text{m/s}^2$ in the direction of $-y$. The discretization size of the arc length is $\Delta s = 5 \cdot 10^{-3}\,\text{m}$.

4.1 Fixing One End

We consider a square cavity of size $0.25\,\text{m} \times 0.25\,\text{m}$ and a curve of length $0.141\,\text{m}$. The curve is fixed at position $(0, 0.125)$ and has a free end at $(0.141, 0.125)$. Initially, the velocities and pressure are set to be zero. Due to gravity, the curve starts bending towards negative y-direction. In first row of figure 2 we have plotted the positions of the curve at initial time $t = 0\,\text{s}$ (left) and at $t = 0.8377\,\text{s}$ (right). In the second row we have plotted the pressure (left), with values varying from $0.29512\,\text{Pa}$ (blue) to $0.84271\,\text{Pa}$ (red) and the velocity field (right) with maximum value $0.059941\,\text{m/s}$ (red) at time $t = 0.8377\,\text{s}$. The time step is chosen as $\Delta t = 5 \cdot 10^{-4}\,\text{s}$.

We consider the above mentioned example to give an impression of the convergence behavior since fixing one end of the sheet leads to a steady state solution. In table 1 the height of the free-end in the steady state is given for different resolutions Δs. The time step is chosen such that $\Delta s/\Delta t = 20\,\text{m/s}$ holds.

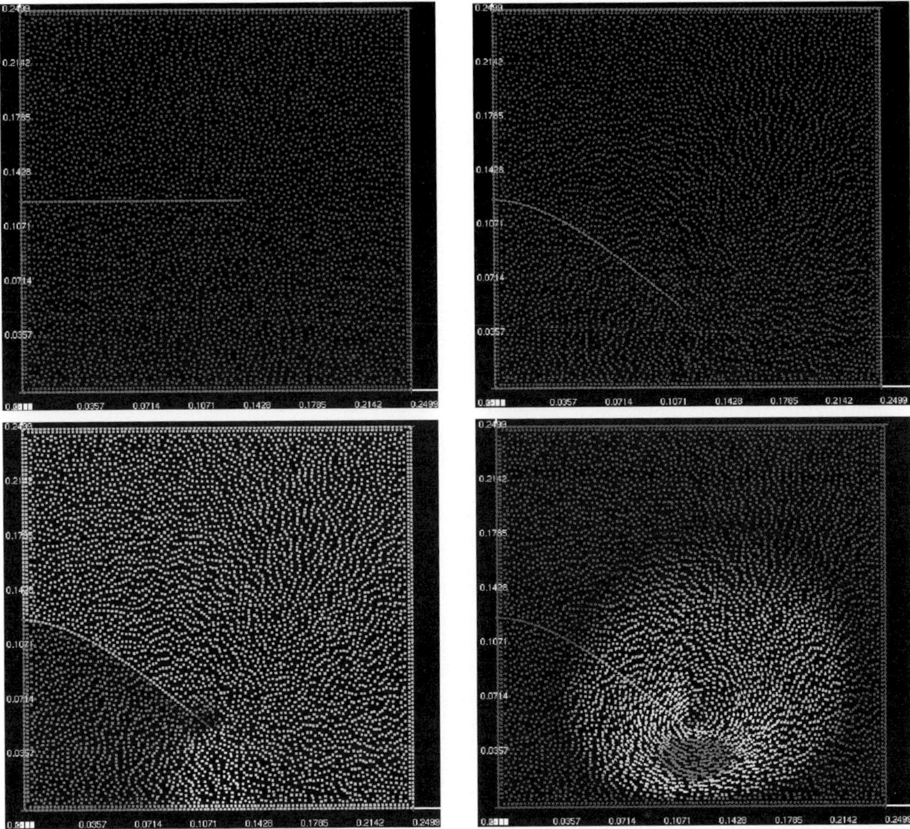

Figure 2. Position of particles and sheet of paper at time $t = 0\,\mathrm{s}$ and $t = 0.8377\,\mathrm{s}$ (top), pressure and velocity field at $t = 0.8377\,\mathrm{s}$ (bottom). See Color Plate 6 on page 303.

Table 1.

Δs	height of free end
0.018 m	0.023756698 m
0.012 m	0.025303442 m
0.009 m	0.026223404 m
0.006 m	0.026814278 m

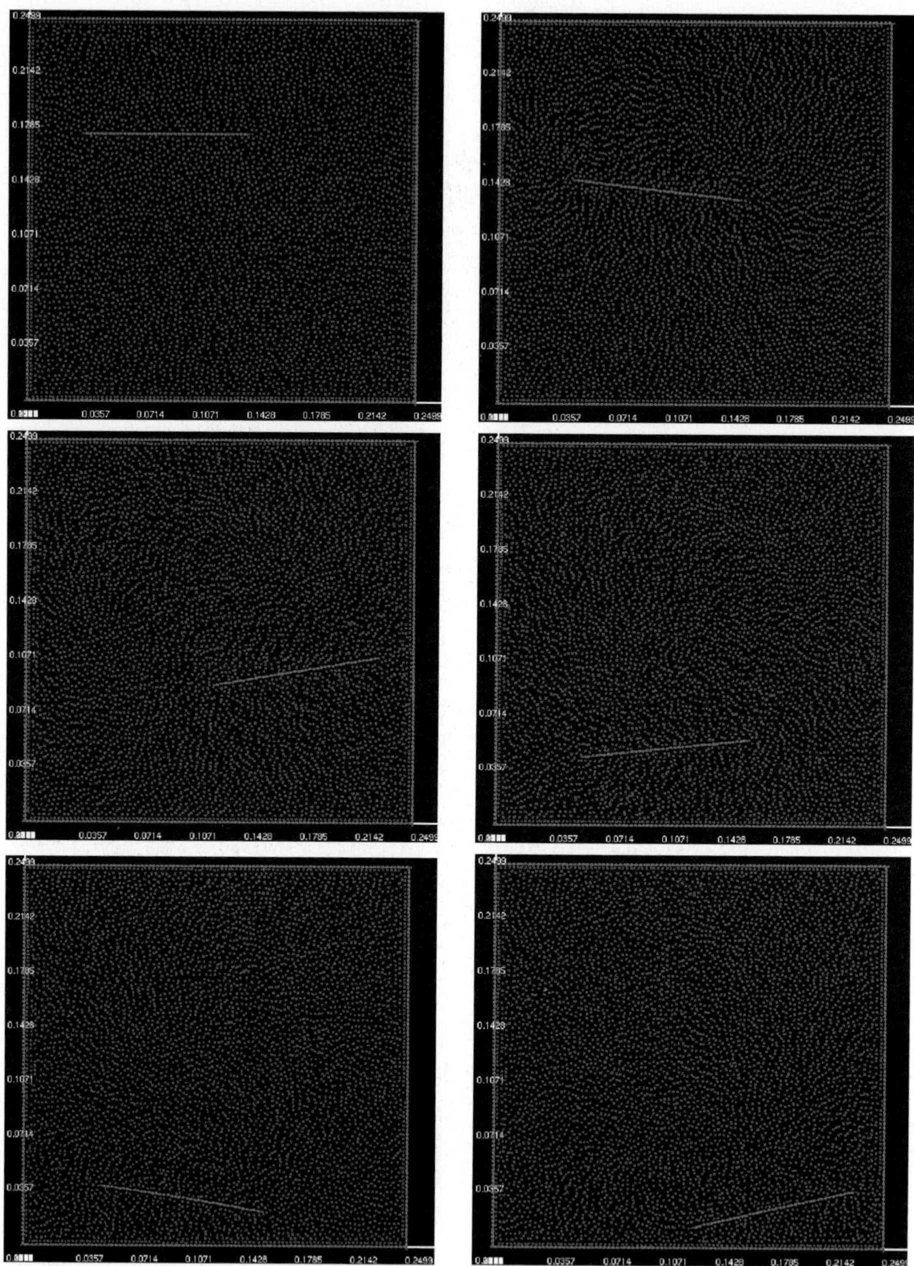

Figure 3. Position of particles and sheet of paper at $t = 0.0, 0.3\,\mathrm{s}$ (top), $t = 0.6, 0.9\,\mathrm{s}$ (center) and $t = 1.2, 1.5\,\mathrm{s}$ (bottom). See Color Plate 7 on page 304.

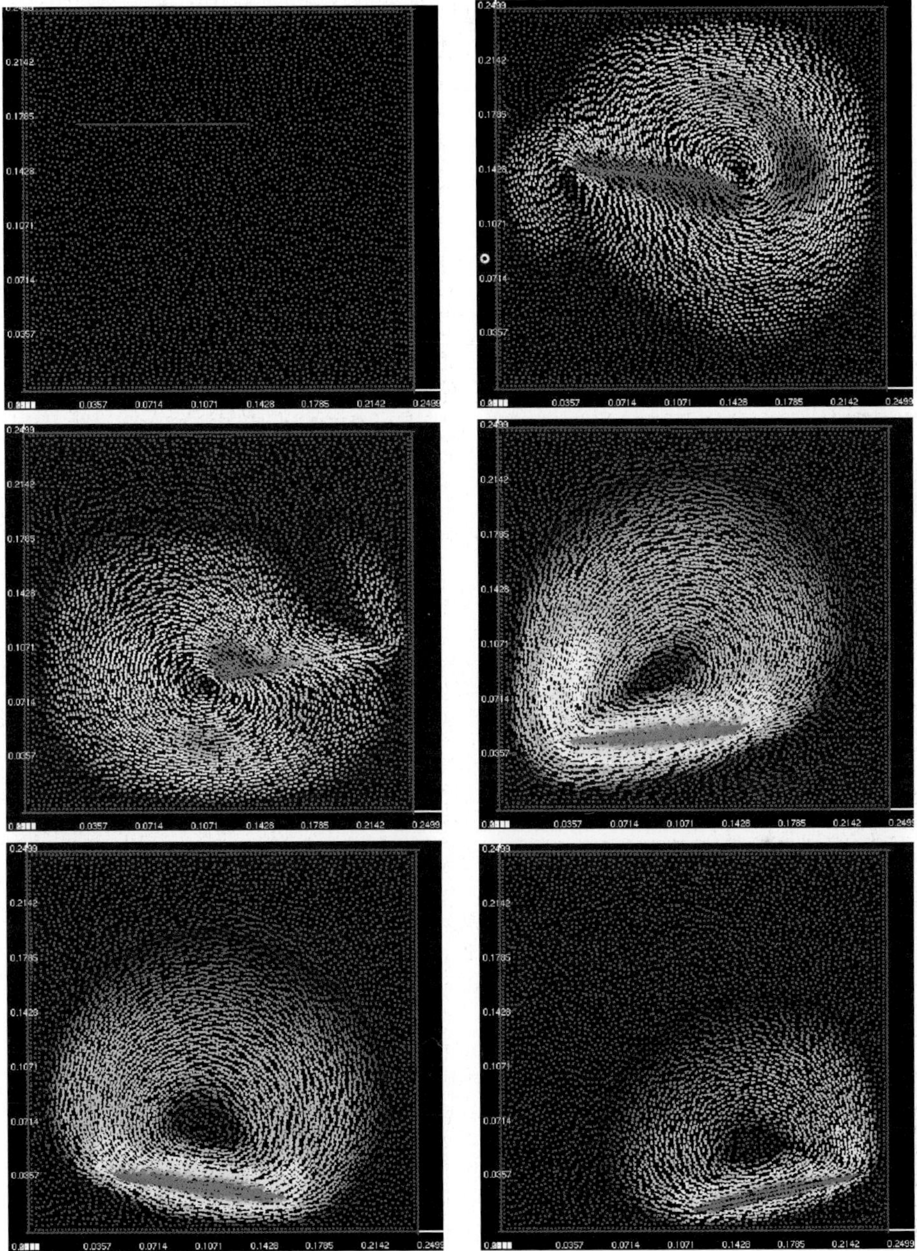

Figure 4. Velocity field at $t = 0.0, 0.3\,\mathrm{s}$ (top), $t = 0.6, 0.9\,\mathrm{s}$ (center) and $t = 1.2, 1.5\,\mathrm{s}$ (bottom). See Color Plate 8 on page 305.

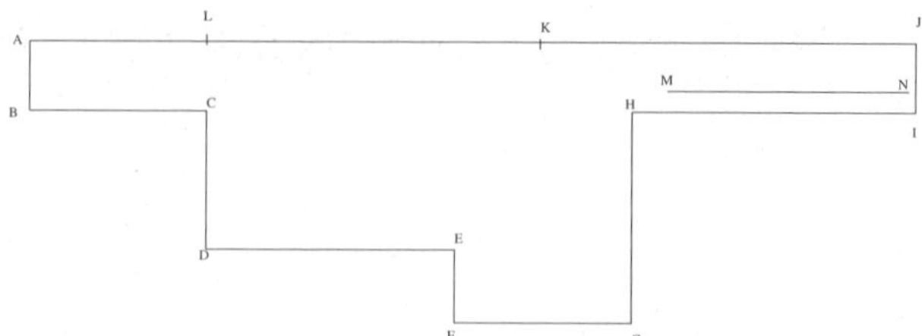

Figure 5. Computational domain

4.2 Free Falling

We again consider the square cavity as above. The curve is positioned at $(0.036, 0.175)$ and $(0.141, 0.175)$ and both ends are considered to be free. The initial velocities and pressure are set to be zero. Due to gravity, the paper starts falling and shows the well-known swinging phenomenon. In figures 3 and 4 we have plotted the positions and the corresponding velocity fields at times $t = 0, 0.3, 0.6, 0.9, 1.2, 1.5\,\mathrm{s}$. The maximum magnitudes of velocity are $0.21310, 0.27685, 0.72154, 0.37865, 0.12890\,\mathrm{m/s}$ at times $t = 0.3, 0.6, 0.9, 1.2, 1.5\,\mathrm{s}$, respectively. The time step was considered the same as in the previous example.

4.3 Flying and Deposition

In this example we consider the computational domain as shown in Figure 5. The boundaries AB and IJ are outflow boundaries. KL is considered as the inflow boundary and the remaining boundaries are solid walls. The sheet of paper MN is pulled at the point M with $4\,\mathrm{m/s}$ in the negative x-direction until M reaches the x-Position of H. Afterwards also this end is left to be free. From the inflow boundary we blow air with $1\,\mathrm{m/s}$ in the negative y-direction. The boundary DE is considered to be the deposition of the sheet of papers. We release the first paper at time $t = 0.01\,\mathrm{s}$ and the others at time $t = 0.05, 0.1$ and $0.15\,\mathrm{s}$. In Figure 6 we have plotted the position, pressure and velocity field at time $t = 0.15\,\mathrm{s}$. At this time the values of the pressure varies from $6.18\,\mathrm{Pa}$ (blue) to $162.16\,\mathrm{Pa}$ (red) and the maximum magnitude of the velocity is $7.1398\,\mathrm{m/s}$. The time step is taken as $\Delta t = 10^{-4}\,\mathrm{s}$. The length of the sheets is exactly the length of the deposition DE. The lengths of the boundaries are $AB = IJ = CD = 0.05\,\mathrm{m}$, $BC = DE = FG = AL = 0.135\,\mathrm{m}$, $EF = 0.03\,\mathrm{m}$, $GH = 0.1\,\mathrm{m}$, $IH = 0.205\,\mathrm{m}$, $KL = 0.202\,\mathrm{m}$, $KJ = 0.272\,\mathrm{m}$ and the length of the paper is $MN = 0.135\,\mathrm{m}$.

5 Conclusion

The Finite Pointset Method (FPM) is suitable to handle a wide range of dynamical fluid structure interactions. In this paper this is demonstrated in a

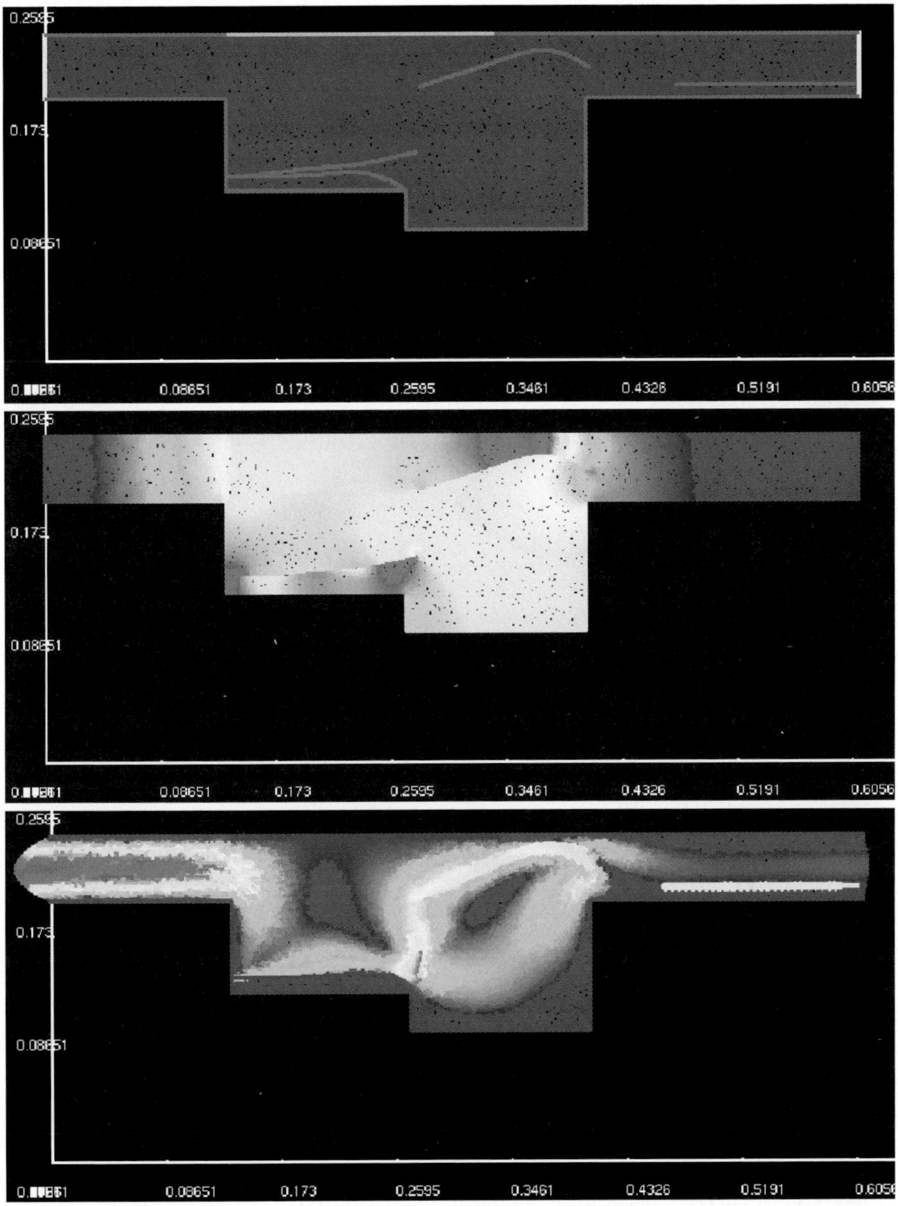

Figure 6. Position of particles and sheets of paper (top), pressure distribution (center) and velocity field (bottom) at $t = 0.15\,\mathrm{s}$. See Color Plate 9 on page 306.

two-dimensional framework that couples FPM with the dynamics of a one-dimensional sheet. Future work will be the extension of the method to three dimensional problems.

Acknowledgement. We would like to thank the "Stiftung Rheinland-Pfalz für Innovation" for the financial support of our research.

References

1. S. S. Antman, *Nonlinear Problems of Elasticity*, Springer Verlag, New York, 1995.
2. F. P. T. Baaijens, *A fictious domain/mortar element method for fluid-structure interaction*, Int. J. Numer. Meth. Fluids, **35** (2001), 743–761.
3. A. Chorin, *Numerical solution of the Navier-Stokes equations*, J. Math. Comput., **22** (1968), 745–762.
4. D. Hietel, M. Junk, J. Kuhnert and S. Tiwari, *Meshless methods for Conservation Laws*, (G. Warnecke, eds.), Analysis and Numerics for Conservation Laws, Springer, 2005, pp. 339–362.
5. O. Iliev and S. Tiwari, *A generalized (meshfree) finite difference discretization for elliptic interface problems*, (I. Dimov, I. Lirkov, S. Margenov, and Z. Zlatev, eds.), Numerical Methods and Applications, Lecture notes in Computer Sciences, Springer 2002, pp. 480–489.
6. N. Marheineke and R. Wegener, *Fiber Dynamics in Turbulent Flows*, Berichte des Fraunhofer ITWM, Nr. 81 (2005), Kaiserslautern, Germany.
7. T. Liszka and J. Orkisz, *The finite difference method on arbitrary irregular grid and its application in applied mechanics* Computers & Structures, **11** (1980), pp. 83–95.
8. S. Tiwari and J. Kuhnert, *Finite pointset method based on the projection method for simulations of the incompressible Navier-Stokes equations*, (M. Griebel and M. A. Schweitzer, eds.), Lecture Notes in Computational Science and Engineering, vol. 26, Springer, 2002, pp. 373–387.
9. S. Tiwari and J. Kuhnert, *Grid free method for solving Poisson equation, Wavelet Analysis and Applications*, New Age International Publishers, 2004, pp. 151–166.
10. S. Tiwari and J. Kuhnert, *A numerical scheme for solving incompressible and low Mach number flows by Finite Pointset Method*, (M. Griebel and M. A. Schweitzer, eds.), Lecture Notes in Computational Science and Engineering, vol. 43, Springer, 2005, pp. 191–206.
11. N. Yamaguchi, K. Yokota and Y. Tsujimoto, *Flutter Limits and Behaviours of a Flexible Thin Sheet in High-Speed Flow I: Analytical Method for Prediction of the Sheet Behaviour*, J. Fluids Eng., **122** (2000), pp. 65–73.

Goal Oriented Error Estimation for the Element Free Galerkin Method

Yolanda Vidal and Antonio Huerta

Laboratori de Càlcul Numèric, Departament de Matemàtica Aplicada III,
Universitat Politècnica de Catalunya, Jordi Girona 1, E-08034 Barcelona, Spain
{yolanda.vidal,antonio.huerta}@upc.edu

Summary. A novel approach for implicit residual-type error estimation in mesh-free methods is presented. This allows to compute upper and lower bounds of the error in energy norm with the ultimate goal of obtaining bounds for outputs of interest. The proposed approach precludes the main drawbacks of standard residual type estimators circumventing the need of flux-equilibration and resulting in a simple implementation that avoids integrals on edges/sides of a domain decomposition (mesh). This is especially interesting for mesh-free methods.

Key words: Element Free Galerkin, mesh-free methods, error estimation, engineering outputs, residual based estimators.

1 Introduction

Assessment of functional outputs of the solution (goal-oriented error estimation) in *computational mechanics* problems is a real need in standard engineering practice. In particular, end-users of finite elements, finite differences or mesh-free codes are interested in obtaining bounds for quantities of engineering interest. Techniques providing these bounds require using error estimators in the energy norm of the solution. Bounds for quantities of interest (functional outputs) are recovered combining upper and lower bounds of the energy error for both the original problem (primal) and a dual problem (associated with the selected functional output) [21, 1, 20].

It is also important to note that bounds for the energy and for quantities of interest are usually obtained with respect to a reference solution (associated with a much larger space of approximation). Bounds for the exact solution of the boundary value problem as presented in [3, 26, 22] are not addressed here.

The need of obtaining reliable upper and lower bounds of the error for quantities of interest has motivated the use of residual error estimators, which are currently the only type of estimators ensuring bounds for the error. Classical residual type estimators, which provide upper bounds of the error, require flux-equilibration procedures (*hybrid-flux* techniques) to properly set boundary conditions for local problems [14, 1]. Flux-equilibration requires a domain decomposition, which is natural in finite elements but not in mesh-free methods. And, moreover, it is performed by a complex algorithm, strongly dependent on the finite element type and requiring a data structure that is not natural in a standard finite element code. Thus highly embedded in the finite element domain decomposition.

The idea of using flux-free estimators, based on the partition-of-the-unity concept and using local subdomains different than elements, has been already proposed in [4, 16, 18, 23] for finite elements. The main advantage of the flux-free approach is the simplicity in the implementation. Obviously, this is especially important in the 3D case.

From the mesh-less point of view, another advantage is the fact that the local subdomains where the error equation is solved are the support of the functions characterizing the partition of unity. This is a concept that also exists in mesh-free methods and thus the extension is possible. Moreover, boundary conditions of the local problems are trivial and the usual data structure of a code is directly employed.

In the last few years, some research has been devoted to develop error estimation procedures for mesh-free methods. Duarte and Oden [7] derived an explicit residual error estimator for the h-p cloud method. Liu et al. [15] used a wavelet solution as an error indicator in an algorithm where multiple-scale adaptive refinement had been introduced. Chung and Belytschko [5] adapted the FEM stress projection technique for error analysis in Element Free Galerkin (EFG). Gavete et al. [10] proposed a sort of recovery-based error estimate, which presents the standard drawbacks of these methods. None of these approaches was able to compute bounds of the energy error. Thus, the assessment of bounds and functional outputs is still an open topic in mesh-free methods.

To the authors knowledge implicit residual based estimators have not been proposed for mesh-free methods. However, these residual based approaches are now standard in finite elements because they are more mathematically sound, more precise and allow to compute upper and lower bounds for energy norms as well as functional outputs.

In this paper the implicit residual-type flux-free error estimator proposed in [23], which has similar efficiency as standard hybrid-flux estimators, is extended to the Element Free Galerkin Method. The remainder of the paper is structured as follows. Section 2, following [23], recalls the basics on output oriented error estimation in finite elements and introduces a flux-free error estimator. Section 3 is devoted to extend step by step the previous concepts to the Element Free Galerkin method: Dirichlet boundary conditions, definition of the reference error, estimation of outputs of interest, numerical integration, local domain for the error equation, and finally the solvability of the local problems. Finally, in Section 4, some numerical examples are shown.

2 Basics on Output Oriented Error Estimation in FEM

2.1 Model Problem

Let $\Omega \subset \mathbb{R}^{\mathtt{n_{sd}}}$ be an open, bounded domain with piecewise linear boundary $\partial\Omega$ and $\mathtt{n_{sd}}$ the number of spatial dimensions. The scalar strong form of the problem is: find u such that

$$\begin{cases} -\boldsymbol{\nabla}^2 u + \sigma u = s & \text{in } \Omega, \\ u = u_D & \text{on } \partial\Omega. \end{cases} \tag{2.1}$$

Only Dirichlet boundary conditions are considered for simplicity. It is trivial to extend these concepts to account also for Neumann-type boundary conditions. The standard weak solution of this problem is $u \in \mathcal{U}$ verifying

$$a(u, v) = l(v) \quad \forall v \in \mathcal{V}, \tag{2.2}$$

where

$$a(u, v) = \int_\Omega \boldsymbol{\nabla} v \cdot \boldsymbol{\nabla} u + \sigma v u \, d\Omega, \quad l(v) = \int_\Omega v s \, d\Omega.$$

The usual solution and test spaces are defined $\mathcal{U} = \{u \in \mathcal{H}^1(\Omega), u|_{\partial\Omega} = u_D\}$ and $\mathcal{V}_0 = \{v \in \mathcal{H}^1(\Omega), v|_{\partial\Omega} = 0\}$, where $\mathcal{H}^1$ is the standard Sobolev space of square integrable functions and first derivatives. The bilinear form $a(\cdot,\cdot)$ induces the energy norm, which is denoted by $\|\cdot\|$, that is, $\|v\|^2 = a(v, v)$.

The finite element interpolation spaces $\mathcal{U}^H \subset \mathcal{U}$ and $\mathcal{V}_0^H \subset \mathcal{V}_0$ are associated with a finite element mesh of characteristic size H and degree p for the complete interpolation polynomial base. The geometric support of the elements for a given mesh are open subdomains denoted by Ω_k, $k = 1 \ldots \mathtt{n_{el}}$, where $\overline{\Omega} = \bigcup_k \overline{\Omega}_k$. It is also assumed that different elements do not overlap, that is, $\Omega_k \cap \Omega_l = \emptyset$ for $k \neq l$. Then, the finite element solution u_H which is an approximation to u, lies in the finite dimensional space $\mathcal{U}^H$ and verifies

$$a(u_H, v) = l(v) \quad \forall v \in \mathcal{V}_0^H.$$

2.2 Error Equations and Reference Error

The goal of a posteriori error estimation is to assess the accuracy of the finite element solution u_H, that is, to evaluate and measure the error, $e := u - u_H$, which belongs to $\mathcal{V}$, either in the energy norm $\|e\|$ or in a quantity of interest as it will be shown next. The equation for the error is recovered from (2.2) replacing the exact solution u by $u_H + e$ and using the linearity of the first argument of $a(\cdot,\cdot)$

$$a(e, v) = l(v) - a(u_H, v) =: R^P(v) \quad \forall v \in \mathcal{V}_0, \tag{2.3}$$

where $R^P(\cdot)$ stands for the weak residue associated to the finite element approximation u_H.

The exact error e is replaced by a reference error, e_h, lying in a finite dimensional space $\mathcal{V}_0^h$ much richer than the original finite element space $\mathcal{V}_0^H$, i.e. $\mathcal{V}_0^H \subset \mathcal{V}_0^h \subset \mathcal{V}_0$. That is, the exact solution u is replaced by the reference (or truth) solution u_h; consequently, $u \approx u_h = u_H + e_h$. Given this definition of e_h it is easy to verify that the reference error is the projection of the exact error into the reference space, that is, $e_h \in \mathcal{V}_0^h$ is the solution of the problem

$$a(e_h, v) = R^P(v) \quad \forall v \in \mathcal{V}_0^h. \tag{2.4}$$

Direct evaluation of e_h is computationally unaffordable because the size of the system of equations is the dimension of $\mathcal{V}_0^h$. The idea behind any implicit residual error estimator is to solve local problems instead of the global problem (2.4). Each of these local problems require proper boundary conditions in order to obtain a good approximation of the error and to ensure solvability.

2.3 Estimation of Outputs of Interest

The actual interest is to bound output quantities $l^{\mathcal{O}}(u)$, where $l^{\mathcal{O}}(\cdot)$ is a linear functional, see for instance [21, 17, 25, 20, 24, 29]. These strategies introduce a dual (or adjoint) problem with respect to the selected output. The weak form of the dual problem reads: find $\psi \in \mathcal{V}_0$ such that

$$a(v, \psi) = l^{\mathcal{O}}(v) \quad \forall v \in \mathcal{V}_0.$$

The finite element approximation of the dual problem is $\psi_H \in \mathcal{V}_0^H$ such that

$$a(v, \psi_H) = l^{\mathcal{O}}(v) \quad \forall v \in \mathcal{V}_0^H.$$

Finally, the dual reference error is $\epsilon_h := \psi_h - \psi_H \in \mathcal{V}_0^h$, such that

$$a(v, \epsilon_h) = l^{\mathcal{O}}(v) - a(v, \psi_H) =: R^D(v) \quad \forall v \in \mathcal{V}_0^h, \tag{2.5}$$

where R^D is the weak residue associated with ψ_H.

If v is replaced by e_h in (2.5), then using Galerkin orthogonality and the parallelogram identity, the following representation of $l^{\mathcal{O}}(e_h)$ can be obtained

$$l^{\mathcal{O}}(e_h) = a(e_h, \epsilon_h) = \frac{1}{4}\|\kappa e_h + \frac{1}{\kappa}\epsilon_h\|^2 - \frac{1}{4}\|\kappa e_h - \frac{1}{\kappa}\epsilon_h\|^2 \tag{2.6}$$

for any arbitrary scalar parameter κ. To simplify the notation the arguments in the squared norms of the r.h.s. in (2.6) are denoted by $z_h^{\pm} = \kappa e_h \pm \frac{1}{\kappa}\epsilon_h$.

In fact, in order to bound the output of the error, $l^{\mathcal{O}}(e_h)$, the r.h.s of (2.6) indicates that it is sufficient to bound the energy norm of z_h^+ and z_h^-, (i.e. the energy norm of linear combinations of e_h and ϵ_h).

Define $\mathrm{E_u}[v]$ and $\mathrm{E_l}[v]$ as the upper and lower bound of $\|v\|^2$, respectively. Note that $\mathrm{E_u}[v]$ and $\mathrm{E_l}[v]$ are not functions; instead, it is a convenient notation of the bounds. Thus, once the bounds for $\|z_h^{\pm}\|^2$ are computed, namely $\mathrm{E_l}[z_h^{\pm}] \leq \|z_h^{\pm}\|^2 \leq \mathrm{E_u}[z_h^{\pm}]$, the output of the error is readily bounded as

$$\frac{1}{4}\mathrm{E}_{\mathrm{l}}[z_h^+] - \frac{1}{4}\mathrm{E}_{\mathrm{u}}[z_h^-] \leq l^{\mathcal{O}}(e_h) \leq \frac{1}{4}\mathrm{E}_{\mathrm{u}}[z_h^+] - \frac{1}{4}\mathrm{E}_{\mathrm{l}}[z_h^-], \tag{2.7}$$

and, obviously, the bounds for the output of interest of the reference approximation, $l^{\mathcal{O}}(u_h)$, are

$$l^{\mathcal{O}}(u_H) + \frac{1}{4}\mathrm{E}_{\mathrm{l}}[z_h^+] - \frac{1}{4}\mathrm{E}_{\mathrm{u}}[z_h^-] \leq l^{\mathcal{O}}(u_h) \leq l^{\mathcal{O}}(u_H) + \frac{1}{4}\mathrm{E}_{\mathrm{u}}[z_h^+] - \frac{1}{4}\mathrm{E}_{\mathrm{l}}[z_h^-].$$

Next section introduces a methodology to obtain both upper and lower bound error estimates in energy norm. This approach is then used to compute $\mathrm{E}_{\mathrm{u}}[z_h^+]$, $\mathrm{E}_{\mathrm{u}}[z_h^-]$, $\mathrm{E}_{\mathrm{l}}[z_h^+]$ and $\mathrm{E}_{\mathrm{l}}[z_h^-]$.

2.4 Upper Bound Estimate of the Reference Error

Let $\boldsymbol{x}^i$, $i = 1, \ldots, \mathtt{n}_{\mathtt{np}}$ denote the vertices of the elements in the computational mesh (thus linked to $\mathcal{U}^H$) and ϕ^i the corresponding linear (or bilinear or trilinear) shape functions, which are such that $\phi^i(\boldsymbol{x}^j) = \delta_{ij}$. The support of ϕ^i is denoted by ω^i and it is called the star centered in, or associated with, vertex $\boldsymbol{x}^i$.

It is important to recall that the linear shape functions based on the vertices are a *partition of unity*. Using this essential property and the linearity of the weak residue $R^P(\cdot)$, defined in (2.3), the residue is decomposed into local contributions over each star

$$R^P(v) = R^P\Big(\sum_{i=1}^{\mathtt{n}_{\mathtt{np}}} \phi^i v\Big) = \sum_{i=1}^{\mathtt{n}_{\mathtt{np}}} R^P(\phi^i v) \quad \forall v \in \mathcal{H}^1(\Omega).$$

Note that $R^P(\phi^i v)$ vanishes if $\operatorname{supp} v \cap \omega^i = \emptyset$, since ω^i is the support of ϕ^i.

The strategy to compute upper bound estimates of the reference error, $\mathrm{E}_{\mathrm{u}}[e_h]$, consist in, first, the evaluation of the finite element solution u_H, which is necessary to compute the residue R^P. And, second, the appraisal of the local estimates $\tilde{e}^{\omega^i} \in \mathcal{V}^h_{\omega^i}$, where $\mathcal{V}^h_{\omega^i} := \mathcal{V}^h_0 \cap \mathcal{H}^1(\omega^i)$, solving problems in each star ω^i

$$a_{\omega^i}(\tilde{e}^{\omega^i}, v) = R^P(\phi^i v) \quad \forall v \in \mathcal{V}^h_{\omega^i}, \tag{2.8}$$

where $a_{\omega^i}(\cdot,\cdot)$ is the restriction of the bilinear form $a(\cdot,\cdot)$ to the star ω^i.

Remark 1. Formally any function $v \in \mathcal{V}^h_{\omega^i}$ is not defined in the whole domain Ω but only in the star ω^i. However, here every $v \in \mathcal{V}^h_{\omega^i}$ is naturally extended to Ω by setting its value outside ω^i equal to zero. Thus, functions in $\mathcal{V}^h_{\omega^i}$ are continuous in ω^i but generally discontinuous across the boundary of the star.

Remark 2. The local restriction $\mathcal{V}^h_0$ to the element Ω_k, $\mathcal{V}^h_{\Omega_k} := \mathcal{V}^h_0 \cap \mathcal{H}^1(\Omega_k)$, is also extended to Ω in the same way. This induces the *broken space*, namely

$$\mathcal{V}^h_{\text{brok}} := \bigoplus_{k=1}^{\mathtt{n}_{\mathtt{el}}} \mathcal{V}^h_{\Omega_k}.$$

Note that functions in $\mathcal{V}^h_{\text{brok}}$ may present discontinuities across the inter-element edges (or faces) and that $\mathcal{V}^h_{\omega^i} \subset \mathcal{V}^h_{\text{brok}}$.

Remark 3. The bilinear form $a(\cdot,\cdot)$ and the energy norm are generalized to accept *broken* functions in its arguments; that is, for v and $w \in \mathcal{V}^h_{\text{brok}}$,

$$a(v,w) := \sum_{k=1}^{\mathtt{n_{el}}} a_{\Omega_k}(v,w) \quad \text{and} \quad \|v\|^2 := \sum_{k=1}^{\mathtt{n_{el}}} \|v\|_k^2,$$

where $a_{\Omega_k}(\cdot,\cdot)$ is the restriction of the bilinear form $a(\cdot,\cdot)$ to the element Ω_k and $\|v\|_k^2 = a_{\Omega_k}(v,v)$.

Finally, adding the local estimates, which have been extended into $\mathcal{V}^h_{\text{brok}}$, a global estimate $\tilde{e} \in \mathcal{V}^h_{\text{brok}}$ is obtained,

$$\tilde{e} := \sum_{i=1}^{\mathtt{n_{np}}} \tilde{e}^{\omega^i},$$

and the upper bound of the energy norm of the reference error is recovered computing the norm of the estimate $\tilde{e}$. See [23] for a detailed description, development and formal analysis (viz. the proof of the next theorem) of this estimator.

Theorem 1. *The estimate* $\tilde{e} = \sum_{i=1}^{n_{np}} \tilde{e}^{\omega^i}$*, where* $\tilde{e}^{\omega^i}$ *is the solution of the local problem given in* (2.8)*, is such that*

$$\mathrm{E_u}[e_h] = \|\tilde{e}\|^2 \geq \|e_h\|^2.$$

3 Extension to Element Free Galerkin

3.1 Reference Error

Similarly to finite elements, in mesh-free methods a finite dimensional space $\mathcal{V}^H \subset \mathcal{H}^1(\Omega)$ is associated with a particle distribution of characteristic size H and degree p for the reproducibility imposed. Thus, the mesh-free solution u_H, which is an approximation to u, belongs to $\mathcal{V}^H$ and verifies

$$a(u_H, v) = l(v) \quad \forall v \in \mathcal{V}^H.$$

However, in mesh-free methods the refined spaces are, in general, not nested, i.e. $\mathcal{V}^H \not\subset \mathcal{V}^h$. The reference error in EFG is directly defined as the projection of the exact error into the reference space, i.e. $e_h \in \mathcal{V}^h$ is the solution of

$$a(e_h, v) = R^P(v) \quad \forall v \in \mathcal{V}^h, \tag{3.9}$$

and, in general, $e_h \neq u_h - u_H$. This weak problem is very similar to (2.4) and only differs in the functional spaces because Dirichlet boundary conditions are imposed differently. This issue will be addressed in Section 3.6. It is important to emphasize that in mesh-free methods the reference error is not anymore $u_h - u_H$ as in finite elements but only the solution of problem (3.9).

3.2 Estimation of Outputs of Interest

When bounds for outputs are sought, as in finite elements —recall equation (2.5)—, a dual reference error is defined as $\epsilon_h \in \mathcal{V}^h$ solution of

$$a(v, \epsilon_h) = l^{\mathcal{O}}(v) - a(v, \psi_H) =: R^D(v) \quad \forall v \in \mathcal{V}^h. \tag{3.10}$$

Note again, that also in this case $\epsilon_h \neq \psi_h - \psi_H$. If v is replaced by e_h in (3.10),

$$l^{\mathcal{O}}(e_h) = a(e_h, \epsilon_h) + a(e_h, \psi_H). \tag{3.11}$$

In finite elements Galerkin orthogonality implies $a(e_h, \psi_H) = 0$ but not in EFG. Nevertheless, the first term in the r.h.s. of (3.11) can be rewritten using the parallelogram identity —as in section 2.3 and equation (2.6)—, namely,

$$l^{\mathcal{O}}(e_h) = \frac{1}{4}\|\kappa e_h + \frac{1}{\kappa}\epsilon_h\|^2 - \frac{1}{4}\|\kappa e_h - \frac{1}{\kappa}\epsilon_h\|^2 + a(e_h, \psi_H).$$

Thus in EFG the output of the error, see (2.7), is bounded by

$$\frac{1}{4}\mathrm{E_l}[z_h^+] - \frac{1}{4}\mathrm{E_u}[z_h^-] + a(e_h, \psi_H) \leq l^{\mathcal{O}}(e_h) \leq \frac{1}{4}\mathrm{E_u}[z_h^+] - \frac{1}{4}\mathrm{E_l}[z_h^-] + a(e_h, \psi_H). \tag{3.12}$$

Several alternatives are possible to cope with the extra term, $a(e_h, \psi_H)$, in EFG. The simplest one is to neglect it because intuitively it is expected to be small. Another alternative is to compute bounds for it. Here, however, it will be evaluated using a computable high-order approximation, thus introducing an error which is negligible in front of the other terms.

3.3 Upper Bound Estimate of the Reference Error

In EFG the upper estimate of the reference error can also be computed with the same rationale as in finite elements. The partition of unity is naturally induced by the *moving least squares* interpolating function, which are also denoted as ϕ^i (i now being the index of each particle). Similar restriction of the functional spaces are defined, that is

$$\mathcal{V}^h_{\omega^i} := \mathcal{V}^h \cap \mathcal{H}^1(\omega^i), \quad \mathcal{V}^h_{\Omega_k} := \mathcal{V}^h \cap \mathcal{H}^1(\Omega_k), \text{ and } \mathcal{V}^h_{\text{brok}} := \bigoplus_{k=1}^{\mathrm{n_{el}}} \mathcal{V}^h_{\Omega_k}.$$

And the estimate also verifies equation (2.8), namely, find $\tilde{e}^{\omega^i} \in \mathcal{V}^h_{\omega^i}$ solving the local problems in each star ω^i

$$a_{\omega^i}(\tilde{e}^{\omega^i}, v) = R^P(\phi^i v) \quad \forall v \in \mathcal{V}^h_{\omega^i}. \tag{2.8}$$

Thus the estimate

$$\tilde{e} := \sum_{i=1}^{\mathrm{n_{np}}} \tilde{e}^{\omega^i}, \tag{3.13}$$

and the upper bound of the energy norm of the reference error is recovered computing the norm of the estimate $\tilde{e}$. Because theorem 1 is also applicable here since it is based on the following lemma, which characterizes this broken approximation.

Lemma 1. *Any estimate $\tilde{e} \in \mathcal{V}^h_{\text{brok}}$ verifying the weak error equation*

$$a(\tilde{e}, v) = R^P(v) \quad \forall v \in \mathcal{V}^h, \tag{3.14}$$

is such that the norm of $\tilde{e}$ is an upper bound of the energy norm of the reference error, that is

$$\|\tilde{e}\|^2 \geq \|e_h\|^2.$$

From practical point of view, to reduce drastically the computational effort and in order to simplify the evaluation of the local estimates, equation (2.8), the definition of the star ω^i is modified. Figure 1 illustrates the definition of the star ω^i. Recall that a local problem is solved in each star and that ϕ^i is extended (equal to zero) outside of ω^i.

Definition 1. *A star, ω^i, is the smallest integration sub-grid that includes the support of ϕ^i.*

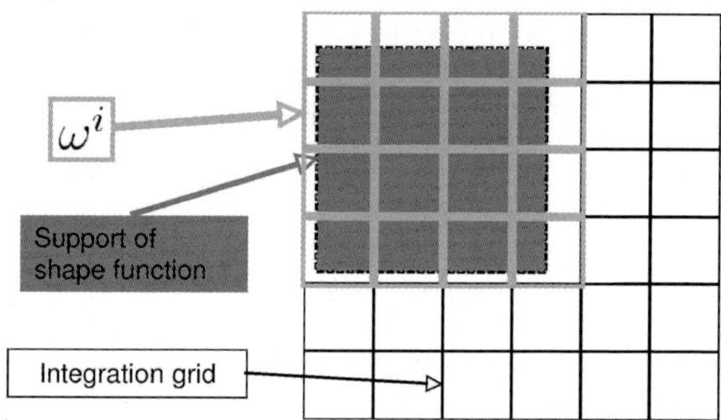

Figure 1. Definition of the star ω^i.

Remark 1. The solvability of the local problem, equation (2.8), is ensured in scalar problems as the kernel of $a_{\omega^i}(w, \cdot)$ for a generic function w includes only constant functions and for v constant the r.h.s is zero by Galerkin orthogonality, i.e. $R^P(\phi^i v) = 0$ for v constant.

3.4 Evaluation of Bounds for the Outputs

As noted in section 3.2 the bounds for the output of interest in EFG introduce a new term $a(e_h, \psi_H)$, which in principle is unknown. Here, a high-order approximation is proposed.

Theorem 1. *The term $a(\tilde{e}, \psi_H)$, where $\tilde{e}$ is given by 3.13 and ψ_H is the coarse solution of the dual (adjoint) problem, is a high-order approximation of $a(e_h, \psi_H)$. In fact,*

$$|a(e_h, \psi_H) - a(\tilde{e}, \psi_H)| \leq C\, h^p H^p. \tag{3.15}$$

Proof. In view of the definition of the reference error (3.9) and the properties of the estimator, see equation (3.14), it is known that,

$$R^P(v) = a(e_h, v) = a(\tilde{e}, v) \quad \forall v \in \mathcal{V}^h,$$

and therefore, $a(e_h - \tilde{e}, \mathcal{T}\!\mathcal{T}^h \psi_H) = 0$. Where the interpolation operator of any function in $\mathcal{V}$ onto the the reference space, $\mathcal{V}^h$, is introduced. That is, $\mathcal{T}\!\mathcal{T}^h : \mathcal{V} \longrightarrow \mathcal{V}^h$ is such that $\mathcal{T}\!\mathcal{T}^h v(\boldsymbol{x}^i) = v(\boldsymbol{x}^i)$ where $\boldsymbol{x}^i$ denote the nodes (particles) of the reference mesh. Thus,

$$a(e_h - \tilde{e}, \psi_H) = a(e_h - \tilde{e}, \psi_H - \mathcal{T}\!\mathcal{T}^h \psi_H)$$

which induces the following bound after using Cauchy-Schwartz inequality

$$|a(e_h - \tilde{e}, \psi_H)| = |a(\tilde{e} - e_h, \psi_H - \mathcal{T}\!\mathcal{T}^h \psi_H)| \leq \|e_h - \tilde{e}\| \, \|\psi_H - \mathcal{T}\!\mathcal{T}^h \psi_H\|.$$

In order to finish the proof the standard interpolation error is employed, i.e. $\|\psi_H - \mathcal{T}\!\mathcal{T}^h \psi_H\| \leq C_1 h^p$, and the bound $\|e_h - \tilde{e}\| \leq C_2 H^p$ is recalled. The latter bound requires that the estimator is efficient (i.e. that exists a constant, $C_3 \leq 1$ such that $C_3 \|\tilde{e}\| \leq \|e_h\| \leq \|\tilde{e}\|$) to obtain the following bound

$$\begin{aligned}
\|e_h - \tilde{e}\|^2 &= \|e_h\|^2 + \|\tilde{e}\|^2 - 2a(e_h, \tilde{e}) \\
&= \|e_h\|^2 + \|\tilde{e}\|^2 - 2R^P(e_h) \\
&= \|e_h\|^2 + \|\tilde{e}\|^2 - 2\|e_h\|^2 \\
&= \|\tilde{e}\|^2 - \|e_h\|^2 \leq \frac{1}{C_3}\|e_h\|^2 - \|e_h\|^2 = (\frac{1}{C_3} - 1)\|e_h\|^2.
\end{aligned}$$

Using now the standard approximation bound $\|e_h\| \leq \|e\| = \|u - u_H\| \leq C_4 H^p$ in the previous equation the desired result $\|e_h - \tilde{e}\| \leq C_2 H^p$ is obtained to end the proof.

□

Thus, in practice, the bounds for the output of the error presented in equation (3.12) are computed using

$$\frac{1}{4}\mathrm{E}_{\mathrm{l}}[z_h^+] - \frac{1}{4}\mathrm{E}_{\mathrm{u}}[z_h^-] + a(\tilde{e}, \psi_H) \leq l^{\mathcal{O}}(e_h) \leq \frac{1}{4}\mathrm{E}_{\mathrm{u}}[z_h^+] - \frac{1}{4}\mathrm{E}_{\mathrm{l}}[z_h^-] + a(\tilde{e}, \psi_H),$$

and, obviously, the bounds for the output of interest of the reference approximation, $l^{\mathcal{O}}(u_h)$, are

$$\begin{aligned}
l^{\mathcal{O}}(u_h) &\geq l^{\mathcal{O}}(u_H) + \frac{1}{4}\mathrm{E}_{\mathrm{l}}[z_h^+] - \frac{1}{4}\mathrm{E}_{\mathrm{u}}[z_h^-] + a(\tilde{e}, \psi_H) \\
l^{\mathcal{O}}(u_h) &\leq l^{\mathcal{O}}(u_H) + \frac{1}{4}\mathrm{E}_{\mathrm{u}}[z_h^+] - \frac{1}{4}\mathrm{E}_{\mathrm{l}}[z_h^-] + a(\tilde{e}, \psi_H).
\end{aligned}$$

3.5 Numerical Integration

Suppose the coarse (global) problem is solved with a given numerical quadrature,

$$a_Q(u_H, v) = l_Q(v) \quad \forall v \in \mathcal{V}^H.$$

Then, as expected, the residue is zero,

$$R_Q^P(v) = l_Q(v) - a_Q(u_H, v) = 0 \quad \forall v \in \mathcal{V}^H.$$

But for a different quadrature,

$$a_q(u_H, v) \neq l_q(v) \quad \forall v \in \mathcal{V}^H,$$

and, thus, Galerkin orthogonality is lost,

$$R_q^P(v) = l_q(v) - a_q(u_H, v) \neq 0 \quad \forall v \in \mathcal{V}^H.$$

Therefore, the same quadrature must be used to compute on the coarse and reference (truth) discretizations because Galerkin orthogonality is needed both (theoretically) to proof the upper bound property and also (practically) to ensure solvability of the local problems, see remark 1. This obviously implies that the so-called "coarse" computation is done with the quadrature of the reference discretization. This is obviously and extra cost which is required to compute the error distribution. This, however, does not preclude any adaptive refinement scheme as it will be seen in the next section.

3.6 Dirichlet Boundary Conditions

In the mesh-free context, shape functions usually do not verify the Kronecker delta property. Therefore, imposing Dirichlet boundary conditions is not as trivial as in the finite element method. In recent years, many specific techniques for the implementation of essential boundary conditions in mesh-free methods have been developed. A general overview on existing techniques is presented in [9]. Of the different techniques discussed in [9] the continuous blending method [12, 13, 8] (i.e. introduce standard finite elements along the boundary and adapt the mesh-free interoplation functions to obtain completeness) and Nitsche's method [19, 27, 2] are the most suitable alternatives. Both alternatives can be used to estimate the error. Here, for compactness of the notation Nitsche's method is used to define the weak form and has been used the examples.

The weak solution of problem (2.1) requires to find $u \in \mathcal{H}^1(\Omega)$ verifying

$$a(u, v) = l(v) \quad \forall v \in \mathcal{H}^1(\Omega), \tag{3.16}$$

where Nitsche's method modifies the forms in the previous equation as follows:

$$a(u,v) = \int_{\Omega} \nabla v \cdot \nabla u + \sigma v u \, d\Omega$$
$$- \int_{\partial\Omega} v \, \boldsymbol{n} \cdot \nabla u \, d\Gamma - \int_{\partial\Omega} u \, \boldsymbol{n} \cdot \nabla v \, d\Gamma + \beta \int_{\partial\Omega} v u \, d\Gamma, \quad (3.17)$$

$$l(v) = \int_{\Omega} v s \, d\Omega - \int_{\partial\Omega} u_D \, \boldsymbol{n} \cdot \nabla v \, d\Gamma + \beta \int_{\partial\Omega} v u_D \, d\Gamma. \quad (3.18)$$

Note that equation (3.16) is identical to (2.2) but the spaces for the approximation and the test functions are different. Note also that as in finite elements, this bilinear form $a(\cdot,\cdot)$ also induces an energy norm.

The last term in (3.17) is required to ensure coercivity of the bilinear form $a(\cdot,\cdot)$ provided that β is large enough. Regarding the choice of the lower bound of β, Nitsche proved that if β is taken as $\beta = \gamma/\ell$, where γ is a large enough constant (independent of h) and ℓ denotes a characteristic measure of the finite element discretization, then the discrete solution converges to the exact solution with optimal order in $\mathcal{H}^1$ and $\mathcal{L}^2$ norms. Moreover, γ can be estimated from the maximum eigenvalue of a generalized eigenvalue problem, see [11]. Similar results are obtained for mesh-free methods and ℓ is related to a measure of the support of the interpolation functions or equivalently to the distance between particles.

Here, two characteristic sizes are used, H is associated to the coarse discretization and h characterizes the reference or truth distribution of particles. As in the previous section *only one* bilinear form is used. Thus $a(\cdot,\cdot)$, and consequently β, is the same for both the global coarse computation in $\mathcal{V}^H$ and the local reference evaluation in $\mathcal{V}^h$. This obviously implies choosing for β the corresponding value associated to the reference mesh (as already done in the numerical integration). Because it will be larger than the one related to $\mathcal{V}^H$ and thus will ensure coercivity of (3.17) and obviously consistency of (3.16).

In summary, as for numerical integration, the parameter β used in the coarse distribution of particles is associated to the reference distribution used later to evaluate the error distribution. Again, this is necessary both for theoretical and practical considerations (Galerkin orthogonality is necessary). Nevertheless, this does not preclude any adaptive scheme, as will be shown on a following publication, because at each refinement step the reference distribution of particles is known a priori (it is directly related to the given "coarse" distribution of particles).

4 Numerical Results

In this section, the proposed estimator is used to evaluate bounds for thermal model problems. Some of the selected examples have been used by other authors to assess the performance of similar techniques [23, 20].

4.1 First Poisson Example

The 2D Laplace problem

$$\begin{cases} \boldsymbol{\nabla}^2 u = 0 & \text{in } (x,y) \in]0,1[\times]0,1[\\ u(x,0) = \sin(\pi x) & \\ u(x,1) = u(0,y) = u(1,y) = 0 & \end{cases}$$

with known analytical solution [28, 9],

$$u(x,y) = \big(\cosh(\pi y) - \coth(\pi)\sinh(\pi y)\big)\sin(\pi x),$$

is considered next. Figure 2 shows this analytical solution and the primal and dual approximations using EFG with bilinear consistency and a uniform distribution of particles of 5×5.

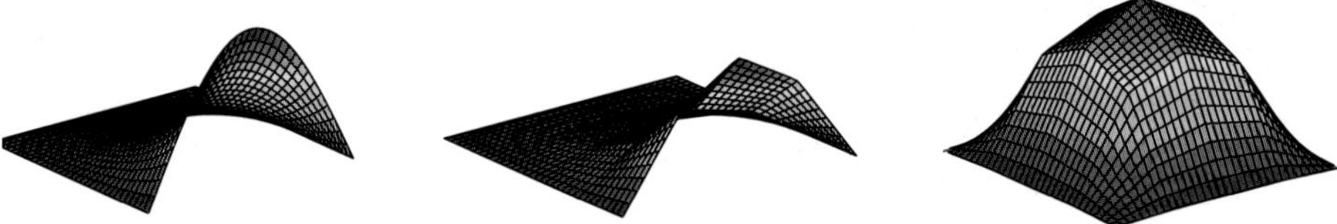

Figure 2. Exact solution (left), EFG primal (center) and dual (right) approximations.

The behavior of the energy norm estimate is compared to the exact energy error norm and the reference error energy norm, see Figure 3. The approximate solution u_H is computed using bilinear consistency and the reference space is associated with a particle distribution $h = H/4$. Uniform distributions of particles have been considered. From a qualitative viewpoint, it is worth noting that the estimated error distribution is in good agreement with the exact error distribution.

Figure 4 shows the spatial distribution of the local effectivity index. It is important to notice that the effectivity index is always close to the optimum value: one. Obviously if the index is evaluated using the reference error, e_h, which is the value actually bounded, the effectivity index is even better (closer to one and more uniform).

Remark 1. Elements of the integration grid with very small contributions to the errors are not considered in the previous plot (areas not plotted) because they are not interesting from an adaptive viewpoint. Moreover, in these areas, small roundoff errors in the error assessment lead to unreasonable effectivity indices (very small absolute error but large relative error). An element is considered to have a "small" contribution to the global error when $\|e_h\|/4n_{el}$ (with n_{el} being the number of elements of the integration grid). This results, in this case, on neglecting 20% of the elements approximately.

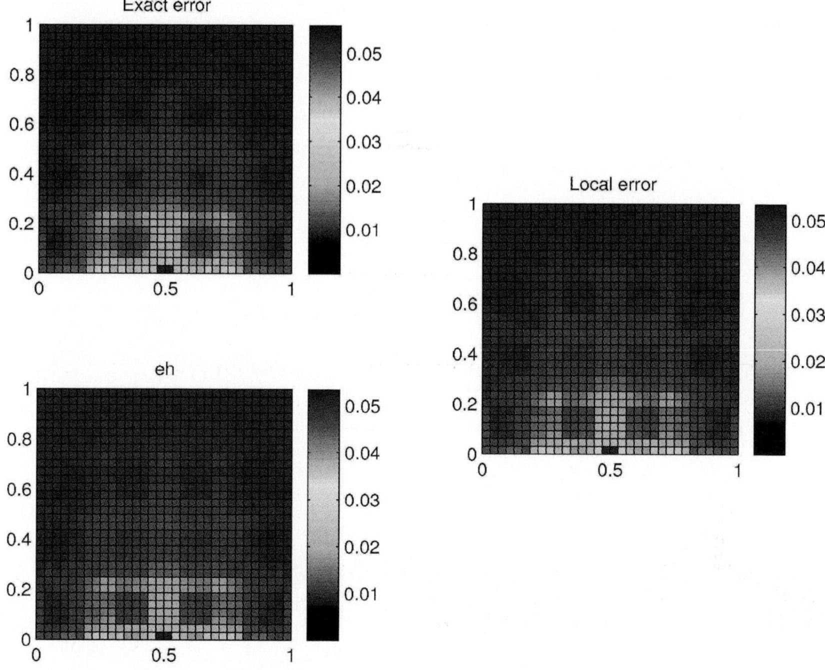

Figure 3. Energy norm of error and estimate.

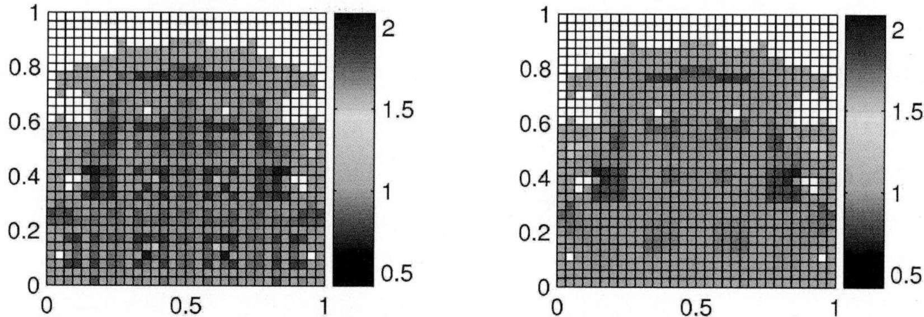

Figure 4. Local effectivity index with respect exact error (left) and with respect the reference error, e_h, (right).

Remark 2. The local energy norm has been represented in the integration grid. From now on, all local representations will be represented in this grid.

Finally, Figure 5 shows the resulting bounds and their convergence. The lower and upper bounds for an approximation to the output of the exact solution are shown as well as the bound average, the EFG approximation and the exact output. As expected they converge to the exact output as the number of particles is refined. Second order convergence is obtained as bilinear consistency is used. For the initial distribution of particles (uniform distribution

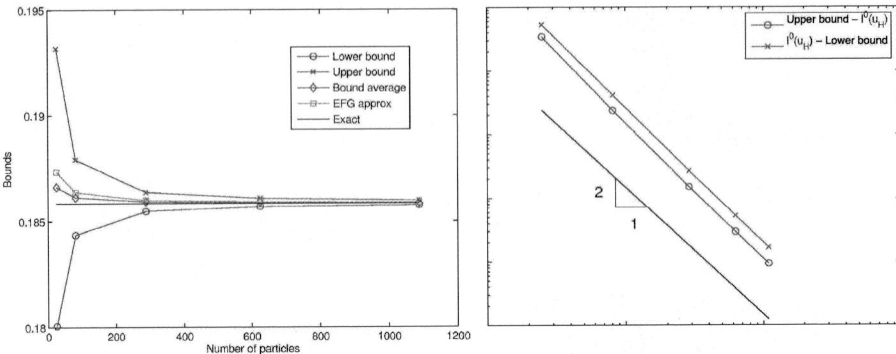

Figure 5. Resulting bounds (left) and their convergence (right).

of 5×5 particles) the output is bounded by, $l^{\mathcal{O}}(u_h) \in [0.1801, 0.1932]$, that is

$$l^{\mathcal{O}}(u_h) = 0.1866 \pm 0.0066 = 0.1866 \pm 3.54\%$$

and for the final distribution (33×33 particles) the output is bounded by $l^{\mathcal{O}}(u_h) \in [0.1858, 0.1860]$

$$l^{\mathcal{O}}(u_h) = 0.1859 \pm 0.0001 = 0.1859 \pm 0.05\%.$$

4.2 Second Poisson Example

A well-known benchmark is solved in this section, see [1, 20, 6, 23]. The problem reads,

$$\begin{cases} -\boldsymbol{\nabla}^2 u = s & \text{in } \Omega, \\ u = 0 & \text{on } \partial\Omega, \end{cases}$$

where the source term is chosen such that the exact solution has the following analytical expression

$$u(x, y) = x^2(1-x)^2(e^{10x^2} - 1)y^2(1-y)^2(e^{10y^2} - 1)/2000.$$

Figure 6 shows this analytical solution and the primal and dual approximations using EFG with bilinear consistency and a uniform distribution of particles of 5×5. Note that the EFG primal solution verifies the Dirichlet boundary condition weakly as expected when it is imposed by Nitsche's method.

The behavior of the energy norm estimate is compared to the exact energy error norm and the reference error energy norm, see Figure 7. The approximate solution u_H is computed using bilinear consistency and the reference space is associated with a particle distribution $h = H/4$. Uniform distributions of particles have been considered. From a qualitative viewpoint, it is worth noting

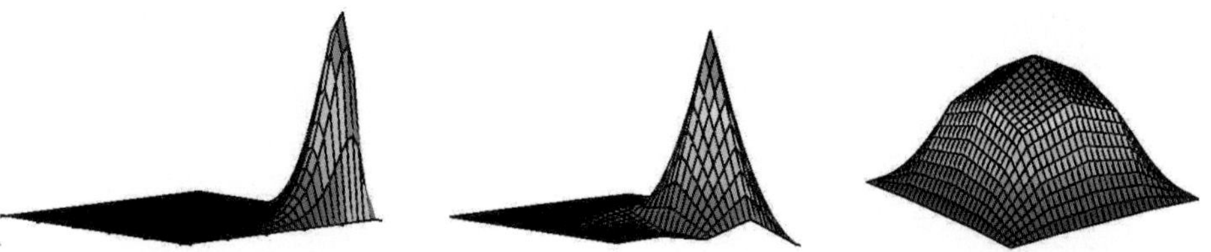

Figure 6. Exact solution (left), EFG primal (center) and dual (right) approximations.

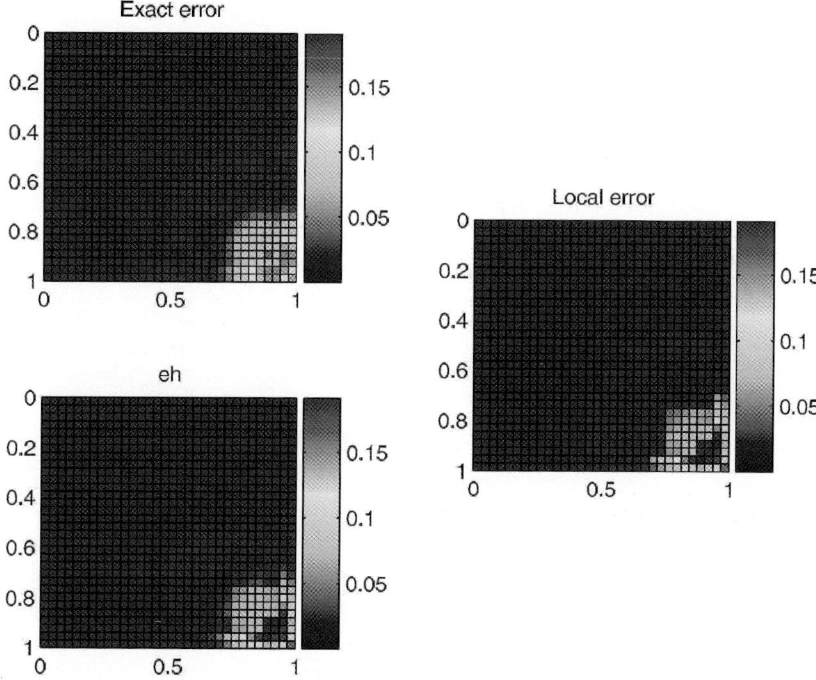

Figure 7. Energy norm of error and estimate.

that also in this example the estimated error distribution is in good agreement with the exact error distribution and almost identical to the reference error distribution. In fact, this comparison is more clear in Figure 8 where the spatial distribution of the local effectivity indexes is plotted.

Finally, Figure 9 shows the resulting bounds and their convergence. The lower and upper bounds for an approximation to the output of the exact solution are shown as well as the bound average, the EFG approximation and the exact output. As expected the optimal rete of converge is obtained as the number of particles is refined.

For the initial distribution of particles (uniform distribution of 5×5 particles) the output is bounded by, $l^{\mathcal{O}}(u_h) \in [-0.0097, 0.0276]$

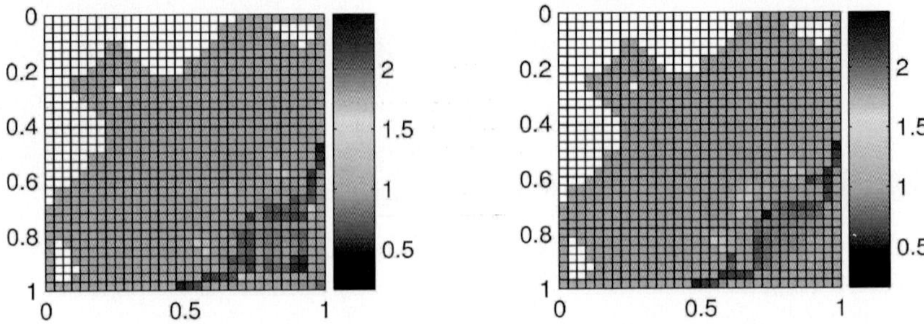

Figure 8. Local effectivity index with respect exact error (left) and with respect e_h (right).

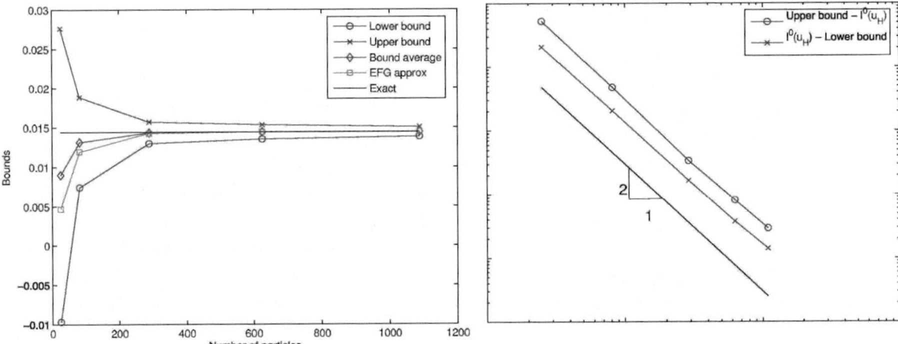

Figure 9. Resulting bounds and their convergence.

$$l^{\mathcal{O}}(u_h) = 0.0090 \pm 0.0187 = 0.0090 \pm 207.78\%$$

and for the final distribution (33×33 particles) the output is bounded by $l^{\mathcal{O}}(u_h) \in [0.0138, 0.0150]$

$$l^{\mathcal{O}}(u_h) = 0.0144 \pm 0.0006 = 0.0144 \pm 4.17\%.$$

5 Concluding Remarks

For the first time in mesh-free methods an implicit residual-based estimation is presented. Moreover, with this strategy bounds for outputs of interest can be computed. The resulting estimate yields upper and lower bounds for the output of the reference error. Moreover, the distribution of local contributions to the error are accurately estimated, both for the energy norm of the error and for the error measured using some functional output. Therefore, this estimate is well suited to guide goal-oriented adaptive procedures.

References

1. M. Ainsworth and J.T. Oden, *A posteriori error estimation in finite element analysis*, John Wiley & Sons, Chichester, 2000.
2. R. Becker, *Mesh adaptation for Dirichlet flow control via Nitsche's method*, Commun. Numer. Methods Eng. **18** (2002), no. 9, 669–680.
3. J. Bonet, A. Huerta, and J. Peraire, *The efficient computation of bounds for functionals of finite element solutions in large strain elasticity*, Comput. Methods Appl. Mech. Engrg. **191** (2002), no. 43, 4807–4826.
4. C. Carstensen and S.A. Funken, *Fully reliable localized error control in the FEM*, SIAM J. Sci. Comput. **21** (1999/00), no. 4, 1465–1484.
5. H.J. Chung and T. Belytschko, *An error estimate in the EFG method*, Comput. Mech. **21** (1998), no. 2, 91–100.
6. P. Diez, N. Pares, and A. Huerta, *Recovering lower bounds of the error by postprocessing implicit residual a posteriori error estimates*, Internat. J. Numer. Methods Engrg. **56** (2003), no. 10, 1465–1488.
7. C.A. Duarte and J.T. Oden, *An h-p adaptive method using clouds*, Comput. Methods Appl. Mech. Eng. **139** (1996), no. 1-4, 237–262.
8. S. Fernandez-Mendez, J. Bonet, and A. Huerta, *Continuous blending of SPH with finite elements*, Comput. Struct. **83** (2005), no. 17–18, 1448–1458.
9. S. Fernandez-Mendez and A. Huerta, *Imposing essential boundary conditions in mesh-free methods*, Comput. Methods Appl. Mech. Eng. **193** (2004), no. 12–14, 1257–1275.
10. L. Gavete, J.L. Cuesta, and A. Ruiz, *A procedure for approximation of the error in the EFG method*, Int. J. Numer. Methods Eng. **53** (2002), 667–690.
11. M. Griebel and M.A. Schweitzer, *A Particle-Partition of Unity Method. Part V: Boundary Conditions*, Geometric Analysis and Nonlinear Partial Differential Equations, (S. Hildebrandt and H. Karcher, eds.), Springer, 2002, pp. 517–540.
12. A. Huerta and S. Fernandez-Mendez, *Enrichment and coupling of the finite element and meshless methods*, Int. J. Numer. Methods Eng. **48** (2000), no. 11, 1615–1636.
13. A. Huerta, S. Fernandez-Mendez, and W.K. Liu, *A comparison of two formulations to blend finite elements and mesh-free methods*, Comput. Methods Appl. Mech. Eng. **193** (2004), no. 12–14, 1105–1117.
14. P. Ladevèze and D. Leguillon, *Error estimate procedure in the finite element method and applications*, SIAM J. Numer. Anal. **20** (1983), no. 3, 485–509.
15. W.K. Liu, S. Jun, D.T. Sihling, Y. Chen, and W. Hao, *Multiresolution reproducing kernel particle method for computational fluid dynamics*, Int. J. Numer. Methods Fluids **24** (1997), 1391–1415.
16. L. Machiels, Y. Maday, and A.T. Patera, *A "flux-free" nodal Neumann subproblem approach to output bounds for partial differential equations*, C. R. Acad. Sci. Paris Sér. I Math. **330** (2000), no. 3, 249–254.
17. Y. Maday, A.T. Patera, and J. Peraire, *A general formulation for a posteriori bounds for output functionals of partial differential equations; application to the eigenvalue problem*, C. R. Acad. Sci. Paris Sér. I Math. **328** (1999), no. 9, 823–828.
18. P. Morin, R.H. Nochetto, and K.G. Siebert, *Local problems on stars: a posteriori error estimators, convergence, and performance*, Math. Comp. **72** (2003), no. 243, 1067–1097.

19. J. Nitsche, *über ein variations zur lösung von dirichlet-problemen bei verwendung von teilräumen die keinen randbedingungen unterworfen sind*, Abh. Math. Se. Univ. **36** (1970), 9–15.
20. J.T. Oden and S. Prudhomme, *Goal-oriented error estimation and adaptivity for the finite element method*, Comput. Math. Appl. **41** (2001), no. 5-6, 735–756.
21. M. Paraschivoiu, J. Peraire, and A.T. Patera, *A posteriori finite element bounds for linear-functional outputs of elliptic partial differential equations*, Comput. Methods Appl. Mech. Eng. **150** (1997), no. 1-4, 289–312.
22. N. Pares, J. Bonet, A. Huerta, and J. Peraire, *The computation of bounds for linear-functional outputs of weak solutions to the two-dimensional elasticity equations*, Comput. Methods Appl. Mech. Eng. **194** (2006), no. 4–6, 406–429.
23. N. Pares, P. Diez, and A. Huerta, *Subdomain-based flux-free a posteriori error estimators*, Comput. Methods Appl. Mech. Eng. **195** (2006), no. 4–6, 297–323.
24. A.T. Patera and J. Peraire, *A general Lagrangian formulation for the computation of a posteriori finite element bounds*, Error estimation and adaptive discretization methods in computational fluid dynamics, Lect. Notes Comput. Sci. Eng., vol. 25, Springer, Berlin, 2003, pp. 159–206.
25. S. Prudhomme and J.T. Oden, *On goal-oriented error estimation for elliptic problems: application to the control of pointwise errors*, Comput. Methods Appl. Mech. Eng. **176** (1999), no. 1-4, 313–331.
26. A.M. Sauer-Budge, J. Bonet, A. Huerta, and J. Peraire, *Computing bounds for linear functionals of exact weak solutions to Poisson's equation*, SIAM J. Numer. Anal. **42** (2004), no. 4, 1610–1630.
27. R. Stenberg, *On some techniques for approximating boundary conditions in the finite element method*, J. Comput. Appl. Math. **63** (1995), no. 1-3, 139–148.
28. G.J. Wagner and W.K. Liu, *Application of essential boundary conditions in mesh-free methods: a corrected collocation method*, Int. J. Numer. Methods Eng. **47** (2000), no. 8, 1367–1379.
29. Z.C. Xuan, N. Pares, and J. Peraire, *Computing upper and lower bounds for the J-integral in two-dimensional linear elasticity*, Comput. Methods Appl. Mech. Eng. **195** (2006), no. 4–6, 430–443.

Bubble and Hermite Natural Element Approximations

J. Yvonnet[1], P. Villon[2], and F. Chinesta[3]

[1] Université de Marne-la-Vallée, 5 Bd Descartes, 77454 Marne-la-Vallée Cedex, France
jyvonnet@univ-mlv.fr
[2] Université de Technlogies de Compiègne, BP 20529, 60205 Compiègne cedex, France
pierre.villon@utc.fr
[3] École Nationale Supérieure d'Arts et Métiers, 151 boulevard de l'Hôpital, F-75013 Paris, France
francisco.chinesta@paris.ensam.fr

Summary. In this paper, new natural element approximations are proposed, in order to address issues associated with incompressibility as well as to increase the accuracy in the Natural Element Method (NEM). The NEM exhibits attractive features such as interpolant shape functions or auto-adaptive domain of influence, which alleviates some of the most common difficulties in meshless methods. Nevertheless, the shape functions can only reproduce linear polynomials, and in contrast to moving least squares methods, it is not easy to define interpolations with arbitrary approximation consistency. In order to treat mechanical models involving incompressible media in the framework of mixed formulations, the associated functional approximations must satisfy the well known inf-sup, or LBB condition. The first proposed approach constructs richer NEM approximation schemes by means of bubbles associated with the topological entities of the underlying Delaunay tessellation, allowing to pass the LBB and to remove pressure oscillations in the incompressible limit. Despite of its simplicity, this approach does not construct approximation with higher order consistency. The second part of the paper deals with a discussion on the construction of second-order accurate NEM approximations. For this purpose, two techniques are investigated : (a) the enrichment in the MLS framework of the bubbles with higher-order polynomials and (b) the use of a new Hermite-NEM formulation.

Key words: Natural Element Method; Bubble functions; Mixed formulations; Incompressible media; LBB condition.

1 Introduction

It is well known that the solution of mechanical problems involving incompressible media using the standard displacement-based finite element technique may yield solutions that are grossly in error [2]. The difficulty is that the computed displacement field needs to satisfy the constraint of very small volumetric strains (which become zero as the condition of total incompressibility is approached) while the pressure is of the order of the boundary tractions. The displacement approximation space is not rich enough to accommodate this constraint without a drastic reduction in the rate of convergence, also known as *locking* [2].

For the analysis of such problems, one solution is to use a mixed formulation in which different approximation spaces are used for the displacement and pressure fields interpolation. Although numerous mixed formulations may be developed, only those that are stable are useful in practice [12, 11]. The solvability, stability and optimality of mixed formulations are related to a compatibility condition, the so-called LBB (or inf-sup) condition [4]. The analytical proof whether the inf-sup condition is satisfied for a specific formulation is, however, difficult, and this has spurred the use of a numerical inf-sup test [17, 4, 9].

Accounting incompressibility in meshless methods is still an open topic. Until recently, it was stated that meshfree methods are immune to locking [3, 33]. Furthermore, the EFG has been actually proposed for treating isochoric elastoplasticity by considering the shape functions support large enough [1]. In the context of the RKPM, a similar claim was made in the context of large deformation of nearly incompressible hyperelastic [6] and elastoplastic materials [16]. Recently, it has been reported that meshfree methods are in fact not locking-free in the incompressibility limit [10]. In a recent paper [14] this issue is clarified determining the influence of the EFG shape functions support on the locking behaviour. The main conclusion was that by increasing the shape functions support the locking can be attenuated, but never suppressed. Several attempts have been proposed to avoid locking in the context of meshfree methods. Huerta *et al.* [27] developed a so-called pseudo-divergence free approximation, consisting in using approximation functions that verify approximately the divergence-free constraint for a given discretization in a diffuse sense. Dolbow and Belytschko [10] have proposed a mixed displacement/pressure formulation and selective reduced integration to alleviate locking. Chen *et al.* note that the use of large support size is computationally expensive and, moreover, cannot remove pressure oscillations [7]. They proposed a pressure projection combined with a reduced integration to remove pressure oscillations in nearly incompressible elasticity problems.

In this paper, we focus on the treatment of incompressibility in the context of the natural element method (NEM). The NEM is a novel meshfree method. Its attractive features are: (a) interpolant character of the shape functions; (b) strict linearity of the shape functions over the boundaries[§]; and (c) the support of the shape functions is based on the Delaunay spheres of the surrounding nodes, and automatically adapts to the local nodal density. The properties (a) and (b) allow direct enforcement of the essential boundary conditions and guarantee conforming approximations in presence of interfaces [29]. Property (c) allows simple refinement strategies [31]. However, unlike the moving least square techniques, it is not possible to directly enrich the basis in order to improve the reproducing conditions (approximation consistency). In the context of incompressible media, Sukumar was the first to propose a mixed NEM interpolation in [26] using constant piecewise shape functions for the pressure approximation, and the standard NEM for the approximation of the displacements. In [13], González *et al.* proposed an enrichment of the NEM in the context of the partition of unity paradigm [18] to construct richer approximations, in order to verify the inf-sup condition. In [28], Chen et al. proposed to use a stabilized nodal integration to avoid locking in near-incompressible elastostatics. We propose in this paper a new approach in the context of the natural element method allowing to define stable mixed formulations for treating mechanical models involving incompressible media. In the proposed technique, additional degrees of freedom, associated with some topological entities of the underlying Delaunay tessellation, i.e. edges, triangles and tetrahedrons, are introduced. The associated *bubble* shape functions are computed from the product of the NEM shape functions related to the generating nodes of the entity. A Hermite-NEM approximation is also proposed, through a discussion dealing with the construction of higher-order NEM approximations.

2 Review of the Natural Element Method

2.1 Natural Neighbor Interpolation

We briefly touch upon the foundation of Sibson's natural neighbor coordinates (shape functions) that are used in the natural element method. For a more in-depth discussion on the Sibson interpolant and its application for solving second-order partial differential equations, the interested reader can refer to Sambridge and Braun [23], and Sukumar *et al.* [25]. The NEM interpolant is constructed on the basis of the Voronoi diagram. The Delaunay tessellation is the topological dual of the Voronoi diagram.

Consider a set of nodes $S = \{n_1, n_2, \ldots, n_N\}$ in $\Re^{dim}$. The Voronoi diagram is the subdivision of $\Re^{dim}$ into regions T_i (Voronoi cells) defined by

$$T_i = \{\mathbf{x} \in \Re^{dim} : d(\mathbf{x}, \mathbf{x}_i) < d(\mathbf{x}, \mathbf{x}_j), \forall j \neq i\}, \quad \forall\, i \tag{2.1}$$

[§] This property is restricted to convex boundaries [25]. However, some techniques have been provided to extend it to non-convex boundaries [30, 8]

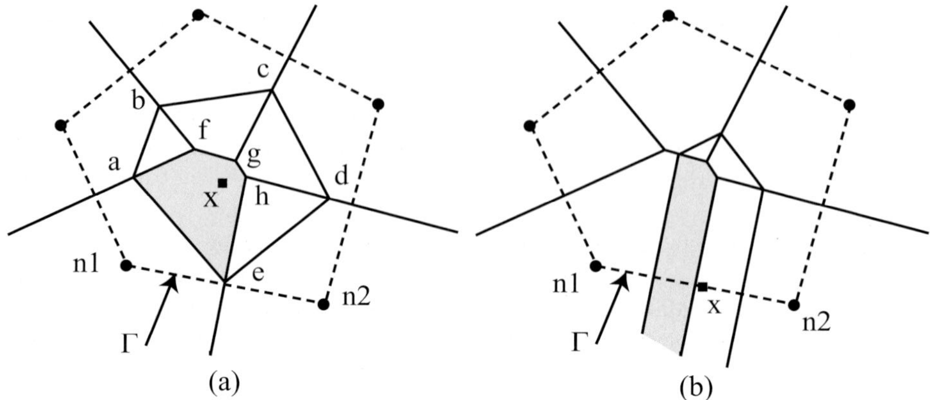

Figure 1. Construction of the Sibson shape functions.

The Sibson coordinates of $\mathbf{x}$ with respect to a natural neighbor n_i (see Fig. 1a) is defined as the ratio of the overlap area (volume in 3D) of their Voronoi cells to the total area (volume in 3D) of the Voronoi cell related to point $\mathbf{x}$. If we consider the 2D example depicted in figure 1(a), we have:

$$\phi_1(\mathbf{x}) = \frac{Area(afghe)}{Area(abcde)} \tag{2.2}$$

If the point $\mathbf{x}$ coincides with the node n_i, i.e. $(\mathbf{x} = \mathbf{x}_i)$, $\phi_i(\mathbf{x}_i) = 1$, and all other shape functions are zero, i.e. $\phi_j(\mathbf{x}_i) = \delta_{ij}$ (δ_{ij} being the Kronecker delta). The properties of positivity, interpolation, and partition of unity are then verified [25]:

$$\begin{cases} 0 \leq \phi_i(\mathbf{x}) \leq 1 \\ \phi_i(\mathbf{x}_j) = \delta_{ij} \\ \sum_{i=1}^{n} \phi_i(\mathbf{x}) = 1. \end{cases} \tag{2.3}$$

The natural neighbour shape functions also satisfy the local coordinate property [24], namely:

$$\mathbf{x} = \sum_{i=1}^{n} \phi_i(\mathbf{x})\mathbf{x}_i \tag{2.4}$$

which combined with Eq. (2.3), implies that the natural neighbour interpolant spans the space of linear polynomials (linear completeness).

Sibson natural neighbour shape functions are C^1 at any point except at the nodes, where they are only C^0. The C^1 continuity everywhere can be obtained by using special classes of natural neighbour shape functions [15].

The support (domain of influence) of a shape function ϕ_i is the union of the Delaunay spheres (circumscribing the Delaunay tetrahedrons) containing the node n_i. This support is thus not radial and automatically adapts to the relative position of n_i and its neighbours, whether is the density or the regularity of the nodal distribution.

Another important property of this interpolant is its strict linearity over the boundary of convex domains. The proof can be found in Sukumar *et al.* [25]. An illustration is depicted in Fig. 1 (b): as the areas associated to points on the boundary become infinite, the contribution of internal nodes vanish in the limit when the point approaches the convex boundary, and the shape functions associated with nodes n_1 and n_2 become linear on the segment $(n_1 - n_2)$. This is not true in the case of non convex boundaries, and an appropriate treatment must be introduced to preserve this property in non-convex domains [30, 8]. Essential boundary conditions can thus be enforced directly, as in the finite element method. This property also guarantees strict continuity of the approximation across material interfaces [29], which is an issue in most meshfree methods.

Consider an interpolation scheme for a vector-valued function $\mathbf{u}(\mathbf{x}) : \Omega \subset \Re^2 \rightarrow \Re$, in the form

$$\mathbf{u}^h(\mathbf{x}) = \sum_{i=1}^{n} \phi_i(\mathbf{x})\ \mathbf{u}_i \tag{2.5}$$

where $\mathbf{u}_i$ are the nodal values of the field at the n natural neighbour nodes, and $\phi_i(\mathbf{x})$ are the shape functions associated with each neighbour node. It is noted that Eq. (2.5) defines a local interpolation scheme. Thus, the trial and test functions used in the discretization of the variational formulation describing the problems treated in this paper take the form of Eq. (2.5).

One of the drawbacks of the NEM is that natural neighbour shape functions can only reproduce at best linear fields, which induces difficulties to construct mixed formulations, where the different fields must be approximated in different approximation spaces in order to avoid numerical locking (LBB condition [4]). In the next section, two new approaches are proposed to enrich the NEM approximation.

3 Hierarchical Bubble Functions in the Natural Element Method

Consider an open bounded domain $\Omega \in \Re^{dim}$ with boundary Γ, dim being the space dimension. Assume that Ω is discretized by a set of nodes S. Let $D(S)$ be the simplicial complex associated with the Delaunay tessellation of S. A simplicial complex K in $\Re^{dim}$ is a collection of simplices (hypertetrahedra) in $\Re^{dim}$ such that:

(i) Every face of a simplex of K is in K;
(ii) The intersection of any two simplices of K is a face of each of them [19];

If we denote F_k the set of $k - simplices$ $(0 \leq k \leq 3)$, in R^3 the Delaunay tessellation $D(S)$ will be defined as the simplicial complex defined by the tetrahedra in F_3, the triangles in F_2, the edges in F_1, and the vertices in F_0. We denote these collections $T(S)$, $F(S)$, $E(S)$ and $V(S)$, respectively.

In order to construct richer approximations, new shape functions can be associated with the different $k - simplices$. The case $1 < k < 3$ is related to the concept of hierarchical methods [34]. The concept of hierarchical bubble shape functions is a very simple way to construct richer approximations. The extension to meshfree methods is not an easy matter in general, in the absence of topology related to some elements. In the natural element, the underlying Delaunay triangulation allows the use of such an approach.

The key idea is to associate new shape functions to the $k - simplices$ of the Delaunay tessellation, i.e. tetrahedra $T' \in T(S)$, triangular facets $F' \in F(S)$ and edges connecting two nodes in the Delaunay triangulation $E' \in E(S)$.

3.1 b-NEM Approximation

A k-simplex (K-S) (vertex, edge, triangular facet or tetrahedron) is generated by $K = k + 1$ vertices ($k = 0$, 1, 2 and 3, respectively). The bubble shape function of an entity χ_j generated by K vertices is computed like

$$\phi_j^*(\mathbf{x}) = \prod_{p=1}^{K} \phi_p(\mathbf{x}) \tag{3.6}$$

where $\phi_p(\mathbf{x})$ is the NEM shape function (Eq. 2.2) associated with node n_p computed at point $\mathbf{x}$.

The support (domain of influence) of a K-S generated by K vertices (nodes) in S is the union of the Delaunay spheres containing the K nodes. It results, in 2D:

(i) if χ_j is a Delaunay triangle $(\chi \in F(S))$ $(k = 2)$, the support of χ_j is composed with one circle containing the 3 generating nodes of the triangle (see fig. 2 (a));
(ii) if χ_j is an edge of a Delaunay triangle $(\chi \in E(S))$ $(k = 1)$, the support of χ_j is composed of the union of two circles (see figure 2 (b)) (if $\chi_j \notin \Gamma$), or one circle if $\chi_j \in \Gamma$ (see figure 2 (a)), containing the 2 generating nodes of χ_j;

We now consider the following approximation scheme

$$\mathbf{u}^h(\mathbf{x}) = \sum_{i=1}^{n} \phi_i(\mathbf{x})\ \mathbf{u}_i + \sum_{j=1}^{m} \phi_j^*(\mathbf{x})\ \gamma_j \tag{3.7}$$

where n is the number of natural neighbours of point $\mathbf{x}$, $\phi_i(\mathbf{x})$ is the NEM shape function related to node $n_i \in S$ computed at point $\mathbf{x}$, $\phi_j^*(\mathbf{x})$ is the

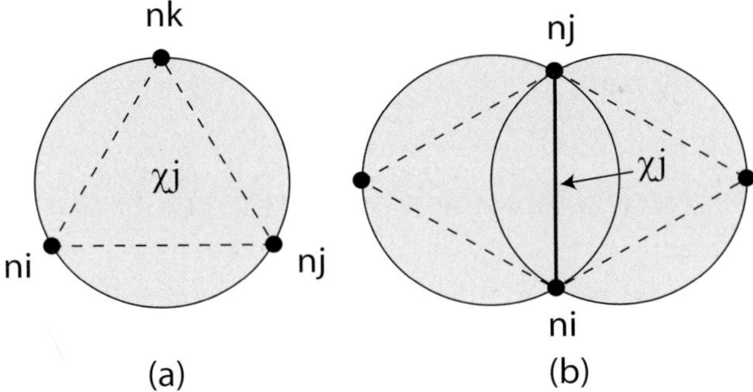

Figure 2. Supports of the bubble shape functions associated with the Delaunay k-simplex; (a) support of a Delaunay triangle χ_j $n_i - n_j - n_k$; (b) support of a Delaunay edge χ_j $n_i - n_j$.

bubble shape function defined in Eq. (3.6) associated with the m influent K-S, and γ_j are additional degrees of freedom.

Remarks.

(i) Different combinations can be chosen for enriching the approximation, i.e. using only bubble functions associated with the edges, with the Delaunay triangles, or both.
(ii) The evaluation of the bubble shape functions associated with the K-S is not costly as it only requires the product of available NEM shape functions computed at point $\mathbf{x}$.
(iii) Despite that the approximation scheme defined in Eq. (3.7) is richer than standard NEM approximation, it does not satisfy any reproducing property other than the linear consistency.

In [32], two variant of the b-NEM approximation have been investigated: (a) the enrichment of the approximation using one bubble function associated with each Delaunay triangle, called b1-NEM, and (b) the enrichment of the approximation using one bubble function associated with each Delaunay edge, called b2-NEM.

3.2 b-NEM with Reproducing Properties : b-NEM$^+$

In this section we proceed to correct the shape functions previously constructed defining the approximation scheme (3.7) within a standard moving least squares framework, in order to evaluate the benefits provided by the higher approximation consistency. We briefly summarize the MLS procedure [20, 3]. Let $w_i(\mathbf{x})$ be some weight function either associated with a standard or a bubble-NEM shape function, computed at point $\mathbf{x}$. Let the following approximation scheme defined by

$$\mathbf{u}^h(\mathbf{x}) = \mathbf{p}^T(\mathbf{x})\mathbf{a}(\mathbf{x}) \tag{3.8}$$

with a polynomial basis $\mathbf{p}^T(\mathbf{x})$, i.e. $\mathbf{p}^T(\mathbf{x}) = [1, x, y, xy]$ and $\mathbf{p}^T(\mathbf{x}) = [1, x, y, xy, x^2, y^2]$ for a bilinear and quadratic basis, respectively, in 2D, and $\mathbf{a}(\mathbf{x})$ a vector of unknown coefficients. In order to determine $\mathbf{a}(\mathbf{x})$, the functional J defined by eq. (3.9) has to be minimized with respect to $\mathbf{a}(\mathbf{x})$:

$$J = \frac{1}{2}\sum_{i=1}^{n} w_i(\mathbf{x}) \left[\mathbf{p}^T(\mathbf{x}_i)\mathbf{a}(\mathbf{x}) - u_i\right]^2 \tag{3.9}$$

where u_i are the nodal unknown associated with neighbours of point $\mathbf{x}$. The minimization of J with respect to the unknown coefficient $a_j(\mathbf{x})$ leads to:

$$\frac{\partial J}{\partial a_j(\mathbf{x})} = \sum_{k=1}^{n} a_k \left[\sum_{i=1}^{n} w_i(\mathbf{x}) p_j(\mathbf{x}_i) p_k(\mathbf{x}_i)\right] - \sum_{i=1}^{n} w_i(\mathbf{x}) p_j(\mathbf{x}_i) u_i = 0 \tag{3.10}$$

which leads to the usual linear system

$$\mathbf{A}(\mathbf{x})\mathbf{a}(\mathbf{x}) = \mathbf{B}(\mathbf{x})\mathbf{u}. \tag{3.11}$$

Substituting $\mathbf{a}(\mathbf{x})$ in Eq. (3.8), results in

$$u^h(\mathbf{x}) = \mathbf{p}^T(\mathbf{x})\mathbf{A}^{-1}(\mathbf{x})\mathbf{B}(\mathbf{x})\mathbf{u}. \tag{3.12}$$

By identification, the new shape functions are given by

$$\boldsymbol{\psi}^T(\mathbf{x}) = \mathbf{p}^T(\mathbf{x})\mathbf{A}^{-1}(\mathbf{x})\mathbf{B}(\mathbf{x}). \tag{3.13}$$

The reproducing b-NEM shape functions are computed by setting $w_i(\mathbf{x}) = \{\phi_i(\mathbf{x});\ \phi_j^*(\mathbf{x})\}$, $\phi_i(\mathbf{x})$ and $\phi_j^*(\mathbf{x})$ being the shape functions defined in (2.2) and (3.6).

In the following, $\boldsymbol{\psi}(\mathbf{x})$ is a vector containing the shape functions associated with influent nodes or K-S at point $\mathbf{x}$.

Remark. The main difference between the reproducing-b-NEM and the b-NEM without additional reproducing properties is that physical coordinates must be associated with each K-S shape function, in order to evaluate the terms $p_j(\mathbf{x}_i)$ and $p_k(\mathbf{x}_i)$ in Eq. (3.10). A simple solution is to consider the K-S centroid coordinates.

In the following, the b1-NEM and b2-NEM schemes described in the previous section are corrected using the MLS procedure just described. In the most unfavourable case a point $\mathbf{x}$ is influenced by four shape functions in the b1-NEM (3 NEM shape functions, and 1 bubble shape function associated with the Delaunay triangle). As these weight functions are independent, the method is stable if the basis $\mathbf{p}^T(\mathbf{x})$ contains 4 monomials. We call b1-NEM$^+$

the enrichment of the b1-NEM from $\mathbf{p}^T(\mathbf{x}) = \{1, x, y, xy\}$. Following similar assumptions, b2-NEM$^+$ results from the enrichment of the b2-NEM using $\mathbf{p}^T(\mathbf{x}) = \{1, x, y, xy, x^2, y^2\}$.

We have shown in [32] that essential boundary conditions can be enforced directly in all the proposed approximation schemes, as the bubble-NEM shape functions vanish over all external boundaries. For further details, see the proofs for the different schemes in the referred paper.

3.3 Natural Element Discretization

We consider the usual mixed variational formulation of the incompressible linear elastostatics problem where displacement trial and test functions are interpolated using the same shape functions, as the same for the pressure trial and test functions. In the following, the pressure is interpolated using the standard (Sibson) NEM shape functions, while the displacements are interpolated using the b-NEM or the b-NEM$^+$ shape functions previously defined.

b-NEM Displacements Interpolation

In the context of the b-NEM, the following approximation scheme is used for the displacements interpolation:

$$\mathbf{u^h}(\mathbf{x}) = \sum_{i=1}^{n} \phi_i(\mathbf{x})\mathbf{u}_i + \sum_{j=1}^{m} \phi_j^*(\mathbf{x})\boldsymbol{\gamma}_j \ , \quad \boldsymbol{\delta}\mathbf{u^h}(\mathbf{x}) = \sum_{i=1}^{n} \phi_i(\mathbf{x})\boldsymbol{\delta}\mathbf{u}_i + \sum_{j=1}^{m} \phi_j^*(\mathbf{x})\boldsymbol{\delta\gamma}_j \tag{3.14}$$

$$p^h(\mathbf{x}) = \sum_{i=1}^{n} \phi_i(\mathbf{x})p_i, \quad \delta p(\mathbf{x}) = \sum_{i=1}^{n} \phi_i(\mathbf{x})\delta p_i \tag{3.15}$$

where $\phi_i(\mathbf{x})$ is the usual (Sibson) NEM shape function related to node n_i computed at point $\mathbf{x}$, $\phi_j^*(\mathbf{x})$ is the bubble shape function associated with the K-S χ_j, being γ_j the degree of freedom associated with χ_j, n the number of neighbour nodes related to point $\mathbf{x}$ and m the number of influent K-S at point $\mathbf{x}$ (number of K-S shape functions whose support contains $\mathbf{x}$).

b-NEM$^+$ Interpolation

In the context of the b-NEM$^+$, the following approximation scheme is used for the displacements interpolation:

$$\mathbf{u}^h(\mathbf{x}) = \sum_{i=1}^{n+m} \psi_i(\mathbf{x})\mathbf{u}_i \ , \quad \delta u^h(\mathbf{x}) = \sum_{i=1}^{n+m} \psi_i(\mathbf{x})\delta\mathbf{u}_i \tag{3.16}$$

$$p^h(\mathbf{x}) = \sum_{i=1}^{n} \phi_i(\mathbf{x})p_i \ , \quad \delta p^h(\mathbf{x}) = \sum_{i=1}^{n} \phi_i(\mathbf{x})\delta p_i \tag{3.17}$$

where $\psi_i(\mathbf{x})$ are the corrected shape functions computed at point $\mathbf{x}$ using the MLS technique described in section 3.2, $n+m$ the number of influent shape functions, including nodes and K-S shape functions, $\mathbf{u}_i$ contains both displacements associated with the nodes, and nodes that have been associated to the K-S shape functions in the MLS procedure.

3.4 Numerical Test for the inf-sup Condition

In order to perform the inf-sup test a sequence of successive refined meshes is considered (uniform distributions). The test is defined in details in [9]. The objective is to monitor the inf-sup values, β_{min} (minimum eigenvalue related to the discrete LBB condition [9]), when h decreases. If $log(\beta_{min})$ decreases with $log(h)$ the approximation scheme does not pass the LBB numerical test, which requires that $log(\beta_{min})$ remains bounded by a positive constant when $log(h)$ decreases.

Figure 3 shows numerical test comparing some mixed NEM approximation schemes, i.e.: b-NEM/NEM, NEM/Thiessen [26] (NEM approximation for the displacements and constant pressure within each Voronoi cell), and the P1/P0 and P2/P1 mixed FEM approximation schemes. The FEM computations are carried out using directly the Delaunay triangles. As claimed in other previous works [13], the mixed NEM/Thiessen approximation scheme does not pass the numerical inf/sup test. The mixed FEM P1/P0 also violates the LBB condition [9]. All the bubble-NEM schemes are clearly LBB compliant, being the results similar to the ones computed by using the P2/P1 FEM, which satisfy the LBB condition.

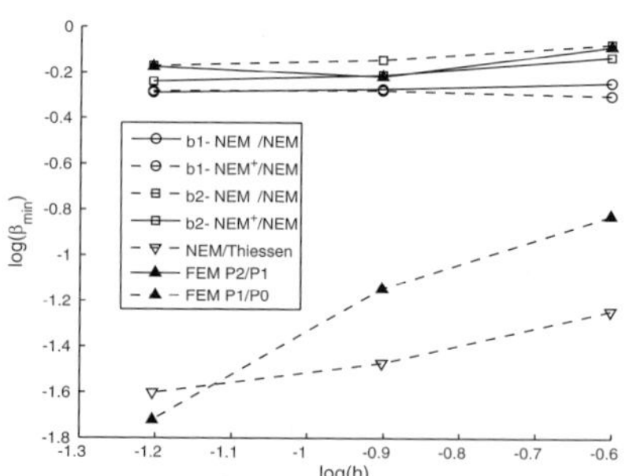

Figure 3. inf-sup numerical test.

More examples in the context of incompressible elasticity can be found in [32], in which we have shown that the b-NEM allows to remove pressure oscillations in the incompressible limit. As noted previously, the NEM shape functions only possess linear completeness [24]. In [32], we have noticed that the enrichment of bubble in the context of MLS does not seem to increase the convergence rate with standard integration despite the proved increase in the approximation consistency, probably due to the highly oscillating character of the obtained shape functions. In the following, a different technique is analyzed with respect to that limitation.

4 Hermite-Natural Element Formulation

4.1 Hermite-NEM Approximation

In this section, quadratic approximation consistency is achieved through a diffuse Hermite-NEM interpolation [22], by using natural neighbour weights in the moving least square approximation. Compared to standard moving least square method, the minimization is performed both with respect to the primary variable, and the diffuse spatial derivatives. For this purpose, we consider an interpolation scheme in the form

$$u^h(\mathbf{x}) = \sum_{i=1}^{n} \psi_i(\mathbf{x}) u_i + \sum_{i=1}^{n} \psi_i^x(\mathbf{x}) \frac{\partial u_i}{\partial x} + \sum_{i=1}^{n} \psi_i^y(\mathbf{x}) \frac{\partial u_i}{\partial y} \tag{4.18}$$

where $\psi_i(\mathbf{x})$ are the shape function associated with the unknown variable u_i, $\psi_i^x(\mathbf{x})$ and $\psi_i^y(\mathbf{x})$ are the shape function associated with the spatial diffuse derivative of u_i with respect to x and y, respectively. In the above framework, u_i, $\frac{\partial u_i}{\partial x}$ and $\frac{\partial u_i}{\partial y}$ are unknown (degrees of freedom). In order to construct the shape functions, we consider the minimization of the functional

$$u^h(\mathbf{x}) = \mathbf{p}(\mathbf{x})^T \mathbf{a} \tag{4.19}$$

where $\mathbf{p}(\mathbf{x})$ is a polynomial basis, i.e. $\mathbf{p}(\mathbf{x}) = \{1, x, y, xy, x^2, y^2\}$ and $\mathbf{a}$ is a vector of unknown coefficients. In order to determine $\mathbf{a}$, we consider the functional

$$J = \frac{1}{2} \sum_{i=1}^{n} w_i(\mathbf{x}) \left\{ \left[\mathbf{p}^T(\mathbf{x})\mathbf{a} - u_i\right]^2 + \alpha \left[\frac{\partial \mathbf{p}^T}{\partial x}(\mathbf{x})\mathbf{a} - \frac{\partial u_i}{\partial x}\right]^2 + \right.$$
$$\left. + \alpha \left[\frac{\partial \mathbf{p}^T}{\partial y}(\mathbf{x})\mathbf{a} - \frac{\partial u_i}{\partial y}\right]^2 \right\} \tag{4.20}$$

where n is the number of natural neighbours of point x, $w_i(\mathbf{x})$ are the natural neighbour shape functions computed at point x, $\frac{\partial \mathbf{p}^T}{\partial x}(\mathbf{x})$ and $\frac{\partial \mathbf{p}^T}{\partial y}(\mathbf{x})$ represent the derivative of the basis $\mathbf{p}(\mathbf{x})$ with respect to x and y, respectively. α is a dimensional parameter which is fixed to 1 in our simulations. Minimizing J with respect to $\mathbf{a}$ leads to the following system of equations:

$$\mathbf{A}\mathbf{a}(\mathbf{x}) = \mathbf{B}\mathbf{q} \tag{4.21}$$

with $\mathbf{q} = \left\{u_1, \frac{\partial u_1}{\partial x}, \frac{\partial u_1}{\partial y}, u_2, \frac{\partial u_2}{\partial x}, \frac{\partial u_2}{\partial y}, ..., u_N, \frac{\partial u_N}{\partial x}, \frac{\partial u_N}{\partial y}\right\}$.

Derivatives of the shape functions are obtained through a standard procedure [3], involving the derivative of the weight functions $w_i(\mathbf{x})$. Closed form of Sibson shape functions derivatives can be found in [21]. The obtained shape functions are depicted in figure 4.

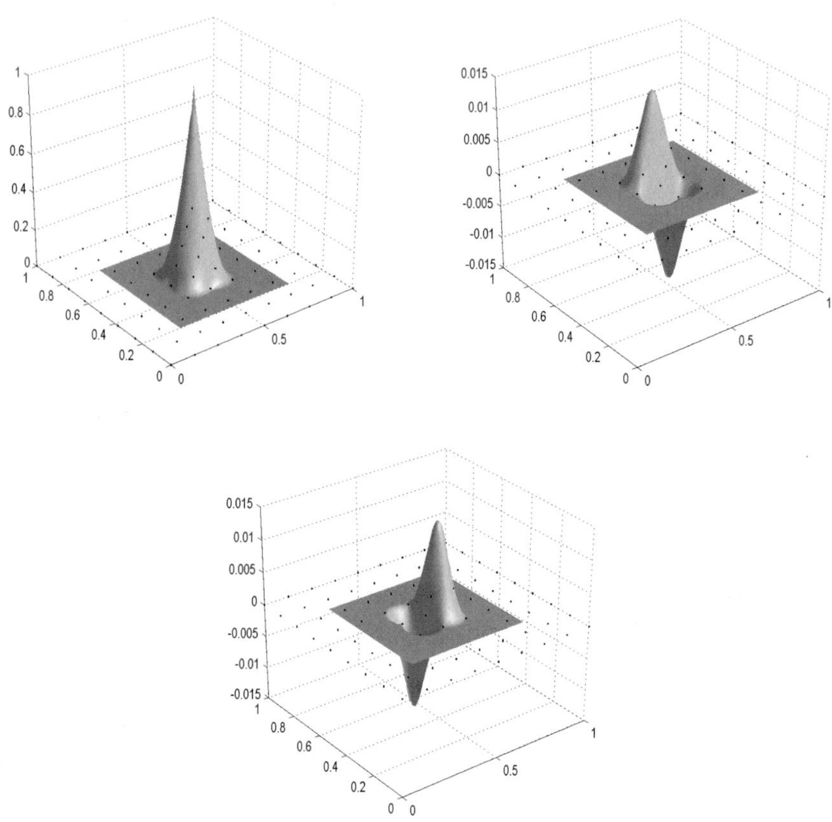

Figure 4. Hermite natural neighbour shape functions.

This approach provides smoother shape functions than the ones obtained in the context of enriched bubbles computed in the MLS framework. Then we assume that for a given number of integration points, the Hermite-NEM scheme will lead to more accurate results than the MLS-bubble NEM. Nevertheless, as in the MLS-bubble NEM there are no issues associated with the boundary conditions, it is not so obvious in the context of Hermite-NEM. According to Eq. (4.20), the new degrees of freedom associated with the derivatives can be interpreted like pseudo-derivatives which do not coincide with the real derivatives. Thus, imposition of essential boundary conditions becomes delicate. Nevertheless, in order to investigate the accuracy of the technique without being polluted by this issue, we consider in next section a Poisson's problem whose solution and its derivatives on the boundary vanish.

4.2 Numerical Example Involving H-NEM Approximation

The boundary value problem is defined by

$$\begin{cases} -\Delta u = f & in\ \Omega =]0,1[\times]0,1[\\ u = \bar{u} & on\ \Gamma_u. \end{cases} \tag{4.22}$$

We consider from now on

$$\begin{cases} \bar{u} = 0, \\ f = 4\pi^2 \left\{ 2cos(2\pi x)cos(2\pi y) - cos(2\pi x) - cos(2\pi y) \right\} \end{cases} \tag{4.23}$$

whose exact solution results in

$$u^{ex}(\mathbf{x}) = \left\{1 - cos(2\pi x)\right\} \left\{1 - cos(2\pi y)\right\} \tag{4.24}$$

The weak form associated with Eq. (4.22) is given by:
Find $u \in H_0^1(\Omega)$ such that

$$\int_\Omega \nabla u \cdot \nabla \delta u d\Omega \int_{\Gamma_u} f \delta u d\Gamma, \quad \forall\ \delta u\ \in\ H_0^1(\Omega) \tag{4.25}$$

where $H_0^1(\Omega)$ is the usual Sobolev functional space. The Hermite-NEM interpolation just described is used to approximate the trial and test functions $\mathbf{u}$ and $\delta\mathbf{u}$, respectively, which are built with the only contribution of internal nodes.

The error in energy norm is computed according to

$$\left\|\mathbf{u} - \mathbf{u}^h\right\|_{E(\Omega)} = \left(\frac{1}{2} \int_\Omega (\nabla u^{ex} - \nabla u^h)^T (\nabla u^{ex} - \nabla u^h) \right)^{1/2}. \tag{4.26}$$

For the evaluation of both Eqs. (4.25) and (4.26), the Voronoi cells associated with each node are triangulated and a Gauss quadrature scheme is applied in each subtriangle, with 3, 6 and 12 points. Figure 5 compares the accuracy of the Hermite-NEM (H-NEM) approximation with the standard NEM. If three Gauss points quadrature scheme is used, the accuracy of the H-NEM exceeds the accuracy of the NEM, but the difference in the convergence rate is not significant. If a fine enough quadrature scheme is applied (6 points), the H-NEM reaches, as expected, a second-order convergence rate.

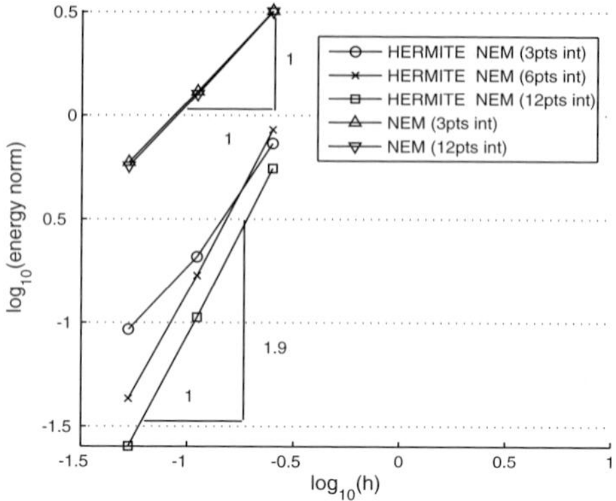

Figure 5. Convergence in energy norm for the two-dimensional Poisson's problem.

5 Conclusion

In this paper, new NEM approximation schemes have been proposed. On one hand, we have proposed a bubble-NEM scheme which does not increase the approximation consistency of the original NEM but allows to construct a richer approximation, which can be used in a mixed formulation to account for incompressibility. On the other hand, we have tried to construct NEM approximations with higher-order consistency. Firstly, we have proposed to enrich the bubbles in the context of the MLS scheme. Despite that this scheme has a consistency of second order in approximation, the order of convergence using standard integration remains of first order, probably due to the highly oscillatory character of the resulting shape functions. Secondly, we have proposed a Hermite-NEM scheme, in which new degrees of freedom are associated with the original nodes. The resulting shape functions are smooth enough to reach second-order convergence in absence of the difficulties related to the boundary conditions prescription.

To circumvent the difficulties related to the imposition of boundary conditions in the H-NEM framework different possibilities are presently examined: (i) the use of Lagrange multipliers; (ii) the assignment of the Hermite degrees of freedom (diffuse derivatives) to the bubble nodes instead to the original nodes, except at the bubbles located on the domain boundary where the approximation does not involve the derivatives degrees of freedom. This approach should allow to enforce essential boundary conditions, but if the resulting shape functions are not smooth enough integration difficulties could subsist being the order of convergence lower than the expected one. This analysis constitutes a work in progress.

References

1. H. Askes, R. de Borst, O. Heeres , *Conditions for locking-free elastoplastic analyses in the element-free Galerkin method*, Computer Methods in Applied Mechanics and Engineering **173** (1999), 99–109.
2. K.J. Bathe, *Finite Element procedures* Cambridge University Press, 2001.
3. T. Belytschko, Y.Y. Lu, L. Gu, *Element-free Galerkin methods*, International Journal for Numerical Methods in Engineering **37** (1994), 229–256.
4. F. Brezzi, M. Fortin, *Mixed and Hybrid Finite Element Methods* New-York, Springer-Verlag, 1991.
5. D. Chapelle, K.-J. Bathe, *The inf-sup test*, Comput. Struct., **48**(5) (1993), 745–760.
6. J.S. Chen, C. Pan, C.T. Wu, *Large deformation analysis of rubber based on a reproducing kernel particle method*, Computer methods in applied mechanics and engineering, **173** (1997), 99–109.
7. J.S. Chen, S. Yoon, H.P. Whang, W.K. Liu, *An improved reproducing kernel particle method for nearly incompressible finite elasticity solids.* Computer methods in applied mechanics and engineering, **181** (2000), 117–145.
8. E. Cueto, M. Doblaré, L. Gracia, *Imposing essential boundary conditions in the natural elements method by means of density-scaled alpha-shapes*, International Journal for Numerical Methods in Engineering **49** (2000), 519–546.
9. D. Chapelle, K.-J. Bathe, *The inf-sup test*, Comput. Struct., **48**(5) (1993), 745–760.
10. J. Dolbow, T. Belytschko, *Volumetric locking in the element-free Galerkin method*, International Journal for numerical methods in engineering **46** (1999), 925–942.
11. J. Donea, A. Huerta, *Finite Element methods for flow problems*, John Wiley and sons, Chichester, 2003.
12. V. Girault, P.-A. Raviart, *Finite element methods for Navier-Stockes equations. Theory and algorithms*, Berlin, Springer-Verlag, 1986.
13. D. González, E. Cueto, M. Doblaré, *Volumetric locking in natural neighbour Galerkin methods*, International Journal for Numerical Methods in Engineering **61** (2004) 611–632.
14. A. Huerta, S. Fernández-Méndez, *Locking in the incompressible limit for the element free Galerkin method* Int. J. Numer. Methods Eng. **51**(11) (2001) 1361–1383.
15. H. Hiyoshi, K. Sugihara, *Improving continuity of Voronoi-based interpolation over Delaunay spheres*, Computational Geometry **22** (2002) 167–183.
16. S. Li, W. Hoa, W.K. Liu, *Mesh-free simulations of shear banding in large deformation*, Int. Journ. Solids. and Sruct. **37** (2000), 7183–7206.
17. D.S. Malkus, *Eigenproblems associated with the discrete LBB condition for the incompressible finite elements*, Int. J. Eng. Sci. **10** (1981), 1299–1310.
18. J.M. Melenk, I. Babuška, *The partition of unity finite element method: basic theory and applications*, Computer Methods in Applied Mechanics and Engineering **4** (1996) 289–314.
19. J.R. Munkres, *Elements of algebraic topology* Perseus Press, 7–14, 1993.
20. B. Nayroles, G. Touzot, P. Villon, *Generalizing the finite element method: diffuse approximation and diffuse elements* Computational mechanics **10** (1992) 307–318.

21. B. Piper, *Properties of local coordinates based on Dirichlet tessellations*, Computing Suppl. **8** (1993) 227–239.
22. A. Rassineux, P. Villon, J.M. Savignat, O. Stab, *Surface remeshing by local Hermite diffuse interpolation*, International Journal for Numerical Methods in Engineering **49**(1-2) (2000) 10–20.
23. M. Sambridge, J. Braun, M. McQueen, *Geophisical parameterization and interpolation of irregular data using natural neighbors*, Geophysi. J. Int. **122** (1995) 837–857.
24. R. Sibson, *A vector Identity for the Dirichlet tessellations*, Math. Proc. Camb. Phil. Soc. **87** (1980) 151–155.
25. N. Sukumar, B. Moran, T. Belytschko, *The natural elements method in solid mechanics*, International Journal for Numerical Methods in Engineering **43** 1998, 839–887.
26. N. Sukumar, The natural element method in solid mechanics. *Ph.D. Dissertation.* Northwestern University, Evanston, Illinois, 1998.
27. Y. Vidal, P. Villon, A. Huerta, *Locking in the element free Galerkin*, Communications in Numerical Methods in Engineering **19**(9) (2003) 725–735.
28. J., Yoo , Moran B., Chen J.-S., *Stabilized conforming nodal integration in the natural-element method*, International Journal for Numerical Methods in Engineering **60** (1998) 861–890.
29. J. Yvonnet, F. Chinesta, P. Lorong, D. Rynckelynck, *The constrained natural element method (C-NEM) for treating thermal models involving moving interfaces*, International Journal of Thermal Sciences **44** (2005), 559–569.
30. J. Yvonnet, D. Ryckelynck, P. Lorong, F. Chinesta, *A new extension of the natural element method for non convex and discontinuous domains : the constrained natural element method (C-NEM)*, International Journal for Numerical Methods in Engineering **60** (2004) 1451–1474.
31. J. Yvonnet, P. Lorong, D. Ryckelynck, F. Chinesta, *Simulating dynamic thermo-elastoplasticity in large transformations with adaptive refinement in the natural element method : application to shear banding*, International Journal of Forming Processes, **8**, (2005) 346–363.
32. J. Yvonnet, P. Villon, F. Chinesta, *Natural Element Approximations involving bubbles for treating incompressible media*, International Journal for Numerical Methods in Engineering, in press.
33. T. Zhu, S.N. Atluri, *A modified collocation method and a penalty formulation for enforcing the essential boundary conditions in the element free Galerkin method*, Computational Mechanics **21** (1998) 211–222.
34. O.C. Zienkiewicz, J.P. De, S.R. Gago, D.W. Kelly, *The hierarchical concept in finite element analysis*, Computers and Structures, **16** (1983) 53–65.

Appendix

Color Plates

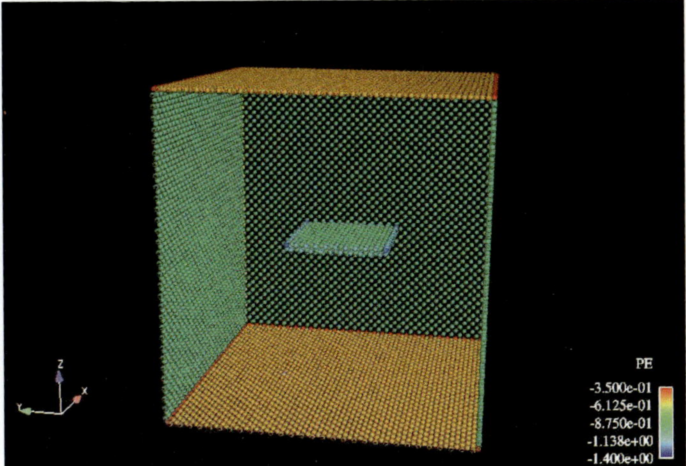

Plate 1. (Figure 3 on page 200) Initial pre-crack for dynamic crack propagation example. Contours of potential energy shown. Only those atoms with potential energy greater than ninety percent of the equilibrium value are shown.

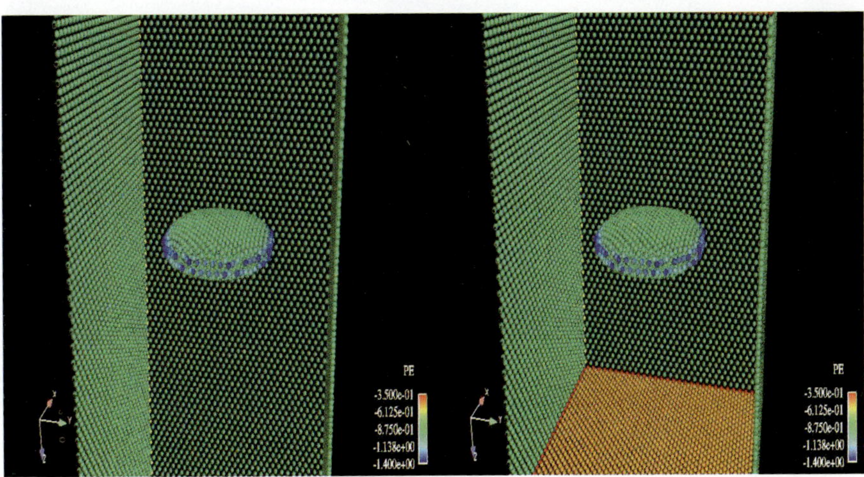

Plate 2. (Figure 4 on page 201) Onset of crack growth for (left) full MD simulation and (right) bridging scale simulation.

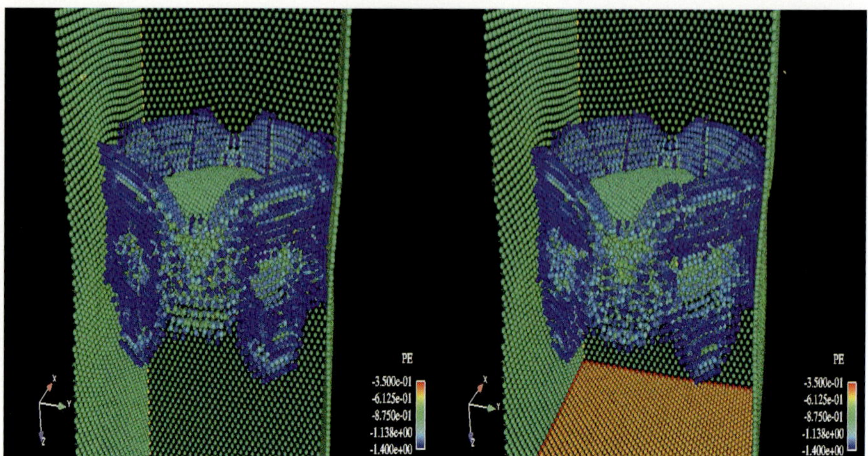

Plate 3. (Figure 5 on page 201) Out of plane crack branching for (left) full MD simulation and (right) bridging scale simulation.

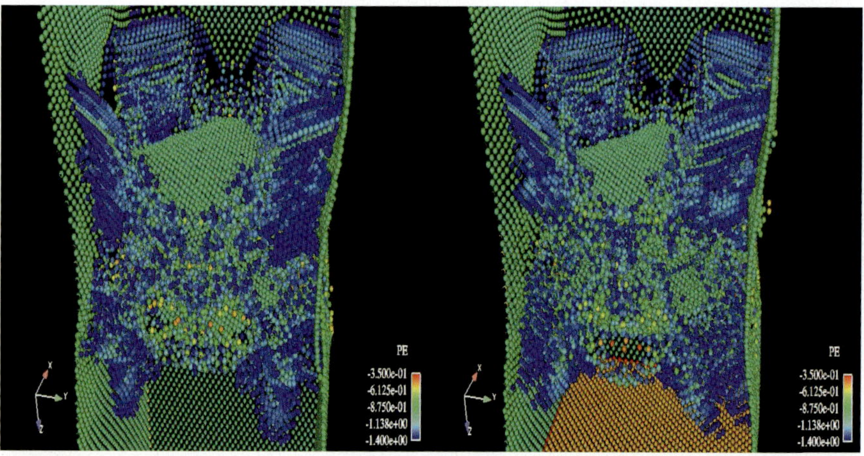

Plate 4. (Figure 6 on page 202) Final configuration in (left) full MD simulation and (right) bridging scale simulation.

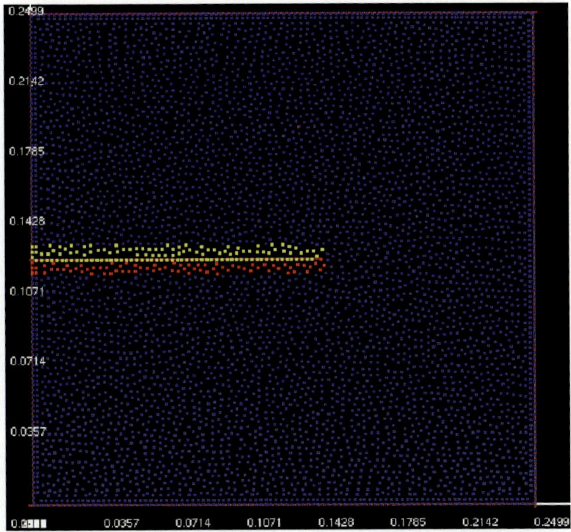

Plate 5. (Figure 1 on page 257) Orientations of the particles around a sheet of paper, blue indicates orientation 0, red +1 and yellow −1

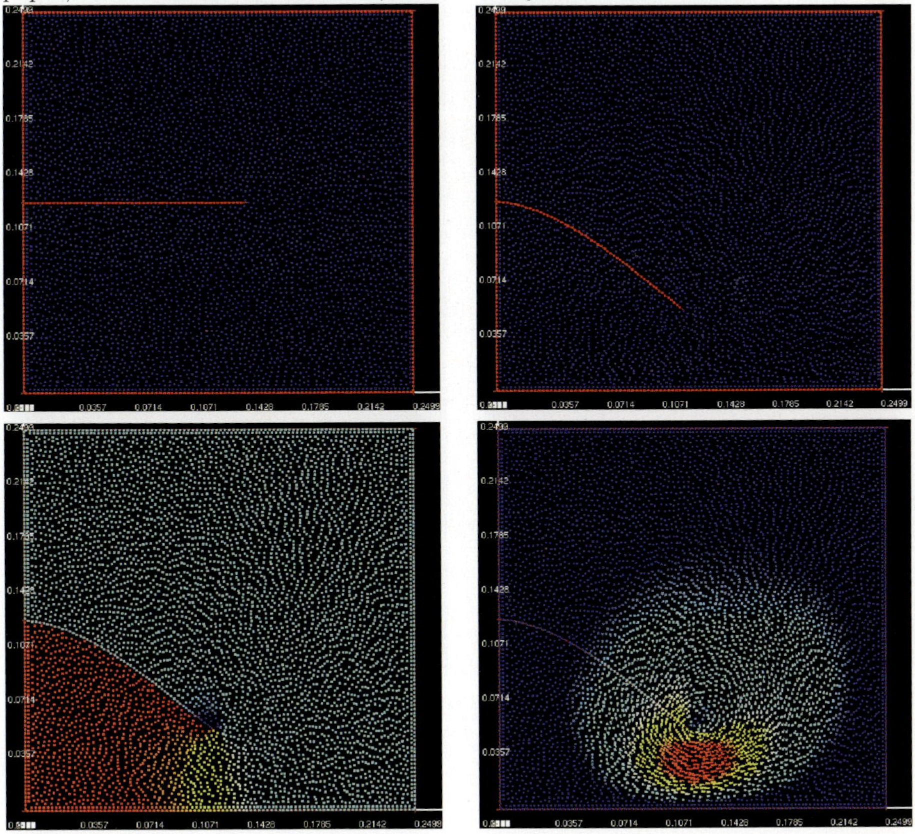

Plate 6. (Figure 2 on page 259 Position of particles and sheet of paper at time $t = 0\,\mathrm{s}$ and $t = 0.8377\,\mathrm{s}$ (top), pressure and velocity field at $t = 0.8377\,\mathrm{s}$ (bottom)

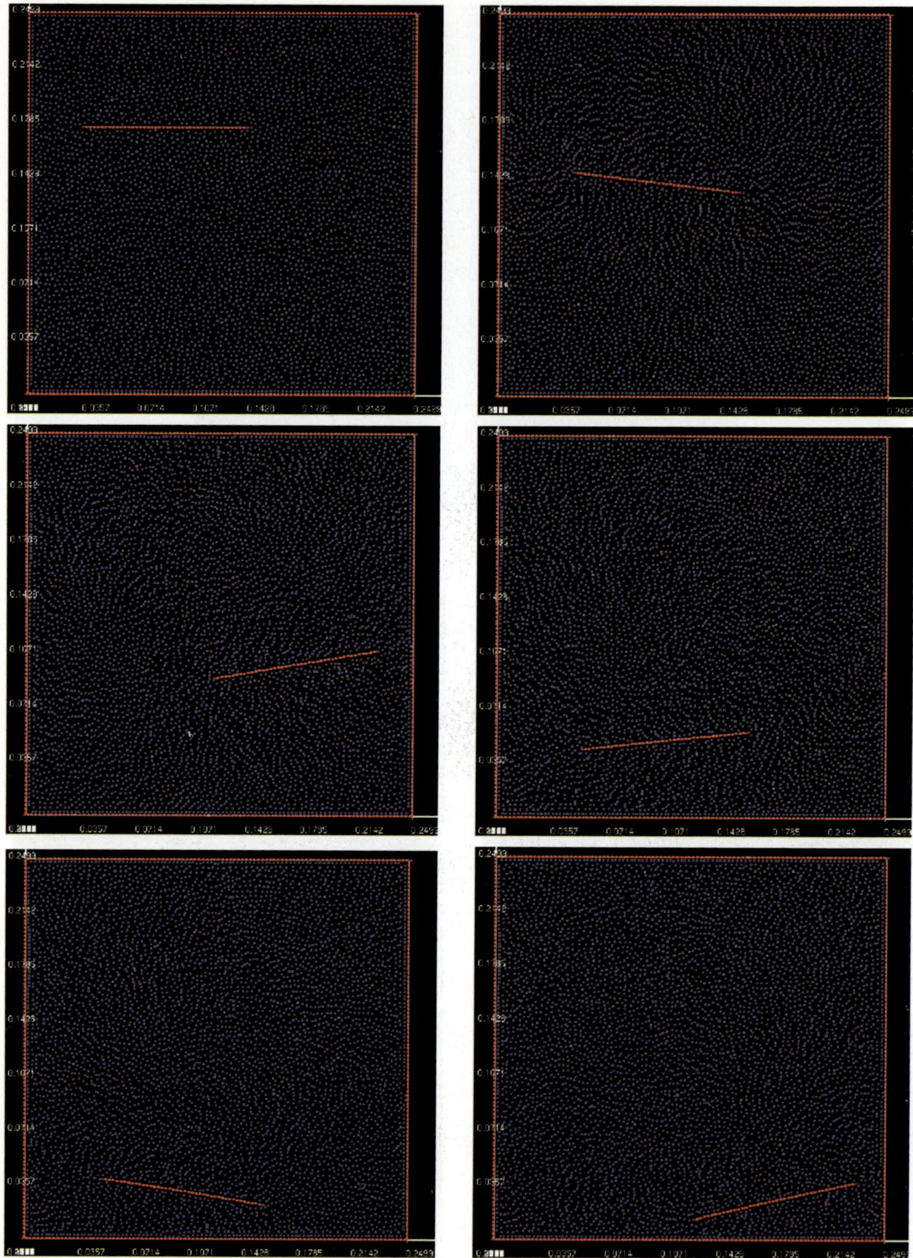

Plate 7. (Figure 3 on page 260 Position of particles and sheet of paper at $t = 0.0, 0.3\,\mathrm{s}$ (top), $t = 0.6, 0.9\,\mathrm{s}$ (center) and $t = 1.2, 1.5\,\mathrm{s}$ (bottom)

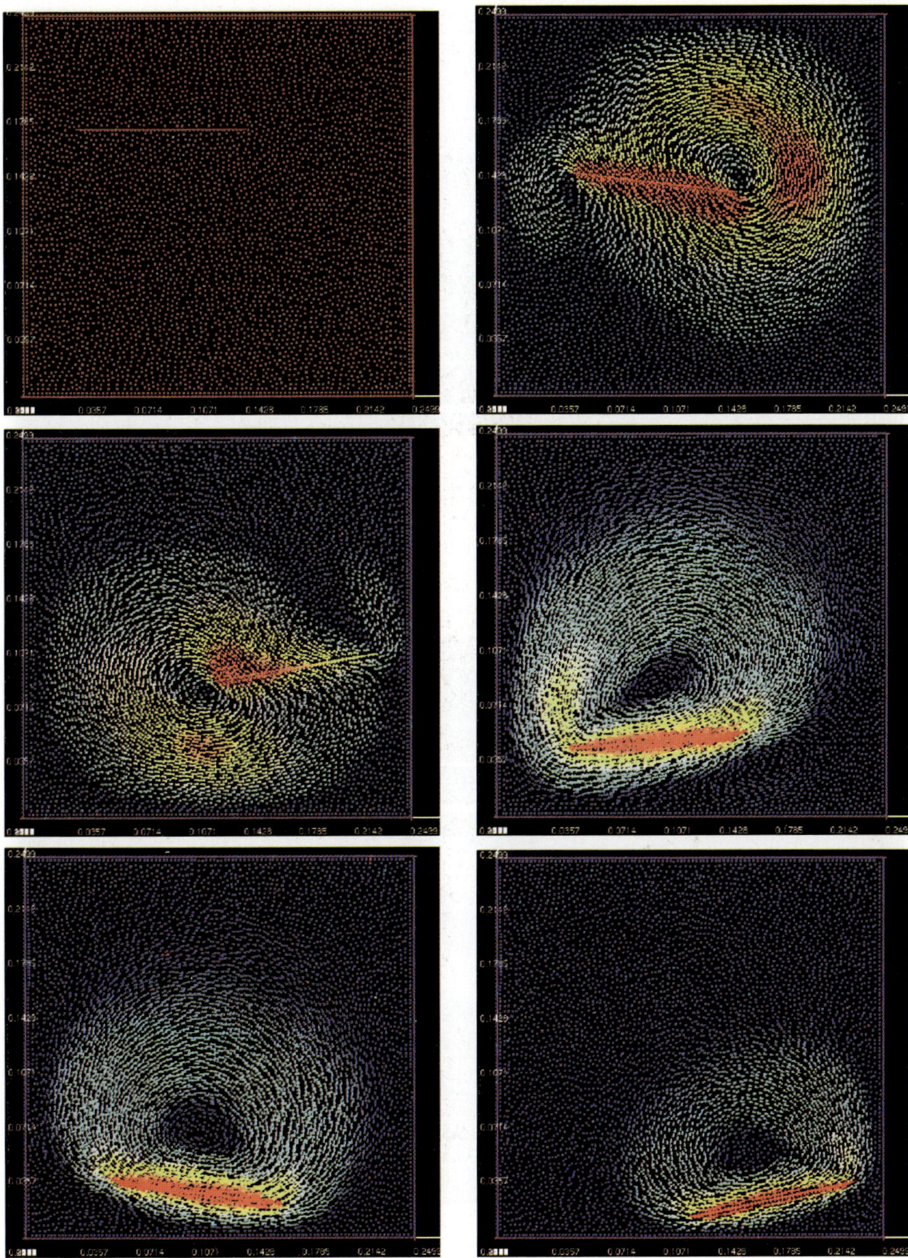

Plate 8. (Figure 4 on page 261 Velocity field at $t = 0.0, 0.3\,\mathrm{s}$ (top), $t = 0.6, 0.9\,\mathrm{s}$ (center) and $t = 1.2, 1.5\,\mathrm{s}$ (bottom)

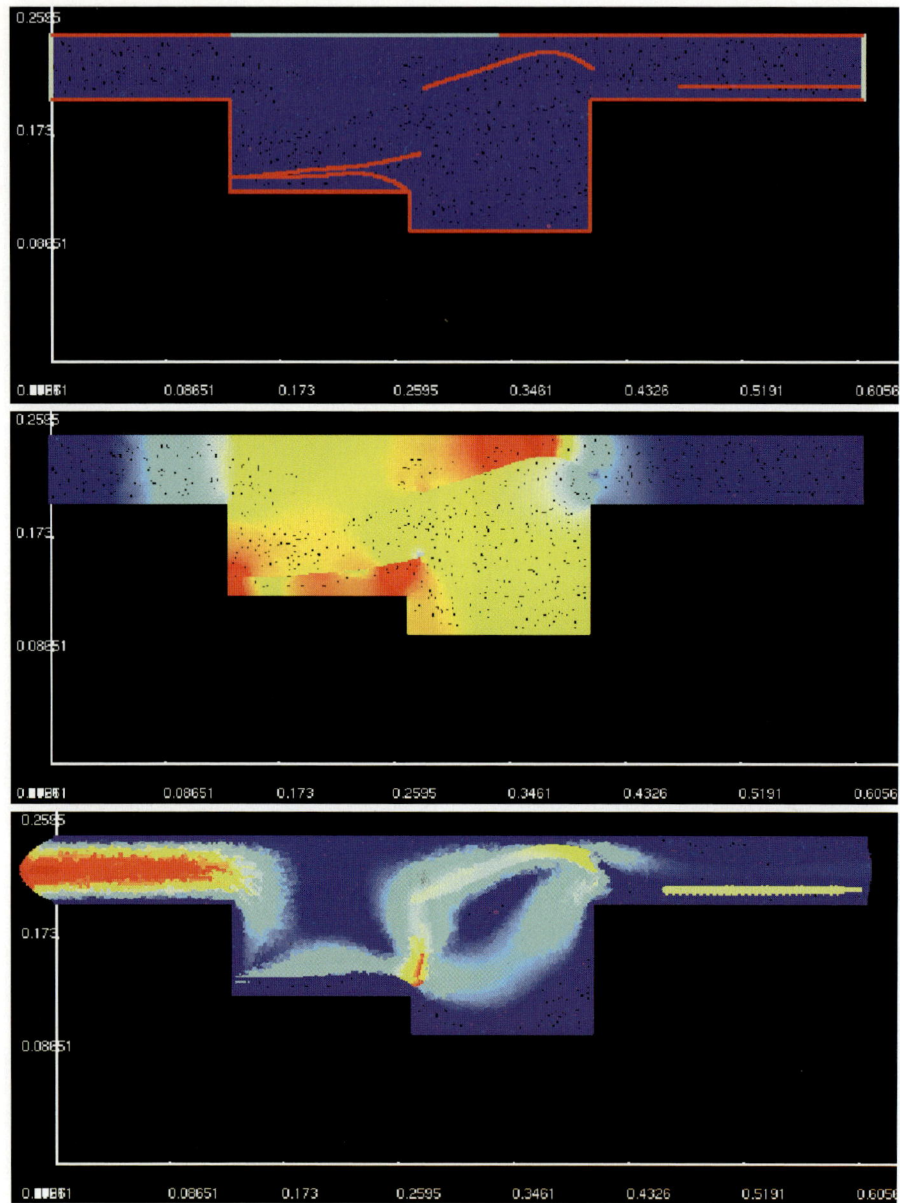

Plate 9. (Figure 6 on page 263 Position of particles and sheets of paper (top), pressure distribution (center) and velocity field (bottom) at $t = 0.15\,\mathrm{s}$

Editorial Policy

1. Volumes in the following three categories will be published in LNCSE:

i) Research monographs
ii) Lecture and seminar notes
iii) Conference proceedings

Those considering a book which might be suitable for the series are strongly advised to contact the publisher or the series editors at an early stage.

2. Categories i) and ii). These categories will be emphasized by Lecture Notes in Computational Science and Engineering. **Submissions by interdisciplinary teams of authors are encouraged.** The goal is to report new developments – quickly, informally, and in a way that will make them accessible to non-specialists. In the evaluation of submissions timeliness of the work is an important criterion. Texts should be well-rounded, well-written and reasonably self-contained. In most cases the work will contain results of others as well as those of the author(s). In each case the author(s) should provide sufficient motivation, examples, and applications. In this respect, Ph.D. theses will usually be deemed unsuitable for the Lecture Notes series. Proposals for volumes in these categories should be submitted either to one of the series editors or to Springer-Verlag, Heidelberg, and will be refereed. A provisional judgment on the acceptability of a project can be based on partial information about the work: a detailed outline describing the contents of each chapter, the estimated length, a bibliography, and one or two sample chapters – or a first draft. A final decision whether to accept will rest on an evaluation of the completed work which should include

- at least 100 pages of text;
- a table of contents;
- an informative introduction perhaps with some historical remarks which should be accessible to readers unfamiliar with the topic treated;
- a subject index.

3. Category iii). Conference proceedings will be considered for publication provided that they are both of exceptional interest and devoted to a single topic. One (or more) expert participants will act as the scientific editor(s) of the volume. They select the papers which are suitable for inclusion and have them individually refereed as for a journal. Papers not closely related to the central topic are to be excluded. Organizers should contact Lecture Notes in Computational Science and Engineering at the planning stage.

In exceptional cases some other multi-author-volumes may be considered in this category.

4. Format. Only works in English are considered. They should be submitted in camera-ready form according to Springer-Verlag's specifications.
Electronic material can be included if appropriate. Please contact the publisher.
Technical instructions and/or LaTeX macros are available via
http://www.springer.com/east/home/math/math+authors?SGWID=5-40017-6-71391-0.
The macros can also be sent on request.

General Remarks

Lecture Notes are printed by photo-offset from the master-copy delivered in camera-ready form by the authors. For this purpose Springer-Verlag provides technical instructions for the preparation of manuscripts. See also *Editorial Policy*.

Careful preparation of manuscripts will help keep production time short and ensure a satisfactory appearance of the finished book.

The following terms and conditions hold:

Categories i), ii), and iii):
Authors receive 50 free copies of their book. No royalty is paid. Commitment to publish is made by letter of intent rather than by signing a formal contract. Springer-Verlag secures the copyright for each volume.

For conference proceedings, editors receive a total of 50 free copies of their volume for distribution to the contributing authors.

All categories:
Authors are entitled to purchase further copies of their book and other Springer mathematics books for their personal use, at a discount of 33,3 % directly from Springer-Verlag.

Addresses:

Timothy J. Barth
NASA Ames Research Center
NAS Division
Moffett Field, CA 94035, USA
e-mail: barth@nas.nasa.gov

Michael Griebel
Institut für Numerische Simulation
der Universität Bonn
Wegelerstr. 6
53115 Bonn, Germany
e-mail: griebel@ins.uni-bonn.de

David E. Keyes
Department of Applied Physics
and Applied Mathematics
Columbia University
200 S. W. Mudd Building
500 W. 120th Street
New York, NY 10027, USA
e-mail: david.keyes@columbia.edu

Risto M. Nieminen
Laboratory of Physics
Helsinki University of Technology
02150 Espoo, Finland
e-mail: rni@fyslab.hut.fi

Dirk Roose
Department of Computer Science
Katholieke Universiteit Leuven
Celestijnenlaan 200A
3001 Leuven-Heverlee, Belgium
e-mail: dirk.roose@cs.kuleuven.ac.be

Tamar Schlick
Department of Chemistry
Courant Institute of Mathematical
Sciences
New York University
and Howard Hughes Medical Institute
251 Mercer Street
New York, NY 10012, USA
e-mail: schlick@nyu.edu

Mathematics Editor at Springer: Martin Peters
Springer-Verlag, Mathematics Editorial IV
Tiergartenstrasse 17
D-69121 Heidelberg, Germany
Tel.: *49 (6221) 487-8185
Fax: *49 (6221) 487-8355
e-mail: martin.peters@springer.com

Lecture Notes in Computational Science and Engineering

Vol. 1 D. Funaro, *Spectral Elements for Transport-Dominated Equations.* 1997. X, 211 pp. Softcover. ISBN 3-540-62649-2

Vol. 2 H. P. Langtangen, *Computational Partial Differential Equations.* Numerical Methods and Diffpack Programming. 1999. XXIII, 682 pp. Hardcover. ISBN 3-540-65274-4

Vol. 3 W. Hackbusch, G. Wittum (eds.), *Multigrid Methods V.* Proceedings of the Fifth European Multigrid Conference held in Stuttgart, Germany, October 1-4, 1996. 1998. VIII, 334 pp. Softcover. ISBN 3-540-63133-X

Vol. 4 P. Deuflhard, J. Hermans, B. Leimkuhler, A. E. Mark, S. Reich, R. D. Skeel (eds.), *Computational Molecular Dynamics: Challenges, Methods, Ideas.* Proceedings of the 2nd International Symposium on Algorithms for Macromolecular Modelling, Berlin, May 21-24, 1997. 1998. XI, 489 pp. Softcover. ISBN 3-540-63242-5

Vol. 5 D. Kröner, M. Ohlberger, C. Rohde (eds.), *An Introduction to Recent Developments in Theory and Numerics for Conservation Laws.* Proceedings of the International School on Theory and Numerics for Conservation Laws, Freiburg / Littenweiler, October 20-24, 1997. 1998. VII, 285 pp. Softcover. ISBN 3-540-65081-4

Vol. 6 S. Turek, *Efficient Solvers for Incompressible Flow Problems.* An Algorithmic and Computational Approach. 1999. XVII, 352 pp, with CD-ROM. Hardcover. ISBN 3-540-65433-X

Vol. 7 R. von Schwerin, ***Multi Body System SIM**ulation.* Numerical Methods, Algorithms, and Software. 1999. XX, 338 pp. Softcover. ISBN 3-540-65662-6

Vol. 8 H.-J. Bungartz, F. Durst, C. Zenger (eds.), *High Performance Scientific and Engineering Computing.* Proceedings of the International FORTWIHR Conference on HPSEC, Munich, March 16-18, 1998. 1999. X, 471 pp. Softcover. ISBN 3-540-65730-4

Vol. 9 T. J. Barth, H. Deconinck (eds.), *High-Order Methods for Computational Physics.* 1999. VII, 582 pp. Hardcover. ISBN 3-540-65893-9

Vol. 10 H. P. Langtangen, A. M. Bruaset, E. Quak (eds.), *Advances in Software Tools for Scientific Computing.* 2000. X, 357 pp. Softcover. ISBN 3-540-66557-9

Vol. 11 B. Cockburn, G. E. Karniadakis, C.-W. Shu (eds.), *Discontinuous Galerkin Methods.* Theory, Computation and Applications. 2000. XI, 470 pp. Hardcover. ISBN 3-540-66787-3

Vol. 12 U. van Rienen, *Numerical Methods in Computational Electrodynamics.* Linear Systems in Practical Applications. 2000. XIII, 375 pp. Softcover. ISBN 3-540-67629-5

Vol. 13 B. Engquist, L. Johnsson, M. Hammill, F. Short (eds.), *Simulation and Visualization on the Grid.* Parallelldatorcentrum Seventh Annual Conference, Stockholm, December 1999, Proceedings. 2000. XIII, 301 pp. Softcover. ISBN 3-540-67264-8

Vol. 14 E. Dick, K. Riemslagh, J. Vierendeels (eds.), *Multigrid Methods VI.* Proceedings of the Sixth European Multigrid Conference Held in Gent, Belgium, September 27-30, 1999. 2000. IX, 293 pp. Softcover. ISBN 3-540-67157-9

Vol. 15 A. Frommer, T. Lippert, B. Medeke, K. Schilling (eds.), *Numerical Challenges in Lattice Quantum Chromodynamics.* Joint Interdisciplinary Workshop of John von Neumann Institute for Computing, Jülich and Institute of Applied Computer Science, Wuppertal University, August 1999. 2000. VIII, 184 pp. Softcover. ISBN 3-540-67732-1

Vol. 16 J. Lang, *Adaptive Multilevel Solution of Nonlinear Parabolic PDE Systems.* Theory, Algorithm, and Applications. 2001. XII, 157 pp. Softcover. ISBN 3-540-67900-6

Vol. 17 B. I. Wohlmuth, *Discretization Methods and Iterative Solvers Based on Domain Decomposition.* 2001. X, 197 pp. Softcover. ISBN 3-540-41083-X

Vol. 18 U. van Rienen, M. Günther, D. Hecht (eds.), *Scientific Computing in Electrical Engineering.* Proceedings of the 3rd International Workshop, August 20-23, 2000, Warnemünde, Germany. 2001. XII, 428 pp. Softcover. ISBN 3-540-42173-4

Vol. 19 I. Babuška, P. G. Ciarlet, T. Miyoshi (eds.), *Mathematical Modeling and Numerical Simulation in Continuum Mechanics.* Proceedings of the International Symposium on Mathematical Modeling and Numerical Simulation in Continuum Mechanics, September 29 - October 3, 2000, Yamaguchi, Japan. 2002. VIII, 301 pp. Softcover. ISBN 3-540-42399-0

Vol. 20 T. J. Barth, T. Chan, R. Haimes (eds.), *Multiscale and Multiresolution Methods.* Theory and Applications. 2002. X, 389 pp. Softcover. ISBN 3-540-42420-2

Vol. 21 M. Breuer, F. Durst, C. Zenger (eds.), *High Performance Scientific and Engineering Computing.* Proceedings of the 3rd International FORTWIHR Conference on HPSEC, Erlangen, March 12-14, 2001. 2002. XIII, 408 pp. Softcover. ISBN 3-540-42946-8

Vol. 22 K. Urban, *Wavelets in Numerical Simulation.* Problem Adapted Construction and Applications. 2002. XV, 181 pp. Softcover. ISBN 3-540-43055-5

Vol. 23 L. F. Pavarino, A. Toselli (eds.), *Recent Developments in Domain Decomposition Methods.* 2002. XII, 243 pp. Softcover. ISBN 3-540-43413-5

Vol. 24 T. Schlick, H. H. Gan (eds.), *Computational Methods for Macromolecules: Challenges and Applications.* Proceedings of the 3rd International Workshop on Algorithms for Macromolecular Modeling, New York, October 12-14, 2000. 2002. IX, 504 pp. Softcover. ISBN 3-540-43756-8

Vol. 25 T. J. Barth, H. Deconinck (eds.), *Error Estimation and Adaptive Discretization Methods in Computational Fluid Dynamics.* 2003. VII, 344 pp. Hardcover. ISBN 3-540-43758-4

Vol. 26 M. Griebel, M. A. Schweitzer (eds.), *Meshfree Methods for Partial Differential Equations.* 2003. IX, 466 pp. Softcover. ISBN 3-540-43891-2

Vol. 27 S. Müller, *Adaptive Multiscale Schemes for Conservation Laws.* 2003. XIV, 181 pp. Softcover. ISBN 3-540-44325-8

Vol. 28 C. Carstensen, S. Funken, W. Hackbusch, R. H. W. Hoppe, P. Monk (eds.), *Computational Electromagnetics.* Proceedings of the GAMM Workshop on "Computational Electromagnetics", Kiel, Germany, January 26-28, 2001. 2003. X, 209 pp. Softcover. ISBN 3-540-44392-4

Vol. 29 M. A. Schweitzer, *A Parallel Multilevel Partition of Unity Method for Elliptic Partial Differential Equations.* 2003. V, 194 pp. Softcover. ISBN 3-540-00351-7

Vol. 30 T. Biegler, O. Ghattas, M. Heinkenschloss, B. van Bloemen Waanders (eds.), *Large-Scale PDE-Constrained Optimization.* 2003. VI, 349 pp. Softcover. ISBN 3-540-05045-0

Vol. 31 M. Ainsworth, P. Davies, D. Duncan, P. Martin, B. Rynne (eds.), *Topics in Computational Wave Propagation.* Direct and Inverse Problems. 2003. VIII, 399 pp. Softcover. ISBN 3-540-00744-X

Vol. 32 H. Emmerich, B. Nestler, M. Schreckenberg (eds.), *Interface and Transport Dynamics.* Computational Modelling. 2003. XV, 432 pp. Hardcover. ISBN 3-540-40367-1

Vol. 33 H. P. Langtangen, A. Tveito (eds.), *Advanced Topics in Computational Partial Differential Equations.* Numerical Methods and Diffpack Programming. 2003. XIX, 658 pp. Softcover. ISBN 3-540-01438-1

Vol. 34 V. John, *Large Eddy Simulation of Turbulent Incompressible Flows.* Analytical and Numerical Results for a Class of LES Models. 2004. XII, 261 pp. Softcover. ISBN 3-540-40643-3

Vol. 35 E. Bänsch (ed.), *Challenges in Scientific Computing - CISC 2002.* Proceedings of the Conference *Challenges in Scientific Computing*, Berlin, October 2-5, 2002. 2003. VIII, 287 pp. Hardcover. ISBN 3-540-40887-8

Vol. 36 B. N. Khoromskij, G. Wittum, *Numerical Solution of Elliptic Differential Equations by Reduction to the Interface.* 2004. XI, 293 pp. Softcover. ISBN 3-540-20406-7

Vol. 37 A. Iske, *Multiresolution Methods in Scattered Data Modelling.* 2004. XII, 182 pp. Softcover. ISBN 3-540-20479-2

Vol. 38 S.-I. Niculescu, K. Gu (eds.), *Advances in Time-Delay Systems.* 2004. XIV, 446 pp. Softcover. ISBN 3-540-20890-9

Vol. 39 S. Attinger, P. Koumoutsakos (eds.), *Multiscale Modelling and Simulation.* 2004. VIII, 277 pp. Softcover. ISBN 3-540-21180-2

Vol. 40 R. Kornhuber, R. Hoppe, J. Périaux, O. Pironneau, O. Wildlund, J. Xu (eds.), *Domain Decomposition Methods in Science and Engineering.* 2005. XVIII, 690 pp. Softcover. ISBN 3-540-22523-4

Vol. 41 T. Plewa, T. Linde, V.G. Weirs (eds.), *Adaptive Mesh Refinement – Theory and Applications.* 2005. XIV, 552 pp. Softcover. ISBN 3-540-21147-0

Vol. 42 A. Schmidt, K.G. Siebert, *Design of Adaptive Finite Element Software.* The Finite Element Toolbox ALBERTA. 2005. XII, 322 pp. Hardcover. ISBN 3-540-22842-X

Vol. 43 M. Griebel, M.A. Schweitzer (eds.), *Meshfree Methods for Partial Differential Equations II.* 2005. XIII, 303 pp. Softcover. ISBN 3-540-23026-2

Vol. 44 B. Engquist, P. Lötstedt, O. Runborg (eds.), *Multiscale Methods in Science and Engineering.* 2005. XII, 291 pp. Softcover. ISBN 3-540-25335-1

Vol. 45 P. Benner, V. Mehrmann, D.C. Sorensen (eds.), *Dimension Reduction of Large-Scale Systems.* 2005. XII, 402 pp. Softcover. ISBN 3-540-24545-6

Vol. 46 D. Kressner (ed.), *Numerical Methods for General and Structured Eigenvalue Problems.* 2005. XIV, 258 pp. Softcover. ISBN 3-540-24546-4

Vol. 47 A. Boriçi, A. Frommer, B. Joó, A. Kennedy, B. Pendleton (eds.), *QCD and Numerical Analysis III.* 2005. XIII, 201 pp. Softcover. ISBN 3-540-21257-4

Vol. 48 F. Graziani (ed.), *Computational Methods in Transport.* 2006. VIII, 524 pp. Softcover. ISBN 3-540-28122-3

Vol. 49 B. Leimkuhler, C. Chipot, R. Elber, A. Laaksonen, A. Mark, T. Schlick, C. Schütte, R. Skeel (eds.), *New Algorithms for Macromolecular Simulation.* 2006. XVI, 376 pp. Softcover. ISBN 3-540-25542-7

Vol. 50 M. Bücker, G. Corliss, P. Hovland, U. Naumann, B. Norris (eds.), *Automatic Differentiation: Applications, Theory, and Implementations.* 2006. XVIII, 362 pp. Softcover. ISBN 3-540-28403-6

Vol. 51 A.M. Bruaset, A. Tveito (eds.), *Numerical Solution of Partial Differential Equations on Parallel Computers* 2006. XII, 482 pp. Softcover. ISBN 3-540-29076-1

Vol. 52 K.H. Hoffmann, A. Meyer (eds.), *Parallel Algorithms and Cluster Computing.* 2006. X, 374 pp. Softcover. ISBN 3-540-33539-0

Vol. 53 H.-J. Bungartz, M. Schäfer (eds.), *Fluid-Structure Interaction.* 2006. VII, 388 pp. Softcover. ISBN 3-540-34595-7

Vol. 54 J. Behrens, *Adaptive Atmospheric Modeling.* 2006. XX, 314 pp. Softcover. ISBN 3-540-33382-7

Vol. 55 O. Widlund, D. Keyes (eds.), *Domain Decomposition Methods in Science and Engineering XVI.* 2007. XXII, 784 pp. Softcover. ISBN 3-540-34468-3

Vol. 56 S. Kassinos, C. Langer, G. Iaccarino, P. Moin (eds.), *Complex Effects in Large Eddy Simulations.* 2007. XII, 440 pp. Softcover. ISBN 3-540-34233-8

Vol. 57 M. Griebel M.A Schweitzer (eds.), *Meshfree Methods for Partial Differential Equations III.* 2007. VIII, 306 pp. Softcover. ISBN 3-540-46214-7

For further information on these books please have a look at our mathematics catalogue at the following URL: `www.springer.com/series/3527`

Monographs in Computational Science and Engineering

Vol. 1 J. Sundnes, G.T. Lines, X. Cai, B.F. Nielsen, K.-A. Mardal, A. Tveito, *Computing the Electrical Activity in the Heart.* 2006. XI, 318 pp. Hardcover. ISBN 3-540-33432-7

For further information on this book, please have a look at our mathematics catalogue at the following URL: `www.springer.com/series/7417`

Texts in Computational Science and Engineering

Vol. 1 H. P. Langtangen, *Computational Partial Differential Equations.* Numerical Methods and Diffpack Programming. 2nd Edition 2003. XXVI, 855 pp. Hardcover. ISBN 3-540-43416-X

Vol. 2 A. Quarteroni, F. Saleri, *Scientific Computing with MATLAB and Octave.* 2nd Edition 2006. XIV, 318 pp. Hardcover. ISBN 3-540-32612-X

Vol. 3 H. P. Langtangen, *Python Scripting for Computational Science.* 2nd Edition 2006. XXIV, 736 pp. Hardcover. ISBN 3-540-29415-5

For further information on these books please have a look at our mathematics catalogue at the following URL: www.springer.com/series/5151